D1629383

Patient Centric Blood Sampling and Quantitative Bioanalysis

Wiley Series on Pharmaceutical Science and Biotechnology: Practices, Applications, and Methods

Series Editor:
Mike S. Lee
New Objective

Mike S. Lee (ed.) • *Integrated Strategies for Drug Discovery Using Mass Spectrometry*

Birendra Pramanik, Mike S. Lee, and Guodong Chen (eds.) • *Characterization of Impurities and Degradants Using Mass Spectrometry*

Mike S. Lee and Mingshe Zhu (eds.) • *Mass Spectrometry in Drug Metabolism and Disposition: Basic Principles and Applications*

Mike S. Lee (ed.) • *Mass Spectrometry Handbook*

Wenkui Li and Mike S. Lee (eds.) • *Dried Blood Spots—Applications and Techniques*

Ayman F. El-Kattan (ed.) • *Oral Bioavailability Assessment: Basics and Strategies for Drug Discovery and Development*

Mike S. Lee and Qin Ji (eds.) • *Protein Analysis using Mass Spectrometry: Accelerating Protein Biotherapeutics from Lab to Patient*

Naidong Wng and Wenying Jian (eds.) • *Targeted Biomarker Quantitation by LC-MS*

Wenkui Li, Wenying Jian, and Yunlin Fu (eds.) • *Sample Preparation in LC-MS Bioanalysis*

Neil Spooner, Emily Ehrenfeld, Joe Siple, and Mike S. Lee (eds.) • *Patient Centric Blood Sampling and Quantitative Bioanalysis: From Ligand Binding to LC-MS*

Patient Centric Blood Sampling and Quantitative Bioanalysis

Edited by

Neil Spooner
Spooner Bioanalytical Solutions
Hertford, UK

Emily Ehrenfeld
New Objective
Cambridge, MA, USA

Joe Siple
New Objective
Cambridge, MA, USA

Mike S. Lee
New Objective
Cambridge, MA, USA

Published by John Wiley & Sons, Inc., Hoboken, New Jersey.
Published simultaneously in Canada.

Library of Congress Cataloging-in-Publication Data applied for
Hardback ISBN 9781119615552

Cover Design: Wiley
Cover Image: © Choun JC/Shutterstock

Set in 9.5/12.5pt STIXTwoText by Straive, Pondicherry, India

Contents

List of Contributors

Catherine E. Albrecht
Labcorp Drug Development
Geneva, Switzerland
catherine.albrecht@labcorp.com

Cecilia Arfvidsson
Integrated Bioanalysis, Clinical
Pharmacology and Safety Sciences,
Biopharmaceuticals, R&D,
AstraZeneca, Gothenburg,
Sweden
Cecilia.Arfvidsson@astrazeneca.com

Christopher Bailey
Integrated Bioanalysis, Clinical
Pharmacology and Safety Sciences,
Biopharmaceuticals, R&D,
AstraZeneca, Cambridge, UK
christopher.bailey@astrazeneca.com

Stephanie Cape
Labcorp Drug Development
Madison, WI, USA
Stephanie.Cape@Labcorp.com

Marc Yves Chalom
Sens Representações Comerciais
Sao Paulo, Sao Paulo, Brazil
mychalom@yahoo.com.br

Bradley B. Collier
Laboratory Corporation of America
Holdings (LabCorp), Center for
Esoteric Testing, Burlington, NC, USA
collib7@labcorp.com

Suzanne K. Cordovado
Centers for Disease Control and
Prevention
Atlanta, GA, USA
snc4@cdc.gov

Carla D. Cuthbert
Centers for Disease Control and
Prevention
Atlanta, GA, USA
ijz6@cdc.gov

Sigrid Deprez
Laboratory of Toxicology
Faculty of Pharmaceutical Sciences
Ghent University, Ghent, Belgium.
sigrid.deprez@ugent.be

Vidar O. Edvardsson
Children's Medical Center
Landspitali–The National University
Hospital of Iceland, Reykjavik, Iceland

Faculty of Medicine, School of Health
Sciences, University of Iceland
Reykjavík, Iceland
vidare@landspitali.is

Amy M. Gaviglio
4ES Corporation
San Antonio, TX, USA
pxd9@cdc.gov

Russell P. Grant
Laboratory Corporation of America Holdings (LabCorp), Center for Esoteric Testing, Burlington, NC, USA
grantr@labcorp.com

Arkady I. Gusev
Biomarker Development
Novartis Institutes of BioMedical Research, Inc.
Cambridge, MA, USA
arkady.gusev@novartis.com

Liesl Heughebaert
Laboratory of Toxicology
Faculty of Pharmaceutical Sciences
Ghent University, Ghent, Belgium
liesl.heughebaert@ugent.be

Rachel Jones
Cheshire, UK
Rachel@10g.co.uk

Carlos Roberto V. Kiffer
Laboratório Especialista de Microbiologia Clínica
Disciplina de Infectologia
Escola Paulista de Medicina
Universidade Federal de São Paulo (UNIFESP), Sao Paulo
Sao Paulo, Brazil
carlos.rv.kiffer@gmail.com

Joseph Loureiro
Disease Area X
Novartis Institutes of Biomedical Research, Inc.
Cambridge, MA, USA
joseph.loureiro@novartis.com

Kristina Mercer
Centers for Disease Control and Prevention
Atlanta, GA, USA
pxd9@cdc.gov

Peyton K. Miesse
Laboratory Corporation of America Holdings (LabCorp), Center for Esoteric Testing, Burlington, NC, USA
pmiesse12@gmail.com

Dmitri Mikhailov
Biomarker Development
Novartis Institutes of BioMedical Research, Inc.
Cambridge, MA, USA
dmitri.mikhailov@novartis.com

Ganesh S. Moorthy
Childrens Hospital of Philadelphia, Perelman School of Medicine, University of Pennsylvania, 3400 Civic Center Boulevard, Building 421, Philadelphia, PA 19104, USA
ganeshsmoorthy@gmail.com

Robert Nelson
Labcorp Drug Development
Geneva, Switzerland
Robert.Nelson@Labcorp.com

Regina V. Oliveira
Núcleo de Pesquisa em Cromatografia (Separare)
Departamento de Química
Universidade Federal de São Carlos
Sao Carlos, Sao Paulo, Brazil
oliveirarv1@gmail.com

Runolfur Palsson
Internal Medicine Services
Landspitali–The National University Hospital of Iceland
Reykjavik, Iceland

Faculty of Medicine
School of Health Sciences
University of Iceland
Reykjavík, Iceland
runolfur@landspitali.is

Konstantinos Petritis
Centers for Disease Control and Prevention
Atlanta, GA, USA
nmo3@cdc.gov

Silvia Alonso Rodriguez
Translational Science and Experimental Medicine Early R&I
Biopharmaceuticals R&D
AstraZeneca, Cambridge, UK
silvia.alonsorodriguez@astrazeneca.com

Jenny Royle
MediPaCe Limited
London, UK
Jenny@medipace.com

James Rudge
Trajan Scientific and Medical
Crownhill Business Centre
Milton Keynes, UK
jrudge@trajanscimed.com

Hrafnhildur L. Runolfsdottir
Internal Medicine Services
Landspitali—The National University Hospital of Iceland
Reykjavik, Iceland
hrafnhr@landspitali.is

Paul Severin
Labcorp Drug Development
Madison, WI, USA
Paul.Severin@Labcorp.com

Christophe P. Stove
Laboratory of Toxicology
Faculty of Pharmaceutical Sciences
Ghent University, Ghent, Belgium
christophe.stove@ugent.be

Veronique Stove
Department of Laboratory Medicine
Ghent University Hospital
Ghent, Belgium

Department of Diagnostic Sciences
Faculty of Medicine and Health Sciences
Ghent University, Ghent, Belgium
veronique.stove@uzgent.be

Unnur A. Thorsteinsdottir
Faculty of Pharmaceutical Sciences
School of Health Sciences
University of Iceland
Reykjavik, Iceland
uth15@hi.is

Margret Thorsteinsdottir
Faculty of Pharmaceutical Sciences
School of Health Sciences
University of Iceland
Reykjavik, Iceland

ArcticMass, Reykjavik, Iceland
margreth@hi.is

Christina Vedar
Childrens Hospital of Philadelphia,
Perelman School of Medicine,
University of Pennsylvania, 3400
Civic Center Boulevard, Building 421,
Philadelphia, PA 19104, USA
cvedar@gmail.com

Nick Verougstraete
Laboratory of Toxicology
Faculty of Pharmaceutical Sciences
Ghent University, Ghent, Belgium

Department of Laboratory Medicine
Ghent University Hospital
Ghent, Belgium
nick.verougstraete@uzgent.be

Alain G. Verstraete
Department of Laboratory Medicine
Ghent University Hospital
Ghent, Belgium

Department of Diagnostic Sciences
Faculty of Medicine and Health
Sciences
Ghent University, Ghent, Belgium
alain.verstraete@ugent.be

Enaksha Wickremsinhe
Lilly Research Laboratories
Eli Lilly and Company
Indianapolis, IN, USA
enaksha@lilly.com

Jinming Xing
Biomarker Development, Novartis
Institutes for BioMedical Research,
Inc., Cambridge, MA, USA
jinming.xing@pfizer.com

Athena F. Zuppa
Childrens Hospital of Philadelphia,
Perelman School of Medicine,
University of Pennsylvania,
Philadelphia, PA, USA
zuppa@email.chop.edu

Foreword

Many of the contributors to this book were authoring their chapters whilst living through a global pandemic, which has changed healthcare and health politics forever, and this book could not have been written at a better time.

Sitting here today, we are facing unprecedented change in international health systems—with spiralling costs, increasing cultural and national disparity in healthcare delivery and acceptance, ageing populations, and infrastructures that are decades old. The world around healthcare is moving at a faster pace than the institution can cope with. In today's world mobile phones and technology are now commonplace in many households. Telehealth and virtual consultancies are, in some cases, gradually replacing traditional face-to-face appointments. And people are starting to accept wellness and prevention as ways that they themselves can tackle the onset of disease.

The boundaries of healthcare are finally beginning to change from the clinic walls and reach out into people's lives. Not only does this result in a healthcare system that is more accessible, but it also brings the possibility of healthcare being culturally tailored for populations and delivered in more sensitive and acceptable ways. Of reaching, and supporting, the most vulnerable in our society.

At the center of this reform is the ability to test and monitor for known illnesses outside of the clinic itself. Patient-centric sampling involves the patient or caregiver taking small amounts of body fluid—blood or saliva as examples—in the comfort of a person's home. Recent technological advances have made this possible, making user friendly, simple, safe, and even painless devices available. These are then packaged as directed and then either posted or collected and sent to a central laboratory for processing. The implementation of this approach, however, is fraught with challenges that need to be overcome before it can be fully integrated as the standard approach that healthcare reaches for first. From the learnings of the pandemic we have real life examples where patient-centric sampling has been successfully implemented within and across countries.

This book shines a light on the whole approach. It presents a balanced look across all aspects—the challenges, technological requirements, assays, processes, delivery ... all the way through to the human behavior and ways to integrate into the norm. It discusses everything we have learnt from the long and rich microsampling history and how this can be used to deliver for the rapidly changing expectations and requirements of future populations and healthcare services. It outlines the unique challenges and opportunities presented through use of patient-centric sampling in the clinical trials that are so essential for developing new medicines.

Every one of the authors in this book wrote their chapters to help others. They represent a diverse background of expertise and share their experiences, insights, and case studies with astonishing honesty, openness, and integrity. This partnership between people is unified by a belief in the welfare of others. It gives a unique insight into the systematic changes that must be undertaken to allow the full potential of patient-centric care to be realized. Throughout, these insights are supported and enhanced by the author's real-life case studies and experiences of using this approach in practice.

It is hoped that by sharing this, you will be the next to add to this wave of change—to join the innovators and make a difference. If we all do this, then together we can play our part in ensuring healthcare becomes accessible and acceptable to those who need it most, the patients.

Jenny Royle
Matthew Barfield

Preface

There is an increasingly broad understanding that the collection of biological samples for the quantitative determination of analytes for healthcare and the support of clinical trials needs to be performed in a manner that puts the needs of the patient at the center. The technologies to enable high quality samples to be collected in a location such as the home, pharmacist, local doctor's surgery, or other locations that are convenient to the patient, rather than at a centralized clinical center, are now readily available. Furthermore, the clamour for this change has gained momentum during the recent COVID-19 pandemic, where all of us were reluctant to go to clinical facilities. Despite this, change is always difficult, particularly for something as well established as the processes for the collection and analysis of blood samples. Thankfully, there is an increasing body of leaders, represented by the authors of the chapters in this book, who realize the benefits of these solutions and understand that by working together across inter- and intra-organizational boundaries we can break down the barriers to their routine adoption and bring benefits to the patient.

A previous book in this series, *Dried Blood Spots*, edited by Wenkui Li and Mike Lee, set the benchmark for our understanding of the benefits of these patient-centric sampling technologies and how they can be implemented in a number of scenarios. This book builds upon those strong foundations, to bring us up to date with how this exciting and fast-moving field has developed. The authors, who are looking to enable the routine use of these technologies for the benefit of their fellow humans, generously share their observations, experiences, concerns, and visions of the future. As such, this book is another benchmark in the continuing change that is healthcare and analytical science. We can say with certainty that there is further change to come in this field and as such we look forward to working as a community to facilitate these important and inevitable changes.

The editors sincerely thank all the distinguished authors for the provision of their wonderful chapters and for their patience and persistence with this project through the difficult events of the pandemic. It has definitely been worth the wait! We also wish to thank the editorial staff at John Wiley & Sons for their unwavering and patient support to this project that is a small part of the phenomenal series edited by Mike Lee.

Neil Spooner
Emily Ehrenfeld
Joe Siple
Mike S. Lee

1

Patient Centric Healthcare – What's Stopping Us?

Jenny Royle[1] *and Rachel Jones*[2]

[1] *MediPaCe Limited, London, UK*
[2] *Cheshire, UK*

1.1 The Evolution of Future Health Systems

The primary aim of healthcare systems around the globe is to improve the well-being of populations, with the World Health Organization defining health as "A state of complete physical, mental and social well-being and not merely the absence of disease or infirmity" (World Health Organisation, 2020). Healthcare is not merely treatments for diseases, it is instead a way to support people to achieve the highest attainable mental and physical and social well-being for themselves. But traditional healthcare systems are not set up in this way, and rather than focusing on integrating all the holistic elements required for promotion of health in an individual, they are orientated toward the treatment of disease and malaise after things have already deteriorated in a person's well-being.

The difference between absence of disease and total well-being is subtle but fundamental. Supporting well-being involves encouraging people to live a healthy lifestyle (both physically and mentally) and providing the tools, systems, and education for each person to aim for the best possible version of themselves often via self-care principles—whether or not they are sick at the current time.

Other factors that have further impacted the tension between the treatment versus a self-care model of health have emerged during the COVID-19 pandemic during 2020, which forced upon us innovations and technologies that were once confined to pilot status. These were mobilized during the 2020 pandemic out of necessity in order to meet the restriction in face-to-face services required to prevent transmission of the virus between people. Many of these rapid accelerations,

Patient Centric Blood Sampling and Quantitative Bioanalysis, First Edition.
Edited by Neil Spooner, Emily Ehrenfeld, Joe Siple, and Mike S. Lee.

particularly in relation to telehealth visits and remote monitoring, are here to stay because, once pushed to try a new approach, it has been found to be efficient and a positive experience for many (Jones et al., 2020; Norman et al., 2020; Wosik et al., 2020). An area that is a fundamental component of this wave of change is the process of microsampling at home.

For the uninitiated, microsampling involves the taking of small droplet amounts of body fluid—blood or saliva as examples—in the comfort of a person's home. The samples themselves are taken by the patient or with assistance from a caregiver and are then stored and packaged as directed (for example by drying on a specialized sample tip or card and sealing into the envelope provided). These are then either posted or collected and sent to a central laboratory for processing, thus negating the need for a patient to visit an outpatient clinic or local surgery (Bateman, 2020). The laboratory assay must be validated and provide sufficient accuracy to support accurate clinical decision-making. This onset of a remote, patient centric approach to sampling brings with it the chance to fundamentally challenge and change the healthcare delivery model. Sampling and appointments can be decentralized, and most routine supportive care can be virtual. This does not mean the end of the hospital or GP visit, but it does mean that the approach used can be fitted to the requirements of the individuals involved and the healthcare decisions that need to be made. Remote sampling and consultations are more time efficient (Ballester et al., 2018; Prasad et al., 2020; Russo et al., 2016) and this means that not only can they be scheduled around peoples' daily lives better but also any face-to-face appointments can be prioritized for people where in-person consultation is truly needed. For overstretched front-line staff and health systems, this is likely to be a very attractive proposition.

Research has also shown that dried blood spot sampling versus conventional blood sampling conferred cost savings across the ecosystem in renal transplant and hemato-oncology patients (Martial et al., 2016). In this study, switching to home sampling was associated with a societal cost reduction of 43% for hemato-oncology patients and 61% for nephrology patients per blood draw. From a healthcare perspective, costs reduced by 7% for hemato-oncology patients and by 21% for nephrology patients due to the replacement of office-based tests with home-based sampling.

So the evidence suggests that virtual care provides a mostly positive patient experience and is more efficient for the health service. Could this also help reduce the number of people not "turning up" for medical appointment (if the consultation comes to them)? Research has shown that the high levels of "no shows" to hospital appointments have a large impact on the organizational structure and cost to a health provider (Dantas et al., 2018; Jefferson et al., 2019; Mohammadi et al., 2018). Although no research has been carried out on this to date, it is possible that the efficient use of home sampling could reduce this "appointment missing," and this more optimized supportive care for patients could offer additional benefits on the downstream impacts to service. Another potential benefit of home sampling could be the time freed up for those more in need of face-to-face contact and better decisions on how to balance the two approaches. There are benefits and limitations to both home-based and in clinic approaches—for example, home care approaches which are decentralized have been shown to give better individualized, immediate care, but along with this, the responsibility for monitoring is largely delegated to technical devices, patients, and their families (Oudshoorn, 2009). Face-to-face appointments in the clinic have been shown to be preferred over telemedicine in specific circumstances such as when patients have low self-management ability and/or depending on the purpose of the consultation (e.g. initial discussions about terminal disease, which may have additional, unspoken support needs; Chudner et al., 2019; Derkson et al., 2020). Designing an integrated approach based on the person's needs may be most beneficial for all. For example, in 2019, Jiang and colleagues found that correctly timing a face-to-face consultation increased a patient's ability to accurately find information digitally and administer self-care post consultation. Integrated approaches also bring the potential to save more face-to-face consultation time for personalized conversations and supportive care, leaving more simple tests and interventions to be carried out at home.

The authors suggest the use of home blood sampling may have positive impacts on a person's overall well-being by allowing intrusive interventions to be carried out within a familiar home environment. A survey was taken of 39 adult kidney transplant patients who underwent both traditional venepuncture and microsampling approaches for monitoring of their condition and the current blood sampling burden was quantified using two measures: anxiety and travel requirements (Scuderi et al., 2020). A third of participants ($n = 13$) reported blood test anxiety and 44% ($n = 17$) spent more than an hour just to travel to the required phlebotomy site for standard of care. Preference between the two approaches was also explored: 85% ($n = 33$) preferred microsampling approaches and 95% ($n = 37$) expressed an interest in collecting their microsample themselves at home. This demonstrates a clear patient preference and willingness to give microsampling a go for monitoring post-transplantation recovery progress.

1.2 Exploring the Barriers to Home Sampling

Given the efficiencies and benefits of home-based care, why is not remote patient centric microsampling more rapidly adopted everywhere? The answers may rest with people and the hurdles involved in fundamentally changing established care pathways and healthcare cultures in which people are already working at maximum capacity to deliver what they know, let alone try something new.

The scientific and technological aspects of patient centric microsampling have accelerated in the past 5 years and are driving the field of healthcare in the home; this chapter aims to focus on many of the key concerns that have been heard through working in the clinic, with patients, and developing technologies. The aim is that by starting a discussion around each of these concerns and by proposing potential solutions, developers and leaders of the future will be able to co-create the approaches with the relevant end users and speed up the realization of benefits that these sampling processes can bring.

1.2.1 Barrier One—The Discord Between Innovation and Practice

Recent events of the pandemic in 2020 have shown us that all healthcare systems run at a finite capacity. To implement change, the very same people who rely on established approaches have to, instead, adopt and implement something brand new, while maintaining their high standard of care in challenging times.

The expertise behind the development of highly sensitive microsampling technology has, up until now, been mostly confined to pharmaceutical companies and private laboratories and was generally not widespread in the labs of front-facing healthcare institutions. Furthermore, existing health systems required to implement microsampling systems are built on the fundamental principle *of primum non nocere* (first, do no harm). Sampling and test results are usually only a small part of a clinical pathway, with many healthcare professionals defaulting to the established pathway approaches and more familiar and trusted in-clinic sampling techniques are automatically selected. This involves the deployment of personnel, for example phlebotomists and nurses in a system that, despite being pushed to its limits, is proven to accurately deliver support to clinical decisions. Convenience to patient and family is sometimes seen as of secondary importance, and any potential increase in decision speed is currently unproven with empirical evidence in the standard front-line healthcare systems. It is now commonly recognized that human decision-making usually relies on a System one (fast, reactive, emotional, and habitual) and System two (slower, higher energy, puzzling out a new challenge) approach (Kahneman, 2003). The vast majority of times, human decisions are ultimately based upon responsive, heuristic, and/or emotions, rather than calculated logic—especially when the individual has many years of experience. It is possible, therefore, that emotional rather than rationale drivers may be slowing the uptake of home sampling—for example—trust. For this established and routine model to be replaced, evidence would need to be gathered that the new pathway provides significant improvement in timeliness and physical and/or mental patient well-being. Importantly, overall cost savings for the healthcare system as a whole (primary, secondary, and social care) across the different healthcare models would also need to be demonstrated. Technological advances have the potential to change elements far beyond use of the device itself—reaching into roles, medic–nurse–patient–carer relationships, behaviors, and healthcare culture and there is understandable caution from front-line decision-makers who are upholding the principle of "do no harm" and have little time to assess the overall healthcare value–benefit versus risk offered by home sampling.

A solution to these problems has been proposed through novel patient centric co-creation and delivery of technology clinical trials that design new or enhanced care pathways with those involved in their routine implementation and then test them empirically in a clinical trial setting. Not only does this mean that the full consequences and potential benefits of an innovation (such as home sampling) that may ripple across the care pathway are considered, but that the proposed ways of leveraging the positives are created by those using the current approach. The subsequent quantitative and qualitative testing of the whole pathway then also provides the empirical evidence as to the impact on people's well-being, ability to deliver the care, and cost. With this approach, barriers preventing uptake are reduced through the intervention design and the drive for change is supported by the robust evidence that healthcare demands (Royle et al., 2021).

1.2.2 Barrier Two—Ethical and Operational Considerations

The ethics involved in home sampling within a new care pathway needs to be thoroughly considered, documented, and discussed with end users before it is put into practice. For example, if the novel innovation is a home sampling test kit then, as well as taking the sample, patients need to accept that they are responsible for taking it. Patients need to store and use the sample kit correctly, dispose of elements in a safe way, and post the sample off in a timely manner. People need to understand why this is necessary and how the data collected from the kit will be communicated to them and what it means—in lay terms. Seamless services are now the norm, but on the odd occasion of a faulty/lost kit, patients should understand how to take action and the provider resolve the problem immediately, so as to maintain both patient and physician confidence in the new system.

There are many areas of healthcare where patients and their families already take responsibility for their own treatment and healthcare. For example, many diabetic patients monitor their own glucose levels and titrate their insulin dose and timings, colorectal cancer screening requires people to send samples for assessment (the affectionately called "poo in the post" screen), all the way through to women taking the responsibility with regard to adherence to the contraceptive pill. The idea that people can take a blood microsample and post it off to a lab should not seem that unusual, and it is logical to assume it will be accepted easily. However, there are very important elements at play in each of these examples. For each, there is a clear and direct benefit for the patient themselves and no "easy" alternative. Each has also been standard of practice for many years and has therefore become the accepted social norm in many societies—it has long been expected and accepted that diabetic patients and their families monitor their daily treatment and therefore from the point of diagnosis patients inevitably have to accept this role.

For new point of care and home sampling approaches, this "norm" does not exist for both physicians and patients. Even if it makes it slightly easier, the "accepted" and "expected" more traditional approach from all perspectives is for their doctor, nurse, or a trained phlebotomist to undertake sampling. A patient given the opportunity to take on the sampling themselves may, in some cases, compare their minimal professional experience with that of their authority figures. In addition to this, when the microsamples are being used to monitor a specific health condition of themselves or those they love—the consequences of unknowingly getting it wrong become even more worrying.

None of these challenges are insurmountable, but steps should be put in place to develop the care pathway with those involved to ensure that each step—even if it is not one directly related to the mechanics of sampling itself, is considered. In this case, there are several clear elements that can easily help to overcome barriers:

Helpful tips for successful adoption	Description
Making it as pleasant and easy as possible	Simple, quick action. Painless if possible. But also consider the timing and routine. Linking in with habits helps people to remember and make it more "normal" to undertake. Try to avoid additional steps such as intricate assembly or the need to refrigerate.
Recommendation by an authority figure	Conviction that this is the best approach to take, being conveyed and supported by their physician and other trusted healthcare anchors a patient may have.
Building up self-efficacy (i.e., someone's belief that they can successfully do what is needed—in most cases, take the sample)	Time for training and questions built into the process. Having a trusted contact point for help if needed. Share support from peers—other patients who have successfully adopted the system and are willing to volunteer as "champions."
Seeing that it matters—their actions have tangible value	Knowing that the results are to be evaluated and will not be lost (and getting confirmation that they have been looked at).
Having a feeling of control and reducing anxiety	Having an action plan for patients that covers all the main predictable things that may happen (e.g., knowing what to do if the system malfunctions or the sample collection goes wrong). Provide patients with support and help to understand and evaluate findings and what these mean for them.

1.2.3 Barrier Three—Where Does the Liability Sit?

Healthcare workers are always cognizant of their legal and ethical responsibilities to protect the patients they care for and to do this to the best of their ability. But in the case of home sampling, the process is moved out of their control because it is no longer being undertaken by professional healthcare colleagues but by the patient or caregiver themselves. How can professionals deliver to these legal responsibilities if they are not personally carrying out or delegating the sampling to their colleagues that have certificates of training? Where does the final legal liability lie, for example if poor or inadequate sampling results in the wrong decision? This area of law and liability is infrequently discussed, but is one to consider as home sampling and other such services become more prevalent in treatment and care pathways. The Royal College of Nursing in the UK states that "*To discharge the legal duty of care, health care practitioners must act in accordance with the relevant* ***standard*** *of care.*" Where the "standard of care" is deemed to be that undertaken by other professionals in a similar situation as a benchmark (Royal College of Nursing, 2020).

Although the legal and ethical requirements which constitute a duty of care to patients may vary across countries, it is important that those who provide and initiate the services of home blood sampling consider their duty of care to patients who may be embarking on a shared or self-care journey. Considerations such as the assessment of the patient and their support network to understand their "health literacy" and capabilities to carry out their tests is an important first step. In addition, a willingness of physicians and nurses to adopt a collaborative shared care approach with patients where support is offered and is gradually decreased as the competence of the patient in this area increases should be evident, in the form of health coaching. In the earlier days of such a culture change, to help build physician–patient trust, this could even mean supplementing the home testing within clinic standard tests at the start of treatment. This would endure until both parties can be confident that the home tests are useful indicators and can be routinely relied on (with in clinic sampling then being for emergency second check only).

Further, the authors believe that the proffering of documented, lay-friendly information coupled with peer support would all allow the physician or nurse to comfortably discharge their duty of care to the patient and caregivers. Indeed, the convenience and possible enhanced quality of life that patients may benefit from is testimony to the ethically robust patient centric decision made within the shared care team.

The authors suggest that standardized protocols for the deployment of home-centered care and sampling would serve to provide guidance, share best practice, and alleviate concerns of healthcare professionals (HCPs). Such a standardized approach in shared care decision-making is discussed by a working party (Elwyn et al., 2012) with *choice, option, and decision* "talks" described as a robust process to ensure a shared care collaboration between HCPs and patients. Such a model could be developed with a focus on home sampling so that any specific operational aspects can be incorporated and standardized.

Similarly, many patients and families who may be comfortable with the more paternalistic relationship they share with their physician may suddenly see this responsibility of self-sampling as a burden, rather than a release from yet another outpatient visit. This may suggest the need to further support more vulnerable communities if home sampling is to become the norm. In practice, many of these concerns can be addressed early at the point of service design and importantly, in collaboration with all parties, so that appropriate support interventions may be put in place. Indeed, it has been suggested that shared care should evolve such that both physician and patients expect and make room for time to discuss shared options and engage in reflective thinking asking "what if clinicians felt just as comfortable asking questions as providing answers? What if patients were allowed more time on their own to reflect on what their clinician explained?" (Pieterse et al., 2019). If home sampling and other shared and self-care measures are to continue to grow and be deployed in a seamless fashion, then clearly behaviors of all parties may need to evolve.

1.2.4 Barrier Four—Addressing the Technology Challenge

As microsampling becomes more prevalent in care pathways, it is likely to be linked to an increasing number of point of care devices for patients and their families to use themselves. As discussed earlier, with a clear partnership and action plans between the patient/patient's family and their physician comes a huge opportunity for timely action and proactive healthcare. However, it also brings technological challenges, since many patients may be overwhelmed by new technology or instructions that are not available in lay language. Should results be provided via a mobile device, many more mature or vulnerable patients may struggle to "play their part" within the system which could lead to enhanced anxiety and a feeling of exclusion, plus the obvious loss of healthcare data. These challenges are surprisingly not confined to patient populations, since previous unpublished work conducted by the authors highlighted the challenges faced by nursing staff when confronted with a novel Bluetooth device to monitor salbutamol inhaler use in patients with chronic obstructive pulmonary disease (COPD). This eventually led to both patient and nurse confusion when using the device and consequently an associated reduction in adherence.

We also need to consider the *inclusivity* of such new microsampling interventions in the home, since in every country there is a proportion of the population that are not native speakers and may rely upon family, friends, and HCPs for their care and navigation within that region's healthcare system. Behavioral science underpins the need to not only provide clear, lay friendly information but also reinforce this with using peer support and coaching to highlight the relevance for patients. Clearly, this is of supreme importance in more vulnerable groups who may live on the periphery of our society.

A further consideration is the technological changes that will need to occur in front-line labs, which are currently not kitted out for performing assays on samples derived from patient centric sampling approaches. For decisions to be timely, the performance of the analytical assays must also be. This means that each healthcare institution will need to make a decision—how can they best link with patient centric sampling approaches? Bring the assays in house? Or outsource? This is likely to be different depending on the type of monitoring, decisions made, and relationships that the institute may already have with a third party. Another consideration is which assays to group together. Even though a huge amount of a venepuncture sample may be thrown away, it is still possible to get many standard test results (a panel) out of a single sample. Standard panels have yet to be established for microsampling, simply because it is so new and early adopters usually have bespoke monitoring considerations that they are trying to address.

1.2.5 Barrier Five—The Human Touch

It is clear that a well-designed home sampling service confers many advantages in terms of efficiencies in resource-constrained environments. But for many patients, attending their outpatient appointment is their only chance of human contact with HCPs, and at times, their peers who they meet at regular appointments. Even with the advent of home sampling and virtual appointments which are fast becoming the norm, many physicians can only truly assess patients' whole care needs, in terms of their mental, physical, and social well-being, by seeing them in the flesh where pallor, gait, and general health function can be truly assessed. It is suggested that focused face-to-face visits are planned into a healthcare pathway to complement home sampling, to provide the most rounded care possible. This is especially true for more vulnerable members of society where safeguarding issues may only be truly addressed via face-to-face assessments. Strategic use of new technologies at home may benefit this approach by freeing up time for less rushed and more valued consultations. As identified by Bender et al. (2014), if limited social support and engagement by the medical team is perceived, and self-management requirements increased, this can have a negative impact on a patient's decisional context. With this warning in mind, the role of a more blended care approach, based on the shared decision-making between the physician and patient, again seems optimal.

Other anxieties may perpetuate among the patient community in terms of a fear of bloodletting and of blood itself, and needle phobic people may prefer to remain in a more passive capacity and delegate procedures such as this to others in authority. Noting these concerns, some adaptations to traditional bloodletting are emerging that may help to address some of these barriers, such as on demand blood sampling kits which aim to deliver clinical-grade blood samples with little pain and via simple instructions. How, or if, these will be taken up by different health systems with varying funding options across the world remains to be seen. Clearly, there is a logical need if people want to avoid needles in practice and to support patients so that they may overcome any fears or barriers to home sampling in a patient centric manner.

Not everyone is ready to become an active guardian of their own health. The way in which results are delivered and acted upon by patients in isolation needs to be considered within any patient centric care pathway. Without integration of the home testing into broader engagement and healthcare pathways, patients may feel increased anxiety and isolation in their own homes. This was assessed during an interview-based study conducted by Tiro et al. (2019) in 46 women (median age 55.5 years) who had received a positive home human papillomavirus self-sampling kit result. After reflecting on their experience, six main themes were identified: the convenience of the test, intense emotions after receiving a positive result, the importance of discussing their results with an expert, seeking information themselves from a wide variety of sources; confusion over the meaning and impact of their positive result, and questioning the accuracy of the test. The authors suggest that supportive and interpretative care is available for patients who require assistance while receiving their test results. Further, it is proposed that in extremely vulnerable circumstances, the results are directed primarily toward physicians who would play an active role within the home sampling pathway and results in dissemination.

1.2.6 Barrier Six—Trust in Data Security

For patients and physicians to fully trust a home-based sampling system, any challenges around data security need to be fully addressed at the outset. There have been several incidences of data breaches within healthcare infrastructures (Bai et al., 2017; McLeod & Dolezel, 2018; Seh et al., 2020), and these may contribute to concerns around the security of personal biological data which is sent to a central location. Postal systems can and do encounter problems, and the use of anonymized and encrypted patient codes used to reference biological samples can help to reassure patients of their data privacy and of the resilience of the supply chain from home to laboratory. Patients should be reassured of their data security at the start of their home sampling journey to assuage any concerns.

1.2.7 Barrier 7—Adherence to Service Change

Adherence to any intervention (therapy, treatment, procedure, etc.) is defined by the World Health Organization as "the degree to which the person's behavior corresponds with the agreed recommendations from a health care provider" (Sabaté & Sabaté, 2003).

Despite people being "told what to do for the good of their own health," rates of adherence to recommendations by patients still vary wildly. In developed countries, it has been estimated that around 50% of patients with chronic conditions do not follow their stipulated treatment requirements correctly (Brown & Bussell, 2011; Kvarnström et al., 2018; Naderi et al., 2012). For extremely involved and multifactorial requirements, such as disease management in patients with chronic kidney disease, such nonadherence estimates vary even wider from 3% up to 80% among patients on hemodialysis (Mechta et al., 2018). Failure to adhere to interventions and medications is not simply due to forgetfulness, but to a complex combination of psychological and lifestyle factors, which vary across different conditions. The barriers to the use of home sampling techniques described above opens up an abundance of challenges to adherence, in terms of knowledge, lack of support, and anxiety to name a few. Similarly, physicians and HCPs may express concerns about a patient's ability to carry out their home sampling with adequate competency and adherence, even though the test samples required are small and the process lacks complexity. This said, service providers should aim to integrate both traditional interventions, such as face-to-face clinic visits, alongside home sampling, domiciliary visits by doctors, or use of community pharmacy services. This could be undertaken with a "phased in approach" to allow a patient and caregiver to gain expertise and confidence or "self-efficacy" in their ability to participate actively.

This does not mean that patients suddenly have to become experts in medicine—their healthcare partner can continue to provide this—but it does mean that their observations and support in collecting and acting on the data are equally valued and are a weighted contribution to the clinical picture (rather than considered as useful "supplementary information"). Even at its simplest level, trusting and supporting patients' self-assessments and creating a mechanism to systematically capture and use them can deliver huge benefits. In a randomized controlled trial involving 766 patients with advanced solid tumors, one group received standard of care follow-up, and the other group had standard of care plus an online symptom assessment form that they completed weekly (after email prompt; Basch et al., 2017). Treating physicians received symptom printouts at visits, and nurses received email alerts when participants reported severe or worsening symptoms. Health-related quality of life (HRQL) improved more frequently among participants in the intervention group than usual care (34 vs. 18%) and worsened among significantly fewer participants (38 vs. 53%; $p = 0.001$). In addition, the mean quality-adjusted overall survival was significantly improved in the intervention group (mean 8.7 vs. 8.0 months, $P = 0.004$) with 75% of the intervention group surviving at 1 year compared with 69% in the standard of care group. The authors draw from this that home sampling could be supported with, for example, internet-based or telephone-based nurse support or indeed advanced peer support, especially at the early stages of home sampling use. Other alternatives could be a phased approach whereby home sampling is punctuated with routine visits to an outpatient clinic as an approach to allow both the patient and physician to grow their confidence in the new model.

1.3 Conclusion: The Changing Role of Home Sampling

The hypothesized *all round* benefits of home sampling are unproven until new care pathways have been objectively tested in the clinical trial setting. But what can be recommended is to design any new care pathway with those involved for maximum patient benefit and ensure that data are gathered to characterize any barriers and alternative solutions that may exist in practice. This detailed knowledge, gathered in practice as real-world evidence, can then be used to drive and share broader patient-involvement approaches to specifically address areas of challenge.

Healthcare has historically had a paternalistic and/or disease-focused culture (Bokhour et al., 2018; Resnik, 2015). In this traditional approach, clinical teams are cast as expert providers, while patients are cast as grateful recipients, with most information exchange occurring when the patient visits their clinician. The advent of home sampling delivers impacts to the healthcare system above and beyond the much-needed upsides of convenience and resource saving. It has the potential to increase the robustness of the data that patients themselves collect iven industries who place their end user or customer at the heart of everything they do, but increasingly, patients are being placed in the driving seat, as the end users and ultimate custodians of healthcare. Growing healthcare costs, a desire for personalized approaches (often facilitated by advancing technologies), along with an increased demand for transparency will continue to drive forward this agenda of co-creation. Home sampling is no exception, and its implementation will further enhance patient centric services which should streamline routine testing and free up more time for more valuable and reflective conversations with physicians at the exact time they are needed most.

Changes that evolved through necessity at the time of writing in 2020 (to avoid face-to-face contact between people during a global pandemic) are here to stay.

With collaboration among the healthcare ecosystem and a willingness to challenge social norms, home sampling can take advantage of the paradigm shifts observed in a year of unprecedented change to become integrated into patient centric pathways of care.

References

Bai, G., Jiang, J. X., & Flasher, R. (2017). Hospital risk of data breaches. *JAMA Internal Medicine*, *177*(6), 878–880.

Ballester, J. M. S., Scott, M. F., Owei, L., Neylan, C., Hanson, C. W., & Morris, J. B. (2018). Patient preference for time-saving telehealth postoperative visits after routine surgery in an urban setting. *Surgery*, *163*(4), 672–679.

Basch, E. M., Deal, A. M., Dueck, A. C., Bennett, A. V., Atkinson, T. M., Scher, H. I., Kris, M. G., Hudis, C. A., Sabbatini, P., Dulko, D., & Rogak, L. J. (2017). Overall survival results of a randomized trial assessing patient-reported outcomes for symptom monitoring during routine cancer treatment. *Journal of Clinical Oncology*, *35*(15_suppl), 197–198.

Bateman, K. (2020). The development of patient-centric sampling as an enabling technology for clinical trials. *Bioanalysis*, *12*(13), 971–976.

Bender, C. M., Gentry, A. L., Brufsky, A. M., Casillo, F. E., Cohen, S. M., Dailey, M. M., Donovan, H. S., Dunbar-Jacob, J., Jankowitz, R. C., Rosenzweig, M. Q., & Sherwood, P. R. (2014, May). Influence of patient and treatment factors on adherence to adjuvant endocrine therapy in breast cancer. In *Oncology nursing forum* (Vol. 41, No. 3, p. 274). NIH Public Access.

Bokhour, B. G., Fix, G. M., Mueller, N. M., Barker, A. M., Lavela, S. L., Hill, J. N., Solomon, J. L., & Lukas, C. V. (2018). How can healthcare organizations implement patient-centered care? Examining a large-scale cultural transformation. *BMC Health Services Research*, *18*(1), 1–11.

Brown, M. T., & Bussell, J. K. (2011, April). Medication adherence: WHO cares?. In *Mayo clinic proceedings* (Vol. 86, No. 4, pp. 304–314). Elsevier.

Chudner, I., Goldfracht, M., Goldblatt, H., Drach-Zahavy, A., & Karkabi, K. (2019). Video or in-clinic consultation? Selection of attributes as preparation for a discrete choice experiment among key stakeholders. *The Patient-Patient-Centered Outcomes Research*, *12*(1), 69–82.

Dantas, L. F., Fleck, J. L., Oliveira, F. L. C., & Hamacher, S. (2018). No-shows in appointment scheduling—a systematic literature review. *Health Policy*, *122*(4), 412–421.

Derksen, F. A., Hartman, T. O., & Lagro-Janssen, T. (2020). The human encounter, attention, and equality: The value of doctor–patient contact. *British Journal of General Practice*, *70*(694), 254–255.

Elwyn, G., Frosch, D., Thomson, R., Joseph-Williams, N., Lloyd, A., Kinnersley, P., Cording, E., Tomson, D., Dodd, C., Rollnick, S., & Edwards, A. (2012). Shared decision making: A model for clinical practice. *Journal of General Internal Medicine, 27*(10), 1361–1367.

Jefferson, L., Atkin, K., Sheridan, R., Oliver, S., Macleod, U., Hall, G., Forbes, S., Green, T., Allgar, V., & Knapp, P. (2019). Non-attendance at urgent referral appointments for suspected cancer: A qualitative study to gain understanding from patients and GPs. *British Journal of General Practice, 69*(689), e850–e859.

Jones, M. S., Goley, A. L., Alexander, B. E., Keller, S. B., Caldwell, M. M., & Buse, J. B. (2020). Inpatient transition to virtual care during COVID-19 pandemic. *Diabetes Technology & Therapeutics, 22*(6), 444–448.

Kahneman, D. (2003). A perspective on judgment and choice: Mapping bounded rationality. *American Psychologist, 58*(9), 697.

Kvarnström, K., Airaksinen, M., & Liira, H. (2018). Barriers and facilitators to medication adherence: A qualitative study with general practitioners. *BMJ Open, 8*(1), e015332.

Martial, L. C., Aarnoutse, R. E., Schreuder, M. F., Henriet, S. S., Brüggemann, R. J., & Joore, M. A. (2016). Cost evaluation of dried blood spot home sampling as compared to conventional sampling for therapeutic drug monitoring in children. *PloS One, 11*(12), e0167433.

McLeod, A., & Dolezel, D. (2018). Cyber-analytics: Modeling factors associated with healthcare data breaches. *Decision Support Systems, 108*, 57–68.

Mechta, N. T., Frøjk, J. M., Feldt-Rasmussen, B., & Thomsen, T. (2018). Adherence to medication in patients with chronic kidney disease: A systematic review of qualitative research. *Clinical Kidney Journal, 11*(4), 513–527.

Mohammadi, I., Wu, H., Turkcan, A., Toscos, T., & Doebbeling, B. N. (2018). Data analytics and modeling for appointment no-show in community health centers. *Journal of Primary Care & Community Health, 9*, 2150132718811692.

Naderi, S. H., Bestwick, J. P., & Wald, D. S. (2012). Adherence to drugs that prevent cardiovascular disease: Meta-analysis on 376,162 patients. *The American Journal of Medicine, 125*(9), 882–887.

Norman, M. L., Malcolmson, J., Armel, S. R., Gillies, B., Ou, B., Thain, E., McCuaig, J. M., & Kim, R. H. (2020). Stay at home: Implementation and impact of virtualising cancer genetic services during COVID-19. *Journal of Medical Genetics, 59*(1), 23–27.

Oudshoorn, N. (2009). Physical and digital proximity: Emerging ways of health care in face-to-face and telemonitoring of heart-failure patients. *Sociology of Health & Illness, 31*(3), 390–405.

Pieterse, A. H., Stiggelbout, A. M., & Montori, V. M. (2019). Shared decision making and the importance of time. *JAMA, 322*(1), 25–26.

Prasad, A., Brewster, R., Rajasekaran, D., & Rajasekaran, K. (2020). Preparing for telemedicine visits: Guidelines and setup. *Frontiers in Medicine*, *7*, 600794.

Resnik, D. B. (2015). Paternalism and utilitarianism in research with human participants. *Health Care Analysis*, *23*(1), 19–31.

Royal College of Nursing. (2020). In *Duty of care*. Retrieved December 1, 2020, from https://www.rcn.org.uk/get-help/rcn-advice/duty-of-care.

Royle, J. K., Hughes, A., Stephenson, L., & Landers, D. (2021). Technology clinical trials: Turning innovation into patient benefit. *Digital Health*, *7*, 205520762110121.

Russo, J. E., McCool, R. R., & Davies, L. (2016). VA telemedicine: An analysis of cost and time savings. *Telemedicine and E-Health*, *22*(3), 209–215.

Sabaté E., & Sabaté E. (Eds.). (2003). *Adherence to long-term therapies: Evidence for action*. World Health Organization. Retrieved December 20, 2022, from https://www.researchgate.net/publication/8967443_Adherence_to_long-term_therapies_evidence_for_action_WHO_WHO_2003.

Scuderi, C. E., Parker, S. L., Jacks, M., John, G., McWhinney, B., Ungerer, J., Mallett, A., Roberts, J. A., Healy, H., & Staatz, C. E. (2020). Kidney transplant recipient's perceptions of blood testing through microsampling and venepuncture. *Bioanalysis*, *12*(13), 873–881.

Seh, A. H., Zarour, M., Alenezi, M., Sarkar, A. K., Agrawal, A., Kumar, R., and Khan, R.A. (2020, June). Healthcare data breaches: Insights and implications. In *Healthcare* (Vol. 8, No. 2, p. 133). Multidisciplinary Digital Publishing Institute.

Tiro, J. A., Betts, A. C., Kimbel, K., Buist, D. S., Mao, C., Gao, H., Shulman, L., Malone, C., Beatty, T., Lin, J., & Thayer, C. (2019). Understanding patients' perspectives and information needs following a positive home human papillomavirus self-sampling kit result. *Journal of Women's Health*, *28*(3), 384–392.

World Health Organisation. (2020). In *Constitution*. Retrieved December 1, 2020, from https://www.who.int/about/who-we-are/constitution.

Wosik, J., Fudim, M., Cameron, B., Gellad, Z. F., Cho, A., Phinney, D., Curtis, S., Roman, M., Poon, E. G., Ferranti, J., & Katz, J. N. (2020). Telehealth transformation: COVID-19 and the rise of virtual care. *Journal of the American Medical Informatics Association*, *27*(6), 957–962.

2

Tips for Successful Quantitative Assay Development Using Mitra Blood Sampling with Volumetric Absorptive Microsampling

James Rudge

Trajan Scientific and Medical, Crownhill Business Centre, Milton Keynes, UK

2.1 What is Volumetric Absorptive Microsampling?

Simply put, VAMS® is a technique where microsamples of a biological fluid are collected on the absorbent tip of a Mitra® device with volumetric precision. Precise volumes of blood and other biofluids are necessary for accurate quantification of the analyte of interest. The purpose of the VAMS technique is to enable high-quality, volumetrically collected microsamples (10, 20, or 30 µL) of capillary blood (or other fluids) samples to be collected by practically anyone, anywhere, and anytime. Mitra devices with VAMS® tips look like pipettes, but the difference is that the Mitra microsampler has a polymeric pad, known as the VAMS tip, for absorbing fluid samples (see Figure 2.1).

For remote sampling, the Mitra microsamplers are housed in protective device cartridges, which allow for ease of handling, air drying, and enclosed transport to laboratories for testing. Once in a laboratory, the samplers are removed from their cartridges and reformatted into 96-well "auto-cartridges," which allow for extractions of 96 samplers simultaneously. This format enables fully manual or automated handling using standard laboratory approaches, without the need to invest in expensive new equipment.

The aim of this chapter is not to reinvent the wheel when it comes to validating assays on dried matrix microsamples. Indeed, in recent years, there have been several fantastic reviews and guides on microsampling, which cover method development and validations. The reader is encouraged to refer to

Patient Centric Blood Sampling and Quantitative Bioanalysis, First Edition.
Edited by Neil Spooner, Emily Ehrenfeld, Joe Siple, and Mike S. Lee.

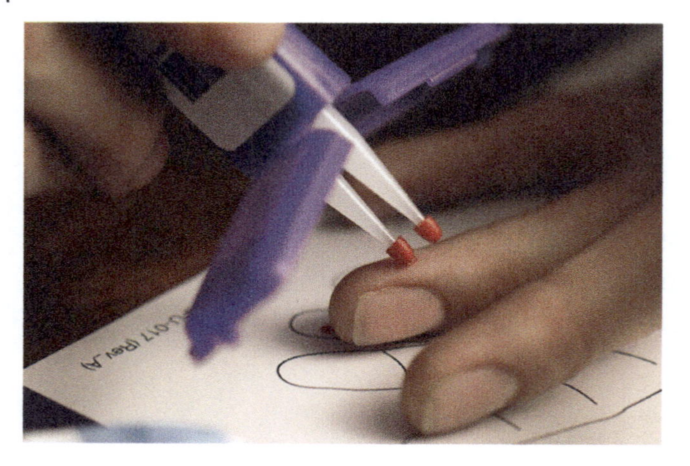

Figure 2.1 Mitra with VAMS microsamplers being used to collect capillary blood from a finger prick.

Spooner et al. (2019), Protti et al. (2019), and Capiau et al. (2019) for guidance. The aim of this chapter is to provide insights and pointers for what to look out for when working with dried specimens—predominantly dried blood—on Mitra devices with VAMS technology, and how these insights help to improve both method development and validation.

2.2 Tip 1—Ensure the Use of a Correct Sampling Procedure to Prevent Volume-Related Biases

Mitra with VAMS samplers are hydrophilic in nature. As shown in Figure 2.1, when placed in contact with a biological fluid from a positive angle (above the blood drop), blood will rapidly (often <2 s) fill the VAMS tips. The tips cannot be overfilled from this angle. However, it is possible to underfill or overfill the tips in other positions, leading to unacceptable biases in analytical results. Overfilling is possible if blood is dropped onto the samplers from above, or if the samplers are fully immersed in the fluid by, for example, dipping the Mitra sampler into a blood tube. It is also advisable to avoid rolling the sampler tips over the blood drop on the finger, which again risks overfilling. Underfilling is seen when the sampler tips are prematurely removed from the fluid, with the tip appearing to remain partially white. If the tip appears to be underfilled, it is fine to reapply the sampler tips to the fluid within 1 min of the first contact event.

The samplers are calibrated to a sampling precision of <5%RSD, which allows for accurate and precise sample collection independent of hematocrit (HCT). This negates the need to precoat internal standard (IS) on the device for liquid chromatography-mass spectrometry (LC-MS)-based assays. The reason for inclusion of an IS (usually a stable isotope labeled form of the analyte) is that it acts to normalize for any analytical or processing errors. One source of error could be inaccurate pipetting of a sample. For example, if the volume pipetted is too low, then without the inclusion of the IS, the results will show a negative bias. However, by taking the area ratio of the analyte and its matched IS, any inaccuracies will be compensated for as it is present in all stages of the analytical process. Any losses, such as those from adsorption of analyte onto walls of well plates, will be normalized by the IS. The nature of the VAMS sampling technique ensures that accurate and precise volumes of blood are collected, so it is sufficient to put the IS in the extractant. The advantage is that this negates the need to produce IS precoated tips, which would be costly and logistically challenging. When good training videos and/or instructions for use (IFUs) are provided, Mitra users routinely report accurate sampling. As an example, a recent paper on extractions to measure methotrexate reported that 825 VAMS capillary blood samples were collected with a 96% accuracy, with only 2% underfilling, and a corresponding 2% overfilling (Brady et al., 2019).

2.3 Tip 2—Working with Wet Whole Blood

For established routine blood tests, when venous blood is collected, either for routine chemistry tests or for measurement of drug concentrations, more often than not, it is not whole blood that is tested but the liquid fraction—serum or plasma. There are, of course, exceptions, such as transplant drugs that partition into red blood cells (RBCs), or hemoglobin and its variants. However, bridging assays from wet plasma or serum onto dried blood VAMS samples raises several considerations when developing assays from whole blood.

2.3.1 Is Your Choice of Assay Biologically Relevant in Blood?

The first thing to consider is the choice of analyte and how it partitions; in other words, where it is found in the blood. A key thought that should be top of mind is whether measuring an analyte in whole blood is clinically relevant or not? Some analytes partition equally between the HCT and plasma portions. In these cases, validations can often be very straightforward. This is explored in a recent paper looking at antiepileptics, where blood to plasma partitioning was examined for each analyte (D'Urso et al., 2019). Also, for a good commentary on blood to plasma partitioning ratio, the reader is encouraged to read work published in 2010 by Rowland and Emmons (2010).

Another consideration is if an analyte is only found in the plasma or serum fraction, then a clinically relevant robust assay can be developed in whole blood, but a negative bias should be expected in the whole blood data. This will be discussed in more detail in Section 2.3.4 on bridging studies. There are, however, examples of analytes that would be impossible to analyze with whole blood and still obtain clinical relevance. The first example is serum potassium. This ion is selectively taken up in cells, including blood cells. For this reason, measurement of serum potassium and not whole blood is clinically relevant as, under normal circumstances, it should be all but absent in serum. The second example is the liver function enzyme aspartate aminotransferase, or AST. AST is found in high concentrations in liver, muscle, heart, kidney, brain, and RBCs, but not in serum or plasma (Burke, 2018). Measuring whole blood would risk a false positive. It must be said, though, that these examples are very much exceptions to the rule. Moreover, with the right analytical technology, a wide range of analyte classes can be successfully developed from whole blood.

2.3.2 Working with Blood as a Matrix

Because blood is a potentially biohazardous matrix, good laboratory practices must be employed when handling it. Moreover, being a wet matrix, blood has limited stability for many analytes, and this is dependent on temperature, analyte, and the type of anticoagulant. For example, Evans et al in 2001 looked at the stability of 19 hormones in the following tubes/conditions: EDTA, lithium heparin, sodium fluoride/potassium oxalate, or tubes without anticoagulant as well as plasma and serum. They measured these at −20, 4, or 30 °C for 24 and 120 hr. Their results showed that not all hormones showed identical stability across all anticoagulants although most of the hormones tested were stable at >120 hr in EDTA and fluoride at 4 °C, but only 13 of these were stable in all anticoagulants. Moreover at 30 °C, as expected, stability was even worse where only three hormones were stable >120 hr in all anticoagulants, although eight were stable in EDTA (Evans et al., 2001).

Most assays have historically been developed for wet serum or plasma, and often the cellular component is not needed for the assay. For this reason, bioanalytical scientists do not typically have to consider the cellular fraction of blood. Blood, however, is a dynamic matrix due to the effect of gravity on the cellular fraction. In other words, the HCT tends to settle without constant mixing. This means that if an analyte is only found in the plasma—such as the metabolized vitamin, 25-hydroxyvitamin D (25(OH)D_3)—once the HCT begins to settle, a time-dependent, dynamic concentration gradient forms. Thus, sampling from the top of a vial with a pipette tip or a Mitra sampler will risk imprecision in the results if care is not taken to ensure the blood is gently mixed throughout the sampling event. For example, slow inversion of the blood tube several times prior to sampling avoiding bubble formation helps to maintain a homogeneous mix. Indeed, the effect of HCT on 25(OH)D_3 was demonstrated by Holly Nicholls and co-workers at the University of East Anglia, United Kingdom (Nichols et al., 2016). During this experiment, non-lysed red cells were added in stages, and it was found that there was a proportional decrease in the analyte concentration. This demonstrated that only trace amounts of the vitamin were found in the cells. Moreover, in unpublished work I conducted (2015) when I was working on developing an assay for drugs of abuse, I noticed that when sampling from the same vial without mixing, the VAMS tips were becoming visibly lighter! This was due to the settling effect of the HCT over time.

Blood, like all proteinaceous matrices, is prone to forming bubbles. This is especially true if the blood has been mixed too vigorously. I have noticed that the HCT levels in bubbles appear to be lower than in the main pool of liquid. It can risk bias in the data if Mitra sampler tips come into contact with bubbles. So, to prevent bubbles from forming, it is important to ensure the blood tubes are mixed, either by tube rollers or by gently inverting them.

2.3.3 Allowing Analytes to Equilibrate *Ex Vivo*

When spiking blood with your analyte for method development and validation, it is important that care is taken to allow analytes to fully equilibrate and partition into the cellular fraction as well as allowing for any binding with plasma proteins. This is less of an issue if calibration lines are developed in whole blood and spiked onto Mitra devices. It is particularly important, though, when plasma is used as a matrix for both calibration and quality control (QC). The reason for this is that if the sample has not fully equilibrated and is allowed to partition then it will not correctly represent the *in vivo* state. This means that biases will be seen in the data if the intracellular concentration has not yet matched the plasma concentration. Furthermore, if a molecule that favors plasma or the cellular component has not equilibrated, then observed biases between plasma and whole blood will not be controlled and will be difficult to correct for. Due to these reasons, it is advisable that calibrators and QCs should match the sample matrix where possible. Finally, often QCs are prepared to coincide with the arrival of samples in the analysis laboratory and are often therefore stored alongside the samples for a period of time. However, calibrants are regularly prepared fresh on the day of analysis. This could give different recovery/stability for calibrators and QCs and should be explored as part of method development.

Both time-dependent and concentration-dependent partitioning needs to be considered and investigated. An excellent case study of the role of RBCs and blood to plasma partitioning was published in 2017 by Saha et al. (2017). They demonstrated clear differences between the blood to plasma partitioning in two related antiviral drugs (Acyclovir and Valacyclovir) over time. In this example, Acyclovir took 30 min to fully equilibrate, whereas Valacyclovir appeared to immediately equilibrate. Therefore, it is advisable to perform such a time course experiment when working with analytes where the blood to plasma partitioning kinetics is not fully understood. Furthermore, analytes may show concentration-dependent partitioning leading to employing a more complicated algorithm to correct for the HCT fraction. This phenomenon was reported on by Chris Baily and colleagues at Astra Zeneca in 2020, and they proposed a mathematical model employing a scaling factor to help compensate for this (Bailey et al., 2020). They are quoted as saying "The model describes the combined influence of B/P and hematocrit on concentration of an analyte in whole blood."

2.3.4 Bridging Between Venous Capillary Blood and the Role of Anticoagulants

One of the key benefits of working with Mitra samplers is that sample acquisition is very straightforward. Capillary sampling with a finger-prick method has been reported to be much preferred over conventional venous draws. In a recent paper that compared samples collected with Mitra samplers to traditional venous draws, it was reported that 82% of patients on immunosuppressive drugs preferred self-collected capillary sampling with Mitra samplers (Mughuni et al., 2020). While it is tempting to move away from venous sampling in favor of the more patient centric capillary sampling method for studies, bridging experiments between venous and capillary blood must be considered. For many studies, the concentrations of analytes in venous and capillary blood are often the same. However, this may not always be the case and should be investigated (Chiou, 1989).

When collecting capillary blood from a finger-prick using a lancet, it is advisable to wipe away the first drop of blood and allow a second drop to form. The main reason for this is the concern that tissue fluid from the first drop will bias the results. This is discussed in Chapter 8 of a recent book called Therapeutic Drug Monitoring Data: A Concise Guide (Dasgupta & Krasowski, 2020). Crawford and workers used this approach when validating an assay for clinical markers comparing plasma to fingerpick blood (Crawford et al., 2020). Care should also be taken not to exacerbate the issue by "milking" the lancet wound by squeezing the finger to encourage blood flow. Under these circumstances, it is better to warm up the hands again by washing with warm soap and water and pricking a different finger with a new lancet.

A further reason the second drop should be used is the risk that any contamination on the surface of the finger might interfere with the results. When collecting blood from babies and very young children, contamination of the collection site is of particular concern, as they often put their hands in their mouths. If an assay is for monitoring orally administered medication, such as a syrup, traces of the medication may remain on the patient's fingers. In short, care should be taken to ensure very young children avoid putting their hands in their mouths before a sampling event. I was involved with a project a few years ago, in which we were investigating orally administered medications in babies. Our blood samples showed unacceptable amounts of the orally administered drug on the babies' hands, even when the hands were cleansed before sampling. The solution was to collect blood from a heel prick, which only works for very young babies. Care also is needed to ensure the baby does not put its heel in its mouth, which is possible!

The final reason the second drop of blood is recommended is anecdotal. It seems that the blood bolus forms better with the second drop compared to the first. It is hypothesized that this might be due to the area from which the blood has been wiped. If the area is more hydrophobic, it allows for the blood to bead more effectively. However, further research is needed to prove this.

An area of concern when working with capillary blood compared to venous blood is the absence of anticoagulant, such as potassium EDTA or lithium heparin in the capillary sample, when collected directly onto the VAMS tip of a Mitra device. The United States Food and Drug Administration (USFDA) guidelines do ask that there is no bias in the data seen with the presence of anticoagulant (or change in the data seen with the presence of anticoagulant) when compared to venous blood samples (USFDA, 2018). The reason for this is that calibrators and QCs are often generated from venous blood containing anticoagulant, and when preparing calibrants and QCs, it is virtually impossible to work with fresh blood without the anticoagulant. However, when it comes to microsampling, the patient samples are often collected as capillary blood. After ensuring there is no biological bias seen between the two sources of blood (see above), there are two strategies that can be used to conduct the bridge between blood with and without anticoagulant.

The first and most popular strategy is to compare two samples of the same blood, but with two different anticoagulants (Spooner et al., 2019). The idea here is that it would be unfortunate if both reagents caused identical bias. A variation of this approach would be to change the concentration of the anticoagulant to see if, by increasing its concentration, it causes bias. The second strategy is to preload anticoagulant on VAMS tips and let them dry so that blood mixes with the anticoagulant on contact. This is akin to the use of EDTA-coated capillaries designed to make EDTA blood when collecting wet capillary blood for blood gas measurements.

The procedure for loading EDTA on VAMS tips is straightforward. First, K_2EDTA compatible solvent, such as 50:50 water/methanol (MeOH; v/v), is selected. A stock solution of K_2EDTA is prepared to match the target concentration in the blood. Sample the solution with a VAMS tip for 6 s and allow it to dry for 2 hr. Even with EDTA coating the tip, the blood-wicking performance of the tip is not affected (Neoteryx, 2017). Thus, a successful bridge can be compared by measuring capillary blood from native tips versus precoated tips.

2.4 Tip 3—Working with Dried Whole Blood

2.4.1 Dried Blood Spot Cards

Microsampling using dried blood dates back to the 1960s when Dr. Robert Guthrie pioneered dried blood spot (DBS) testing for the diagnosis and monitoring of inborn errors of metabolism in newborns, such as phenylketonuria or PKU (Guthrie & Susi, 1963). As well as saving the lives of thousands of newborn babies and improving the lives of many others, Dr. Guthrie left a huge legacy to mankind. He made it feasible to conduct mass remote sampling using dried blood from capillary samples with a heel-stick or finger-stick method, which has now become common practice. Indeed, the International Society of Neonatal Screening (ISNS) has more than 470 members from 70 countries worldwide, with many members running full-time testing programs using DBS cards (ISNS Regions, 2021). Despite its success, the humble DBS card has several inherent drawbacks.

The first drawback is sample quality. Applying a blood sample to a DBS card requires a drop of blood to fall onto the card in a precise location, or a hanging drop of blood to make contact with the card in a precise location. The blood must be able to soak through the card's filter paper, so it can be observed on both sides of the card. If the DBS sample is not properly positioned or does not soak through appropriately, then it risks bias in the analytical results due to under- or oversampling. As a result, many health authorities give guidance on how to sample DBS samples correctly. One such example is the guidance produced by the Montana Department of Public Health and Human Services (Blood Spot Specimen Quality Check, 2017). The Department's guidance shows correctly sampled versus incorrectly sampled blood spots. It also provides detailed descriptions of why and how to improve the sampling technique.

Mitra devices with volumetric absorptive microsampling were created to help overcome some of the challenges presented by DBS cards and to simplify the dried blood collection process. The Mitra device was designed with the patient or research volunteer in mind, enabling anyone to easily self-collect an accurate volume of blood on the device's absorbent VAMS tip without assistance from a healthcare worker. Indeed, in 2017, a comparison was made between DBS and VAMS blood sampling techniques among a cohort of 44 patients who were taking the drug hydroxychloroquine and collecting their own blood samples (Qu et al., 2017). In this study, around 20% of the patients failed to produce enough blood to cover a 6-mm diameter spot on the DBS card, compared to 100% of the patients who were able to use the Mitra devices with VAMS tips to sample correctly. For those trained on using DBS cards routinely, the rates of failure can be improved. However, efforts to overcome the inherent challenges of DBS cards to help improve sampling technique and sample quality have yielded some ingenious innovations. These innovations include combining capillary sampling with DBS, lateral flow solutions, and volumetric absorptive microsampling (Nakadi et al., 2020; Protti et al., 2020b).

Other issues observed when working with DBS cards are those associated with quantitation. It has been reported that poor quantitation has been observed when DBS cards have been "squashed in the post." Also, some researchers have reported that the position of a subpunch on a DBS card can govern the size of the analytical signal, which indicates that there is, for some analytes, a spot geography concentration effect (Cobb et al., 2013; George & Moat, 2016; Lawson et al., 2016).

2.4.2 Volumetric Hematocrit Bias—Blood Viscosity

The final but most discussed issue associated with DBS cards is volumetric HCT bias. One has to be careful when considering bias pertaining to differences in packed cell volume or HCT. As discussed in Tip 2, natural bias is always going to be observed when comparing blood to plasma when the ratio between the two matrices is not 1:1. Biases can also be observed from contributions of the matrix to the ion signal, especially observed when using mass spectrometric detection. Moreover, extraction biases have been observed with increasing HCT, and with regard to the age of the matrix, so attention should be made to optimize extraction methodology to mitigate for these phenomena (Xie et al., 2018). However, the volumetric bias that many are more familiar with is due to sample viscosity.

Simply put, cellular components in blood get trapped in fibers of the DBS card, impeding the lateral migration of the blood sample on the card. The higher the percentage of the HCT, the lesser the lateral flow. This means that if subpunching is used to obtain a "fixed volume" sample, those samples with a greater percentage of HCT levels will spread out less and more enriched subpunches are possible. Indeed, this was elegantly demonstrated by Denniff and Spooner in 2010 when a greater than 40% bias was observed across the mean of a range of HCT blood samples (20–80%) when measuring acetaminophen on a Whatman FTA Elute paper (Denniff & Spooner, 2010). Issues with quality and quantitation have driven innovation in dried matrix microsampling. Volumetric absorptive microsampling addresses both DBS sample quality and solves the HCT bias caused by blood viscosity, because this method collects the whole sample. However, dried blood is a complicated matrix and must be fully understood and tested when developing robust assays (Spooner et al., 2015).

2.4.3 Dried Blood is a Complicated Matrix

There is a lot to consider when embarking on developing assays using dried blood and; although VAMS solves issues associated with blood spot migration, there are other biases to consider beyond this and blood to plasma partitioning (as mentioned in Tip 3). The best approach when working with dried blood is to view it as an uncontrolled mixed-mode solid-phase extraction (SPE) sorbent, where every intermolecular integration is theoretically possible. Dried blood changes over time and is a moving bed of interactions, so to speak. What typically seems to happen is that molecules become increasingly difficult to extract over time, so harsher extraction conditions are required. This was observed in a study by Iris Xie et al. (2018) when developing a method on a proprietary drug. This brings into focus the importance of optimizing extraction efficiency, which will be covered in Tip 4.

It seems that a lot of changes in dried blood occur in the first day of drying due to oxidative effects. Work on reflectance spectroscopy of dried blood stains, aimed at aging or establishing a timeline at crime scenes for forensic investigations, has demonstrated that oxyhemoglobin changes to methemoglobin, and then a dimer of hemoglobin, called hemichrome. Rolf Bremmer et al found, in their 2011 study on Biphasic Oxidation of Oxy-Hemoglobin in Bloodstains (Bremmer et al., 2011), that most of these changes in blood occur in the first day or so of drying. Indeed, Christophe Stove's research group at the University of Ghent has used the same technique to predict HCT levels in DBS samples (Capiau et al., 2016). Stove et al observed similar trends with just less than half of the hemoglobin converting to hemichromes in 24 hr. Certainly, oxidative changes to dried hemoglobin have been linked to changes in relative retention times of HbA1c peaks to hemoglobin using a Toso Haas hemoglobin analyzer (Neoteryx, 2020a).

2.4.4 Working with "Aged" Blood

Due to the changes observed with dried blood and its effect on the extractability of many analytes, care must be taken to develop extraction protocols on "aged" dried blood irrespective of the dried matrix. Ideally, it is best to validate the assay on tips that have been stored for periods of time and reflect when a tip has been sampled and when it has been analyzed during a study. Mitra devices and VAMS tips are normally stored at room temperature (dependent on the stability of the analyte in question). For many assays, storage times would be only a few days or 2 weeks at most. Often, there is an urgency around reporting data in the shortest time frame possible for clinical reasons or due to project deadlines.

Thankfully, there are many examples in the literature which show that, with the right extraction procedure, analytes can extract with high recovery showing little to no temporal bias (a change in assay bias over time) after storage. An example of this is where a panel of 16 antiepileptic drugs showed acceptable temporal bias at −20, 4, 10, and 37 °C for at least 10 days (D'Urso et al., 2019). Furthermore, as an extreme example, the benzodiazepine Midazolam has shown no significant temporal bias, even after being stored for 131 days at ambient temperature (Abu-Rabie et al., 2019)! Finally, to accelerate the aging of tips at elevated temperatures such as 40–60 °C, Christophe Stove and co-workers evaluated the stability of first-generation anti-epileptic drugs on VAMS tips. They found that all compounds studied were stable for at least 1 month when stored at room temperature and, interestingly, for at least 1 week when stored at 60 °C (Velghe & Stove, 2018).

If sampled VAMS tips are to be stored longer than the time taken in validation, it is advisable to store tips sealed in a foil specimen bag with desiccant in the freezer, preferably at −80 °C. A key point to mention is to allow tips to equilibrate to ambient temperature before opening the packs to prevent buildup of condensation. An advantage of storing at −80 °C is that, when analytes are unstable in dried blood, freezing at −80 °C confers excellent long-term stability. This was demonstrated by a team at the European Academy of Bozen-Bolzano (EURAC Research) center where researchers looked at the stability of analytes in an untargeted metabolomics study (Volani et al., 2017). It must be noted that, irrespective of the storage conditions, stability experiments must always cover the maximum time the samples are stored. However, this depends on the expected storage time. For example, some biobanks may store samples for years, even decades, so checking sample integrity during the storage time would be desired. However, this could be problematic to achieve, as freeze– thawing cycles may accelerate compound degradation.

By and large, many molecules show excellent stability at room temperature in dried blood samples, especially drugs, hormones, oligonucleotides, peptides, and proteins. A good summary of stability of a number of analytes is included in an excellent review paper by Michele Protti from the University of Bologna (Protti et al., 2019). This list is slightly out of date, but it provides a good example of what is possible.

2.4.5 Temporal Extraction Bias or Degradation?

When embarking on developing and validating an assay on dried blood and a negative temporal bias is being observed, how does one ascertain the difference between compound stability and extractability? One solution is to measure the samples spectroscopically, typically using a mass spectrometer in a full scanning mode. This allows for a range of ions to be measured simultaneously, rather than measuring specific ions. This is what is typically done to give good sensitivity, specificity, and quantitation for an assay (Introduction to MS Quantitation and Modes of LC/MS Monitoring, 2016). This will help to give a temporal snapshot of any compound degradation. When running in whole scan mode, if new peaks are forming over time as the peak of interest is decreasing, this means there is evidence of degradation. If the identity of metabolites and/or degradants is known, then the appropriate SRM transitions can be monitored. Another approach is to perform precursor ion scanning using a product ion that is typical for the analyte and its metabolites/degradants. However, if the mode of analyte detection does not allow for investigation, such as for example enzyme-linked immunosorbent assay (ELISA), then a more indirect method can be used to ascertain if extractability or degradation is causing negative bias, as discussed toward the end of this section. As mentioned above, blood can be viewed as a complicated mixed-mode SPE sorbent. It transpires that for some analytes, the greater the percentage of HCT, the greater the chance that more extreme extraction conditions will be needed to boost absolute recovery (Kushon et al., 2014). This is covered in more detail in Section 2.5.

Figure 2.2 shows that when the analgesic drug acetaminophen is spiked into plasma and blood of increasing HCT levels (20, 45, and 65%), absolute recovery

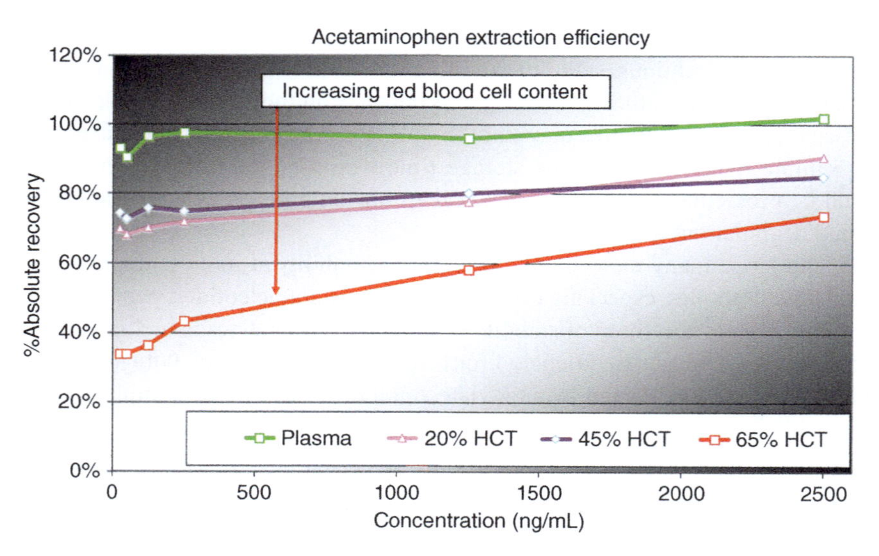

Figure 2.2 Effect of percentage hematocrit (HCT) and analyte concentration on the percentage absolute recovery of acetaminophen from Mitra VAMS tips. Adapted from Kushon et al. (2014).

percentage reduces (Kushon et al., 2014). This has nothing to do with blood to plasma ratio partitioning, because the same amount of drug was spiked into all tubes at equimolar concentrations. Instead, this is because analytes overall become progressively more difficult to extract from the sample and, thus, require harsher extraction conditions. As seen with other analytes, the lower the concentration, the more difficult it is to extract from the dried blood. This gives us tremendous insight into what may be causing this phenomenon. Again, taking the mixed-mode SPE model, it seems there are a limited number of high-affinity sites for analytes to interact with. One such interaction that has been observed is ion exchange. This will be covered in the next Section 2.5 on optimizing extraction methodology.

Ionic interactions are the strongest intermolecular forces, so care is required to ensure high-extraction yields are possible (Intermolecular Forces, 2021). With this in mind, there may be a limited number of ion exchange sites in the dried blood that analytes can access, so analytes preferentially bind to these first. As these sites become saturated, analytes then interact with more abundant, but weaker intermolecular forces, such as Van der Waals. The higher the concentration of analyte in the blood, the greater the fraction of these molecules that interact through weak intermolecular forces. Therefore, the higher the concentration of analyte, the easier it is to extract from in percentage terms. Undoubtedly, the model is a lot more complicated than this, but it acts as a practical explanation and provides a potential approach to improving extraction yields.

Equipped with the understanding that the percentage HCT and analyte concentration have an effect on extraction yield, this can help determine, indirectly, if temporal negative bias is caused by issues with either extraction or compound degradation. The theory is that if a compound is degrading, then this will be seen across all percentage HCT samples equally. However, if extraction bias is the cause, then there will be a greater effect at higher HCT levels which would also be exacerbated at lower analyte concentration, which certainly makes logical sense. This phenomenon is covered in a recent publication on validating methods on dried blood microsamples for therapeutic drug monitoring (Capiau et al., 2019).

2.5 Tip 4—Optimizing Extraction Efficiencies from VAMS

So far, this chapter has discussed how the volumetric nature of VAMS, good sampling technique, bridging between capillary and venous blood, and an understanding of blood to plasma partitioning ratio, all play into developing robust and validated assays. However, we have also learned that not optimizing the extraction conditions can cause validations to fail because of differences in percentage HCT, sample concentration, and time spent in the dried blood matrix. Given these different parameters, it is advisable to optimize absolute extraction efficiencies as high as possible (>80% is recommended) such that any bias in the data caused by changes in the HCT is overcome. An excellent example of this was published by Paul Abu-Rabie and colleagues in Analytical Chemistry (Abu-Rabie et al., 2015). When comparing mean percentage recovery of four drugs from dried blood on DBS cards, drugs (Midazolam and Naproxen) that showed a high-extraction efficiency (>80% at 45% HCT) showed no recovery bias across a range of HCT levels (25, 35, 45, 55, and 65%). Conversely, drugs that extracted with extraction efficiency (<80%) showed a HCT-related extraction bias (see Figures 2.3 and 2.4).

2.5.1 Measuring Percentage Recovery from Mitra Samplers

To measure percentage recovery from the sampled VAMS tips on Mitra devices, or any dried blood sample for that matter, blood at 45% HCT is spiked with the analyte and aliquoted onto a minimum of six tips, dried and extracted with solvent, or buffer containing the IS (condition A). Blank blood (ideally from the same source) is aliquoted onto another six tips, dried and extracted with solvent or buffer containing the IS in the same way (condition B). The analyte standards are then spiked into the extractant at a final concentration, which represents a 100%

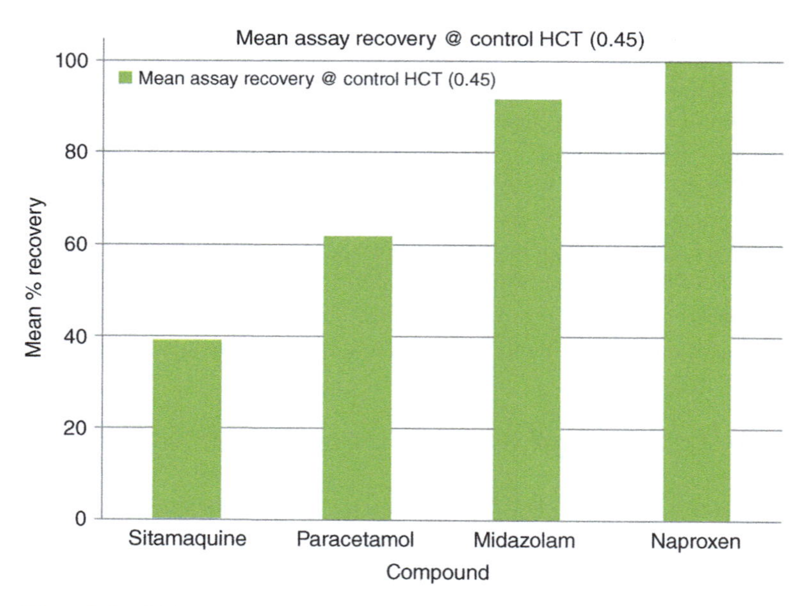

Figure 2.3 Differences in percentage extraction recovery of four drugs from dried blood spot (DBS) samples under the same extraction conditions at a hematocrit of 45 (figure shown by author's permission Paul Abu-Rabie (GSK). Adapted from Abu-Rabie et al. (2015).

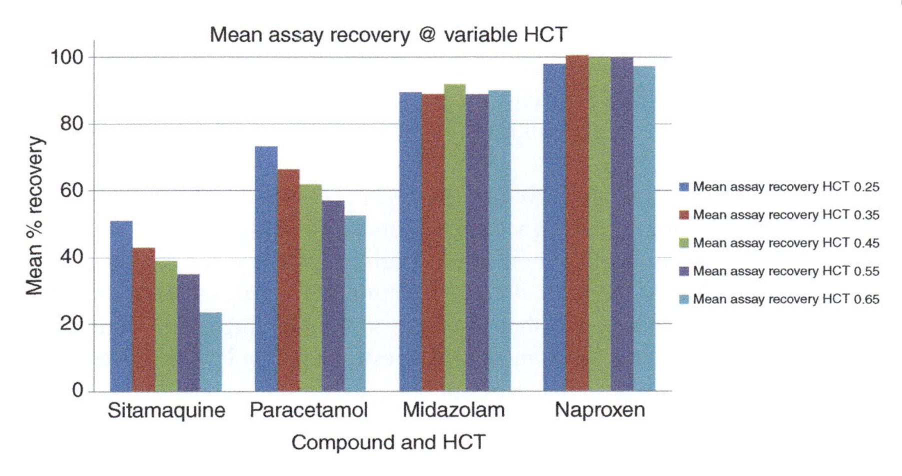

Figure 2.4 Effect of percentage blood hematocrit (HCT) on mean percentage recovery of four drugs extracted from dried blood spot (DBS) samples under the same conditions (figure shown by the author's permission Paul Abu-Rabie (GSK). Adapted from Abu-Rabie et al. (2015).

extraction recovery. Finally, to get a percent recovery, analytical results of A are divided by analytical results of B multiplied by 100.

It is advisable that this experiment is carried out across the HCT range, which the test is intended for (e.g. 25, 45, and 65% HCT). Furthermore, across a range of concentrations (i.e. high QC, medium QC, and Low QC) and under conditions that mimic assumed shipping conditions (e.g. −20, ambient and +40 °C sealed in desiccant packs, stored for a week under each condition). This is a lot of work, especially when trying out different extraction methodologies to see which is most optimal. So, a good compromise when scoping out different extraction conditions is to work on tips that have been dried for at least a full day (1 week is preferable), using >45% HCT blood and working with low concentration standards such as Low QC or five times lower limit of quantitation (LLOQ). The goal here, by using these more challenging conditions, is to aim for >80% extraction recovery. This will minimize the chances of the method failing validation due to unacceptable bias (usually negative bias) in the data.

2.5.2 Extraction Conditions—Where to Start

2.5.2.1 Consulting the Literature and Matching Physicochemical Properties

At the time of writing, there are more than 240 peer-reviewed manuscripts using VAMS and many discussing successful analytical validations when using Mitra with VAMS devices (Neoteryx, 2020b). These manuscripts provide a plethora of extraction methods. One approach to embarking on method development is to consult the literature. If there is not a direct match to the analyte of interest, a good approach is to try and match the structure/physical chemical properties of the analyte(s) of interest to a publication where a similar analyte and analytical equipment was used.

For example, recently there have been several papers on steroid hormones and drugs that can act as a guide to good initial conditions if the intended analyte(s) for assay development are based on a cholesterol backbone (Fragala et al., 2018; Gwenaël et al., 2017; Marshall et al., 2020a; Schmidt et al., 2019). Given that molecules with a steroid backbone show medium-to-high hydrophobicity and there are a lot of resources that report measured or estimated Log *P* values, then these can act as a guide to help compare hydrophobicities to similar analytes reported in the literature (Shoshtari et al., 2008). Therefore, if analytes match the hydrophobicity of those reported, then these may also extract well under the conditions validated from the published methods on VAMS for similar analytes. Please note that Log *P* values act only as a rough guide and must also be considered with other physical or chemical properties, such as pK_a and Log *D*.

2.5.2.2 Adapting Published DBS Methods

Many of the assays developed on DBS cards can often be easily adapted to work on Mitra with VAMS devices, too. The reason is that the dried blood consistency/chemistry should be broadly similar on a DBS compared to a VAMS device, unless the DBS had been treated to change its properties. Due to the longevity of the DBS technique, there are many excellent papers and reviews on DBS cards that can be used as good starting points for new assay development. For example, two reviews of note in the clinical chemistry field are those by Keevil (focusing on LC-MS applications) (Keevil, 2011) and Lehmann, Sylvain and Delaby (focusing on immunoassay) (Lehmann et al., 2013).

If a method is being adapted from DBS samples onto a Mitra device with a VAMS tip, then to keep extraction conditions the same, it is good to match the disk diameter with a suitable tip volume. If this is not possible, then knowing the difference in the two device volumes allows for an adjustment of extraction volume. As discussed in Tip 3, the actual volume of blood obtained from a DBS subpunch is dependent on the viscosity of the blood. However, as an estimate, and knowing that disk volume has a square relationship to the disk diameter, 3-mm subpunches hold around 3 μL of blood, 6-mm subpunches hold around 12 μL of blood, and so on.

2.5.2.3 Converting from a Wet (Whole Blood or Plasma) Method

When a dried method has not been published on an analyte of interest, it is also useful to consult the literature on wet matrices such as whole blood and even plasma or serum. For wet whole blood, conversion to dried methods can be very straightforward indeed. For example, measurement of the immunosuppressive drug tacrolimus is measured in whole blood due to a high accumulation of the drug into the HCT. As a result, more reliable values are obtained from extracting intact blood rather than serum or plasma—where blood concentrations would be underreported in these matrices. Indeed, David Marshall and co-workers demonstrated that tacrolimus can be reliably measured from VAMS extracts using similar extraction conditions to the standard wet method (Marshall et al., 2020b). However, care must be taken to evaluate analyte extractability and stability as well as analytical sensitivity when converting from wet blood methods.

Converting methods from wet plasma or serum is more complicated due to blood to plasma partitioning ratios and if measuring analytes in whole blood versus plasma or serum is physiologically relevant (Rowland and Emmons, 2010). These scenarios are further discussed in Section 2.3. Again, as above, analyte extractability and stability as well as analytical sensitivity must also be considered. Detector compatibility must also be considered. As well as sensitivity, the detection method may require clear samples to allow for meaningful results. For example, colorimetric methods measuring enzymatic kinetics require conversion of a substrate into a pigmented product. Such assays are commonly found in clinical chemistry analyzers. Adding a fully hemolyzed dried blood extract would lead to high analytical backgrounds, making data interpretation difficult if not impossible.

One final point is that if reliable data obtained from whole blood makes analytical and physiological sense, this may be different to data obtained from plasma or serum. It must be noted that this data may not be necessarily inaccurate just simply different and so would require full analytical and clinical validation setting new reference ranges for the new matrix. For many analytes, industry acceptance would also be required.

2.5.2.4 Starting from a Blank Canvas—What to Consider?

One of the questions that is often asked is whether there is a generic but effective extraction procedure that will allow for a straightforward extraction of a range of analytes, using common solvents and equipment. The answer is dependent on the vastly different bioanalytical fields in which Mitra with VAMS devices can be employed. Having a one-size-fits-all extraction procedure is probably too difficult to achieve. However, there are some promising generic methods emerging, which are discussed at the end of this Section 2.5.5. When developing extractions from Mitra, there are several considerations that need to be addressed before choosing the right conditions, including:

- Choice of analyte/analyte panel
- Compatibility of analyte and extraction solvent

- Choice of matrix (e.g. dried blood, plasma, urine, cerebrospinal fluid (CSF), etc.)
- Choice of analytical equipment

2.5.2.4.1 Choice of Analyte or Panel of Analytes Most of the time choosing what to measure is not a choice, but the actual reason for developing an assay in the first place. An example of such an assay is measurement of renal function and antirejection drug trough levels to remotely monitor transplant patients that are isolating safely at home (Marshall et al., 2020b). However, there are occasions when choice of analyte is necessary to prove an emerging technology, such as using VAMS for direct spray MS—a related technique to paper spray (Hecht et al., 2017; Morato et al., 2019). Other key factors to consider are analyte stability when dry and analyte volatility. If a molecule is volatile (e.g. ethanol), then losses will be seen on drying. In this case, dried matrix storage would be not advisable.

2.5.2.4.2 Compatibility of Analyte and Extraction Solvent Analyte compatibility is key when determining the correct choice of solvent and when embarking on method development. There are three main considerations: analyte stability, detector compatibility for a specific solvent, and solvent toxicity.

In terms of analyte/solvent compatibility, it would be unadvisable to use anything other than aqueous extractions when measuring intact proteins. Use of organic solvents would risk denaturing the proteins. Moreover, organic extractions would also interfere with the performance of the immunoassay that was employed. However, in the case of extracting small molecules like drugs, organic extractions intended for LC/MS-MS might be the perfect injection solvent. Care must be taken so that the solvent is not too strong for the chromatography which will result in poor peak shapes. If this is the case, then sample dilution into weaker solvent prior to injection or evaporation and reconstitution into a more compatible injection solvent is recommended. Such an approach was discussed by Tanna et al in reporting a successful validation of cardiovascular drugs. The extraction solvent was MeOH, but water was added before injection of the sample to prevent solvent mismatch in this reversed-phase mobile phase (MeOH:water 40:60 v/v with 0.1% formic acid) (Tanna et al., 2018).

Injection solvent mobile phase mismatch can occur when the injection solvent is stronger than the initial mobile phase conditions leading to the analyte initially not interacting with the stationary phase until the mobile phase has mixed with the injection slug. Diluting the sample in a weaker solvent certainly acts to improve peak shapes by allowing the analytes to interact appropriately with the stationary phase, but the act of dilution can negatively affect sensitivity. One solution to this is to inject a greater volume. If the injection solvent is weaker than the initial mobile phase conditions, then the injection slug will be more likely to preconcentrate on the column and elute with better peak shape.

Solvent pH should also be a consideration in terms of assay compatibility. As discussed in Section 2.5.4, pH can have a dramatic effect on overall extraction efficiency. Nevertheless, some compounds are stable at acidic pHs, but labile under basic conditions, and vice versa.

Some detectors are incompatible with certain solvents. For example, injecting organic solvents into immunoassay systems has the potential to denature reagents in the immunoassay, such as capture antibodies. Moreover, use of MeOH in the injection solvent for hydrophobic interaction liquid (HILIC) chromatography methods would be deleterious for peak shape and retention times.

Some solvents are more toxic than others. Water is considered to be fairly safe, but organic solvents and mobile phase additives like tetrahydrofuran (THF) require adherence to Control of Substances Hazardous to Health (COSHH) or equivalent national regulations (Control of Substances Hazardous to Health, 2002). A risk assessment must always be conducted when dealing with potentially toxic chemicals before embarking on any experimentation, and the reader is advised to follow their institution's health and safety protocols and procedures. Furthermore, certain solvents (i.e. THF) can damage polymers used in LC-MS systems.

2.5.2.5 Choice of Matrix

When it comes to dried matrix assays, most are conducted on blood due to the ease of collection and variety of analytes that can be measured from blood. As discussed in Tip 2—working with wet whole blood, there are some analytes that cannot be meaningfully measured from whole dried blood, so other fluids should be considered. Indeed, matrices such as urine, CSF, and saliva are attractive alternative options for Mitra with VAMS specimen collection and analysis (Delahaye et al., 2018; Marasca et al., 2020; Mercolini et al., 2016; Protti et al., 2020a). One of the issues in working with less viscous samples (such as urine) on DBS cards is that the wet matrix often spreads out with too large diameter to be useful for extraction. VAMS solves this by consistently collecting the same sample volume, regardless of which biological fluid is being sampled. However, it must be noted that there are slight differences in volume collected due to differences in the surface tension of the fluid. Indeed, it is for this reason that on certificate of analyses, the Neoteryx microsampling team at Trajan Scientific and Medical measures average water absorbed across a batch of VAMS tips but then converts this value to a blood volume. The volume in blood is always a little less than water due to the differences in surface tension of the two matrices.

2.5.2.5.1 Choice of Analytical Equipment The final consideration when embarking on a new extraction methodology is the choice of instrumentation. There are primarily two classes of techniques that are compatible with analyzing dried matrix microsamples. The first is chromatography hyphenated with mass spectrometry, where LC-MS/MS predominates, and the second technique is immunoassay. Both these techniques offer the sensitivity and specificity to measure analytes reliably from microsamples. However, it must be noted that various techniques in the molecular diagnostic space are also compatible. An example of a successful implementation of next-generation sequencing from VAMS extracts can be seen from Pirritano et al. (2018).

When considering LC-MS/MS, the validation steps for developing a robust assay from whole blood or plasma samples must also be applied to those from dried blood. Key things to note are (a) that the injection solvent is compatible with the instrument used so that the chromatographic separation is conserved and ions are not suppressed, and (b) that the injection solvent does not interfere or damage the MS detector. Furthermore, care must be taken to explore any matrix-related ion suppression or enhancement that could bias the results. A comprehensive open-access review on the topic was published in the journal, *Therapeutic Drug Monitoring*, by Rani George and colleagues in 2018, which the reader is encouraged to consult (George et al., 2018).

When considering immunoassay, organic solvents should be avoided and care must be taken to understand the effect of dilution steps in the assay. For many immunoassay measurements that have been optimized for serum or plasma, the instrument often injects the neat liquid matrix to begin the process. Aqueous extracts from VAMS, however, are prediluted typically for at least 10 times. If the analyte is found only in the plasma portion of blood, then the expected negative bias in the results would be up to 20 times. Dependent on the method and instrument, this level of dilution could be too much for the sensitivity of the instrument. If this is the case, then care must be taken to minimize dilution steps in the immunoassay method. One solution that has worked for ELISA is to add the VAMS tip directly to the ELISA plate and use the first reagent to desorb the blood from the tips directly in the plate (Wharton et al., 2017). Finally, aqueous blood extracts are more complex in nature and have the potential to interfere with the sensitivity and specificity of an immunoassay. Therefore, care should be taken to ensure enough wash steps are employed to avoid any unwanted side effects.

2.5.3 Aqueous Extraction Conditions

The simplest extraction medium is water. Within minutes of adding a dried Mitra sampler to water, blood will begin to extract off the VAMS tip. Complete extraction depends on the analyte of interest, extraction duration, and whether vortexing and/or sonication is employed. A minimum of 20 min of vortexing would be recommended as a starting point, but vortexing overnight is recommended for certain analytes (Wang et al., 2019; Wharton et al., 2017). Like organic extractions, the minimum recommended volume to extract into is 100 μL of extract for every

10 μL of blood. Others will then use buffers such as phosphate-buffered saline (PBS), surfactants such as tween, and even sodium azide for prolonged extractions (Bloem et al., 2018). If immunoassay is the intended analytical technique, then the blood buffer solution is often all that is needed to prepare the sample of analysis. However, if the intended detection method is LC-MS/MS, then the sample aqueous extract needs to be further processed by employing protein precipitation or liquid/liquid extraction (Keevil, 2011; Schmidt et al., 2019).

2.5.4 Organic Extraction Conditions

When dried blood on a Mitra sampler's VAMS tip comes into contact with water miscible solvents, such as MeOH or acetonitrile (ACN), proteins in the blood precipitate and become trapped on the tip. Small molecules, such as drugs and small biomarkers, begin to elute. Indeed, the extraction is so clean, sometimes no further sample preparation is required before being injected onto an LC-MS (Neoteryx, 2016b). If water is needed to aid the extraction, then unwanted cellular components can co-extract. This necessitates further sample clean-up, such as filtration, centrifugation, or even SPE. Typically, ACN and MeOH are the solvents of choice for organic extraction methods. It is always advisable to screen different solvents to obtain optimal extraction efficiency. Figure 2.5 shows how

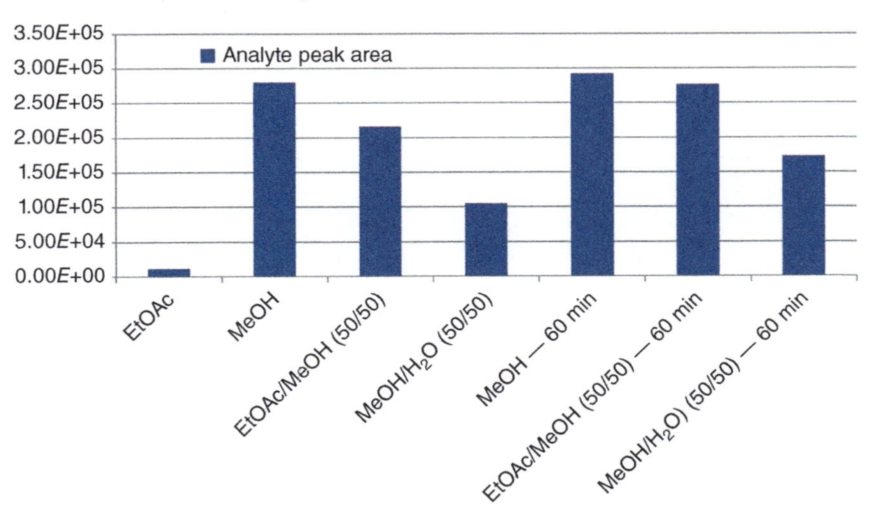

Figure 2.5 Screening the right solvents for maximizing extraction efficiencies of organic extractions from dried blood on Mitra devices with VAMS tips (Neoteryx, 2016a). EtOAc, ethyl acetate; H_2O, water; MeOH, methanol. Courtesy of SciAnalytical Strategies.

different solvents and extraction conditions affect the comparative peak area of acetaminophen extracted from Mitra tips and analyzed by LC-MS/MS.

The pH of the extraction solvent or buffer can also influence the percentage recovery of the analyte of interest from the sampler tips. Using the analogy that dried blood acts like a mixed-mode SPE sorbent, pH can be used to suppress either ionic interactions from a surface or from the analyte. Indeed, using formic acid in organic extractions can act to suppress weak acid groups on analytes, and this has been shown to significantly improve extraction efficiency. The same is also true with suppressible basic moieties on analytes, when using additives which are basic in nature.

Figure 2.6 shows that the addition of acid can significantly improve the percentage recovery of Naproxen by suppressing the carboxylic acid on the molecule. Suppressions of the acid will make the analyte more hydrophobic, but the

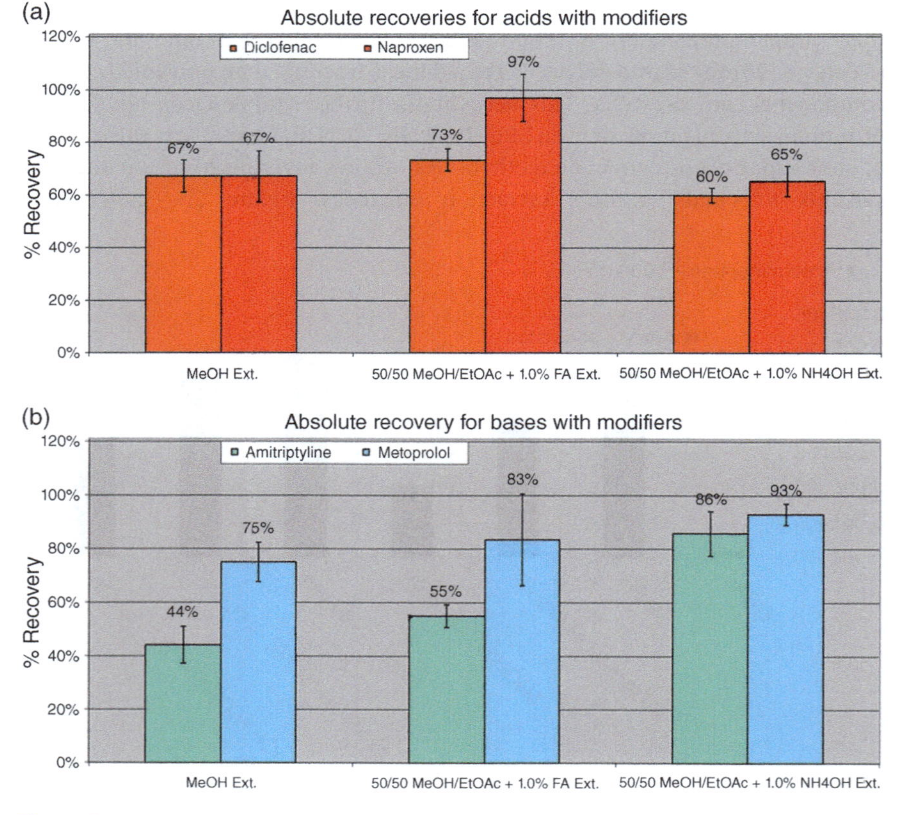

Figure 2.6 Using pH to suppress (a) acid (Diclofenac [pK_a 4.15] and Naproxen [pk_a 4.15]) and (b) basic moieties (Amitriptyline [pK_a 9.7] and Metropolol [pK_a 9.4]) on analytes to aid percentage recoveries.

addition of ethyl acetate helps to break any Van der Waals forces to aid extraction. The same trend can be seen when ammonium hydroxide is used to suppress the tertiary ammonium moiety on amitriptyline.

2.5.5 Generic Extraction Conditions

The Holy Grail for any technique is whether there is a one-size-fits-all approach to extractions that can overcome any percentage or temporal extraction bias. Indeed, there are two promising approaches to this.

The first approach is impact-assisted extraction, where VAMS tips are detached from their Mitra sampler bodies and added to the wells of 96-well plate tissue homogenization equipment (Youhnovski et al., 2017). The vigorous shaking, plus the addition of ball bearings to the wells, acts to add frictional forces to aid with extraction. The tips remain largely intact. However, if organic solvents have been employed, centrifugation of the extract is recommended, as precipitated cellular debris is often observed. This method is highly effective and worth consideration.

Although better extractions may be possible for some analytes using a leave-the-tip-behind approach, one disadvantage of removing the VAMS tip from the Mitra sampler body is that chain of custody for the sample can be lost if barcodes are used on the samplers. Furthermore, by racking up the samplers (tips attached to sampler bodies) in 96-Autoracks™ (see Figure 2.7), 96 samplers can be extracted simultaneously which leads to more efficient extractions from a time perspective and allows for plate automation using liquid-handling robots.

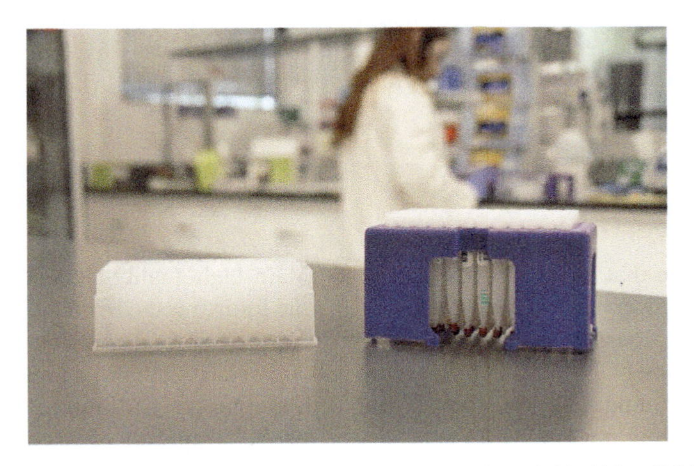

Figure 2.7 96-Autorack™ used for lab processing of dried blood VAMS samples both manually and for automation.

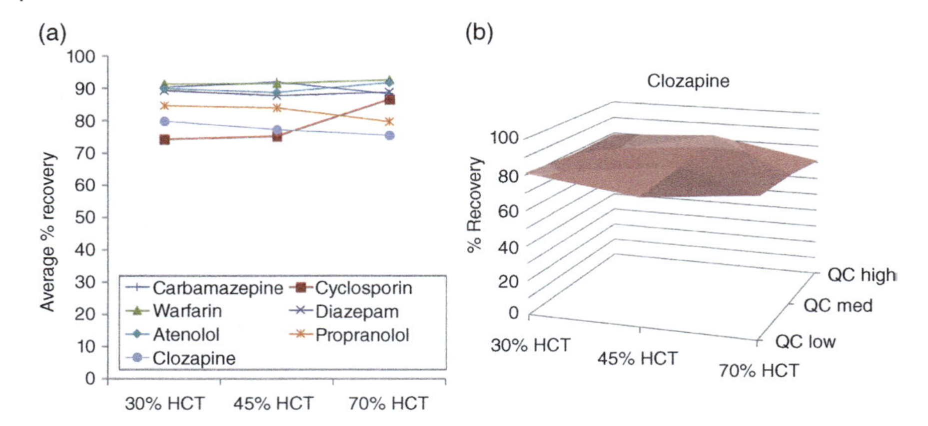

Figure 2.8 Optimized extraction recovery of probe compounds from whole blood dried onto Mitra devices with VAMS tips and extracted after 8 days. The effect of hematocrit (HCT) on average recovery across different analytes (a) and the impact of HCT and concentration on recovery of clozapine (b), % Recovery = Peak area ratio extracted sample/Peak area ratio post spiked extracted blank matrix × 100. Lucey et al. (2018)/with permission of LGC Limited.

The second approach to a generic extraction is to screen and optimize a set of extraction conditions, which covers a wide range of physiochemical properties. Promising work in this area was conducted by Richard Lucey and colleagues at LGC Fordham, United Kingdom. They screened a huge amount of extraction conditions to arrive at a fully optimized set of conditions for a drug panel (Atenolol, Diazepam, Warfarin, Carbamazepine, Propranolol, Clozapine, and Cyclosporin) chosen to represent differing physiochemical space properties. This yielded high recoveries and acceptable recovery bias. The optimized conditions were 5% NH_3 (25:75 ACN:H_2O) mixed at room temperature for 1 hr (see Figure 2.8) (Lucey et al., 2018).

2.6 Conclusions

VAMS is a technique that is enabling new ways to obtain high-quality clinical blood samples and deliver new assays for current tests and future needs. To obtain results that pass analytical validation requirements, there are several points to consider. These include everything from determining how to collect a high-quality sample to understanding the importance of blood to plasma partitioning. Furthermore, it is important to have an appreciation for the complexity of dried blood as a matrix and how this affects extraction of analytes based on percentage HCT and sample age. In terms of extractions, both aqueous and organic extraction

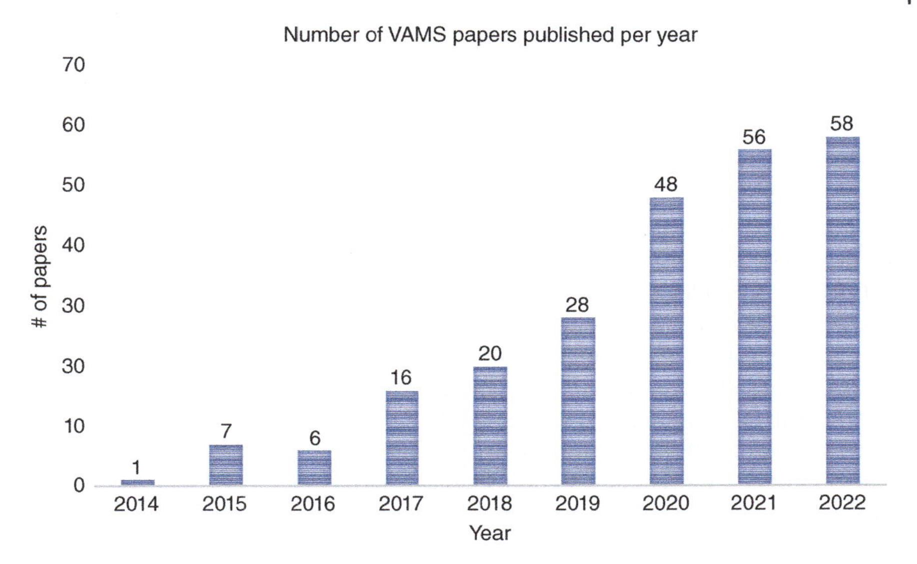

Figure 2.9 Rate of VAMS publications over time (data compiled from various sources including an in-house publication database).

approaches can work equally well. However, it depends on the analyte, matrix, and intended detection technique. Since 2014, more than 240 peer-reviewed manuscripts have been published (see Figure 2.9), exploring assays developed using Mitra devices and the VAMS technique. Indeed, this technique has proven to be a versatile platform for developing next-generation clinical and biological assays.

References

Abu-Rabie, P., Denniff, P., Spooner, N., Chowdhry BZ, Pullen FS. (2015). Investigation of different approaches to incorporating internal standard in DBS quantitative bioanalytical workflows and their effect on nullifying hematocrit-based assay bias. *Analytical Chemistry*, *87*(9), 4996–5003. https://doi.org/10.1021/acs.analchem.5b00908.

Abu-Rabie, P., Neupane, B., Spooner, N., Rudge J, Denniff P, Mulla H, Pandya H. (2019). Validation of methods for determining paediatric midazolam using wet whole blood and volumetric absorptive microsampling. *Bioanalysis*, *11*(19), 1737–1754. https://doi.org/10.4155/bio-2019-0190.

Bailey, C., Arfvidsson C., Woodford L, de Kock M. (2020). Giving patients choices: AstraZeneca's evolving approach to patient-centric sampling. *Bioanalysis*, *12*(13), 957–970. https://doi.org/10.4155/bio-2020-0105.

Bloem, K., Schaap, T., Boshuizen, R., Kneepkens EL, Wolbink GJ, de Vries A, Rispens, T. (2018). Capillary blood microsampling to determine serum biopharmaceutical concentration: Mitra® microsampler vs dried blood spot. *Bioanalysis, 10*(11). https://doi.org/10.4155/bio-2018-0010.

Blood Spot Specimen Quality Check. (2017). Montana newborn screening: bloodspot specimen quality check. Retrieved May 13, 2021, from https://www.babysfirsttest.org/newborn-screening/resources/blood-spot-specimen-quality-check.

Brady, K., Qu, Y., Stimson D, Apilado R, Vezza Alexander R, Reddy S, Chitkara P, Conklin J, O'Malley T, Ibarra C, Dervieux, T. (2019). Transition of methotrexate polyglutamate drug monitoring assay from venipuncture to capillary blood-based collection method in rheumatic diseases. *Journal of Applied Laboratory Medicine, 4*(1), 40–49. https://doi.org/10.1373/jalm.2018.027730.

Bremmer, R. H., de Bruin, D. M., de Joode, M., Buma WJ, van Leeuwen TG, Aalders MCG. (2011). Biphasic oxidation of oxy-hemoglobin in bloodstains. *PLoS ONE, 6*(7), e21845. https://doi.org/10.1371/journal.pone.0021845.

Burke, D. (2018). Aspartate aminotransferase (AST) test. Healthline. Retrieved May 13, 2021, from https://www.healthline.com/health/ast.

Capiau, S., Veenhof, H., Koster, R. A., Bergqvist Y, Boettcher M, Halmingh O, Keevil B G, Koch BCP, Linden R, Pistos C, Stolk LM, Touw DJ, Stove C, Alffenaar JWC. (2019). Official international association for therapeutic drug monitoring and clinical toxicology guideline: Development and validation of dried blood spot-based methods for therapeutic drug monitoring. *Therapeutic Drug Monitoring, 41*(4), 409–430. https://doi.org/10.1097/ftd.0000000000000643.

Capiau, S., Wilk, L. S., Aalders MCG, Stove, C. P., (2016). A novel, nondestructive, dried blood spot-based hematocrit prediction method using noncontact diffuse reflectance spectroscopy. *Analytical Chemistry, 88*(12), 6538–6546. https://doi.org/10.1021/acs.analchem.6b01321.

Chiou, W. L. (1989). The phenomenon and rationale of marked dependence of drug concentration on blood sampling site. Implications in pharmacokinetics, pharmacodynamics, toxicology and therapeutics (Part I). *Clinical Pharmacokinetics, 17*(3), 175–199. https://doi.org/10.2165/00003088-198917030-00004.

Cobb, Z., de Vries, R., Spooner, N., Williams S, Staelens L, Doig M, Broadhurst R, Barfield M, van de Merbel N, Schmid B, Siethoff C, Ortiz J, Verheij E, van Baar B, White S, Timmerman P. (2013). In-depth study of homogeneity in DBS using two different techniques: Results from the EBF DBS-microsampling consortium. *Bioanalysis, 5*(17), 2161–2169. https://doi.org/10.4155/bio.13.171.

Control of Substances Hazardous to Health (COSHH). (2002). Retrieved May 17, 2021, from https://www.hse.gov.uk/coshh/.

Crawford, M. L., Collier BB, Bradley MN, Holland PL, Shuford CM, Grant RP. (2020). Empiricism in microsampling: Utilizing a novel lateral flow device and intrinsic normalization to provide accurate and precise clinical analysis from a finger stick. *Clinical Chemistry*, *66*(6), 821–831. https://doi.org/10.1093/clinchem/hvaa082.

Dasgupta, A., & Krasowski, D. (2020). *Therapeutic drug monitoring data: A concise guide* (4th ed., pp. 91–98). Cambridge, MA: Academic Press.

Delahaye, L., Dhont, E., De Cock, P., De Peape P, Stove, C. (2018). Volumetric absorptive microsampling as an alternative sampling strategy for the determination of paracetamol in blood and cerebrospinal fluid. *Analytical and Bioanalytical Chemistry*, *411*, 181–191. https://doi.org/10.1007/s00216-018-1427-6.

Denniff, P., & Spooner, N. (2010). The effect of hematocrit on assay bias when using DBS samples for the quantitative bioanalysis of drugs. *Bioanalysis*, *2*(8), 1385–1395. https://doi.org/10.4155/bio.10.103.

D'Urso, A., Rudge, J., Patsalos, P., & de Grazia, U. (2019). Volumetric absorptive microsampling: A new sampling tool for therapeutic drug monitoring of antiepileptic drugs. *Therapeutic Drug Monitoring*, *41*(5), 681–692. https://doi.org/10.1097/FTD.0000000000000652.

Evans, M., Livesey, J., & Ellis, M. (2001). Effect of anticoagulants and storage temperatures on stability of plasma and serum hormones. *Clinical Biochemistry*, *34*(2), 107–112. https://doi.org/10.1016/S0009-9120(01)00196-5.

Fragala, M. S., Goldman, S. M., Goldman MM, Mildred M; Bi C, Colletti J D, Arent S M, Walker A J, Clarke, N. J. (2018). Measurement of cortisol and testosterone in athletes: Accuracy of liquid chromatography-tandem mass spectrometry assays for cortisol and testosterone measurement in whole-blood microspecimens. *Journal of Strength and Conditioning Research*, *32*(9), 2425–2434. https://doi.org/10.1519/jsc.0000000000002726.

George, R., Haywood, A., Khan, S., Radovanovic M, Simmonds, J, Norris, R. (2018). Enhancement and suppression of ionization in drug analysis using HPLC-MS/MS in support of therapeutic drug monitoring: A review of current knowledge of its minimization and assessment. *Therapeutic Drug Monitoring*, *1*(40), 1–8. https://doi.org/10.1097/ftd.0000000000000471.

George, R. S., & Moat, S. J. (2016). Effect of dried blood spot quality on newborn screening analyte concentrations and recommendations for minimum acceptance criteria for sample analysis. *Clinical Chemistry*, *62*(3), 466–475. https://doi.org/10.1373/clinchem.2015.247668.

Guthrie, R., & Susi, A. (1963). A simple phenylalanine method for detecting phenylketonuria in large populations of newborn infants. *Pediatrics*, *32*, 338–343.

Gwenaël, N., Gallez, A., Kok MGM, Cobraiville G, Servais A-C, Piel G, Pequeux C, Fillet, M. (2017). Whole blood microsampling for the quantitation of estetrol without derivatization by liquid chromatography-tandem mass spectrometry. *Journal of Pharmaceutical and Biomedical Analysis*, *140*, 258–265. https://doi.org/10.1016/j.jpba.2017.02.060.

Hecht, M., Evard, H., Takkis, K., Veigure R, Aro R, Lohmus R, Herodes K, Leito I, Kipper K. (2017). Sponge spray—reaching new dimensions of direct sampling and analysis by MS. *Analytical Chemistry*, *89*(21), 11592–11597. https://doi.org/10.1021/acs.analchem.7b02957.

Intermolecular Forces. (2021). Retrieved May 13, 2021, from https://chem.fsu.edu/chemlab/chm1046course/interforces.html.

Introduction to MS Quantitation and Modes of LC/MS Monitoring. (2016). Retrieved May 14, 2021, from https://www.ionsource.com/tutorial/msquan/intro.htm.

ISNS Regions. (2021). Retrieved May 13, 2021, from https://www.isns-neoscreening.org/isns-regions/.

Keevil, B. G. (2011). The analysis of dried blood spot samples using liquid chromatography tandem mass spectrometry. *Clinical Biochemistry*, *44*(1), 110–118. https://doi.org/10.1016/j.clinbiochem.2010.06.014.

Kushon, S. Bischofberger A, Carpenter A, Denniff P, Guo Y, Rahn P, Rudge J, Spooner N, Welch E. (2014). Novel dried matrix microsampling device that eliminates the volume based hematocrit bias associated with DBS sub-punch workflows. Phenomenex. Retrieved May 14, 2021, from https://www.future-science.com/userimages/ContentEditor/1403606247205/A%20Novel%20Dried%20Matrix%20Microsampling%20Device%20that%20Eliminates%20the%20Volume%20Based%20Hematocrit%20Bias%20Associated%20with%20DBS%20Sub-punch%20Workflows_phenomenex.pdf.

Lawson, A. J., Bernstone, L., & Hall, S. K. (2016). Newborn screening blood spot analysis in the UK: Influence of spot size, punch location and haematocrit. *Journal of Medical Screening*, *23*(1), 7–16. https://doi.org/10.1177/0969141315593571.

Lehmann, S., Delaby, C., Vialaret, J., Ducos J, Hirtz C. (2013). Current and future use of "dried blood spot" analyses in clinical chemistry. *Clinical Chemistry and Laboratory Medicine*, *51*(10), 1897–1909. https://doi.org/10.1515/cclm-2013-0228.

Lucey, R., Munday, C., Cobb, Z., & Wright, M. (2018). Extraction of drugs from Mitra® micro-sampling devices: Overcoming a new haematocrit effect. *2018 European Bioanalytical Conference*, Barcelona (21–23 November 2018). https://www2.lgcgroup.com/l/31922/2020-10-01/ncnnj8.

Marasca, C., Protti, M., Mandrioli, R., Atti AR, Armirotti A, Cavalli A, De Ronchi D, Mercolini L. (2020). Whole blood and oral fluid microsampling for the monitoring of patients under treatment with antidepressant drugs. *Journal of Pharmaceutical and Biomedical Analysis*, *188*, 113–384. https://doi.org/10.1016/j.jpba.2020.113384.

Marshall, D. J., Adaway, J. E., Hawley, J. M., & Keevil, B. G. (2020a). Quantification of testosterone, androstenedione and 17-hydroxyprogesterone in whole blood collected using Mitra microsampling devices. *Annals of Clinical Biochemistry*, *57*(5), 351–359. https://doi.org/10.1177/0004563220937735.

Marshall, D. J., Kim, J. J., Brand, S., Bryne C, Keevil BG. (2020b). Assessment of tacrolimus and creatinine concentration collected using Mitra microsampling devices. *Annals of Clinical Biochemistry: International Journal of Laboratory Medicine*, *57*(5), 389–396. https://doi.org/10.1177%2F0004563220948886.

Mercolini, L., Protti, M., Catapanoa, M. C., Rudge J, Sberna AE. (2016). LC-MS/MS and volumetric absorptive microsampling for quantitative bioanalysis of cathinone analogues in dried urine, plasma and oral fluid samples. *Journal of Pharmaceutical and Biomedical Analysis*, *123*, 186–194. https://doi.org/10.1016/j.jpba.2016.02.015.

Morato, N. M., Pirro, V., Fedick, P. W., Cooks G. (2019). Quantitative swab touch spray mass spectrometry for oral fluid drug testing. *Analytical Chemistry*, *91*(11), 7450–7457. https://doi.org/10.1021/acs.analchem.9b01637.

Mughuni, MM., Stevens, MA., Langman, LJ., Kudva YC, Sanchez W, Dean PG, Jannetto, PJ. (2020). Volumetric microsampling of capillary blood spot vs whole blood sampling for therapeutic drug monitoring of Tacrolimus and Cyclosporin A: Accuracy and patient satisfaction. *The Journal of Applied Laboratory Medicine*, *5*(3), 516–530. https://doi.org/10.1093/jalm/jfaa005.

Nakadi, F. V., Garde, R., da Veiga, M., Cruces, J., & Resano, M. (2020). A simple and direct atomic absorption spectrometry method for the direct determination of Hg in dried blood spots and dried urine spots prepared using various microsampling devices. *Journal of Analytical Atomic Spectrometry*, *1*, 136–144. https://doi.org/10.1039/C9JA00348G.

Neoteryx. (2016a). Application note: Evaluation of Mitra microsampling device for in vivo pharmacokinetic studies. Retrieved May 17, 2021, from https://hubs.ly/H0zCsmv0.

Neoteryx. (2016b). Tech brief: Guidelines for the efficient extraction of dried blood from the Mitra microsampling device. Retrieved May 17, 2021, from https://hubs.ly/H0d4t4J0.

Neoteryx. (2017). Protocol to pre-absorb anticoagulant onto the Mitra microsampling device. Retrieved May 13, 2021, from https://app.hubspot.com/documents/1806452/view/24541032?accessId=87cf14.

Neoteryx. (2020a). Technical brief: Using remote blood collection with VAMS to accurately measure HbA1c in diabetes. Retrieved May 14, 2021, from https://www.neoteryx.com/hubfs/Content/Technical%20Briefs/HbA1c-Tech-Review-v5.pdf.

Neoteryx. (2020b). *VAMS publication list*. Retrieved May 14, 2021, from https://hubs.ly/H0bG1Z-0.

Nichols, H., Tang JCY, Dutton J, Rudge J. (2016). Evaluation of the Mitra™ microsampling device against dried blood spot cards for measurement of 25-hydroxy vitamin D3 by LC-MS/MS. Poster presented at the 2016 Mass Spectrometry: Applications to the Clinical Laboratory Conference EU Salzburg (September 2016). http://dx.doi.org/10.13140/RG.2.2.20843.75043.

Pirritano, M., Fehlmann, T., Laufer T, Ludwig N, Gasparoni G, Li Y, Meese E, Keller A, Simon, M. (2018). Next generation sequencing analysis of total small noncoding RNAs from low input RNA from dried blood sampling. *Analytical Chemistry*, *90*(20), 11791–11796. https://doi.org/10.1021/acs.analchem.8b03557.

Protti, M., Mandrioli, R., & Mercolini, L. (2019). Tutorial: Volumetric absorptive microsampling (VAMS). *Analytica Chimica Acta*, *1046*, 32–47. https://doi.org/10.1016/j.aca.2018.09.004.

Protti, M., Mandrioli, R., & Mercolini, L. (2020a). Microsampling and LC-MS/MS for antidoping testing of glucocorticoids in urine. *Bioanalysis*, *12*(11), 769–782. https://doi.org/10.4155/bio-2020-0044.

Protti, M., Mandrioli, R., & Mercolini, L. (2020b). Quantitative microsampling for bioanalytical applications related to the SARS-CoV-2 pandemic: Usefulness, benefits and pitfalls. *Journal of Pharmaceutical and Biomedical Analysis*, *191*, 113597. https://doi.org/10.1016/j.jpba.2020.113597.

Qu, Y., Brady, K., Apilado R, O'Malley T, Reddy S, Chitkara P, Ibarra C, Alexander RV, Dervieux, T., (2017). Capillary blood collected on volumetric absorptive microsampling (VAMS) device for monitoring hydroxychloroquine in rheumatoid arthritis patients. *Journal of Pharmaceutical and Biomedical Analysis*, *140*, 334–341. https://doi.org/10.1016/j.jpba.2017.03.047.

Rowland, M., & Emmons, G. T. (2010). Use of dried blood spots in drug development: Pharmacokinetic considerations. *The AAPS Journal*, *12*, 290–293. https://doi.org/10.1208/s12248-010-9188-y.

Saha, A., Kumar, A., Gurule SJ, Khuroo A, Srivastava P. (2017). Role of RBC partitioning and whole blood to plasma ratio in bioanalysis: A case study with valacyclovir and acyclovir. *Mass Spectrometry & Purification Techniques*, *3*(2), 1–12.

Schmidt, M., Rauh, M., Schmid MC, Huebner H, Ruebner M, Wachtveitl R, Cordasic N, Rascher W, Menendez-Castro C, Hartner A, Fahlbush, F. B., (2019). Influence of low protein diet-induced fetal growth restriction on the neuroplacental corticosterone axis in the rat. *Frontiers in Endocrinology*, *10*, 124. https://doi.org/10.3389/fendo.2019.00124.

Shoshtari, S. Z., Wen, J., Alany, R. G. (2008). Octanol water partition coefficient determination for model steroids using an HPLC method. *Letters in Drug Design & Discovery*, *5*(6), 394–400. https://doi.org/10.2174/157018008785777333.

Spooner, N., Anderson, K., Siple J, Wickremsinhe ER, Xu Y, Lee, M. (2019). Microsampling: Considerations for its use in pharmaceutical drug discovery and development. *Bioanalysis*, *11*(10), 1015–1038.

Spooner, N., Denniff, P., Michielsen L, De Vries R, Ji Q C, Arnold ME, Woods K, Woolf EJ, Xu Y, Boutet V, Zane P, Kushon, S., Rudge, J. (2015). A device for dried blood microsampling in quantitative bioanalysis: Overcoming the issues associated blood haematocrit. *Bioanalysis*, *7*(6), 653–659. https://doi.org/10.4155/bio.14.310.

Tanna, S., Alalaqi, A., Bernieh, D., & Lawson, G. (2018). Volumetric absorptive microsampling (VAMS) coupled with high-resolution, accurate-mass (HRAM) mass spectrometry as a simplified alternative to dried blood spot (DBS) analysis for therapeutic drug monitoring of cardiovascular drugs. *Clinical Mass Spectrometry*, *10*, 1–8. https://doi.org/10.1016/j.clinms.2018.08.002.

US Food and Drug Administration. (2018). Bioanalytical method validation: Guidance for industry. Retrieved May 13, 2021, from https://www.fda.gov/files/drugs/published/Bioanalytical-Method-Validation-Guidance-for-Industry.pdf.

Velghe, S., & Stove, C. (2018). Volumetric absorptive microsampling as an alternative tool for therapeutic drug monitoring of first-generation anti-epileptic drugs. *Analytical and Bioanalytical Chemistry, 410*, 2331–2341. https://doi.org/10.1007/s00216-018-0866-4.

Volani, C., Caprioli, G., Calderisi, G., Sigurdsson BB, Rainer J, Gentilini I, Hicks AA, Pramstaller PP, Weiss G, Smarason SV, Paglia G. (2017). Pre-analytic evaluation of volumetric absorptive microsampling and integration in a mass spectrometry-based metabolomics workflow. *Bioanalytical Chemistry, 409*(26), 6263–6276. https://doi.org/10.1007/s00216-017-0571-8.

Wang, J., Li, D., Wiltse, A., Emo, J., Hilchey, SP, Zand MS. (2019). Application of volumetric absorptive microsampling (VAMS) to measure multidimensional anti-influenza IgG antibodies by the mPlex-Flu assay. *Journal of Clinical and Translational Science, 3*(6), 332–343. https://doi.org/10.1017/cts.2019.410.

Wharton, R. E., Feyereisen, M. C., Gonzalez, A. L., Abbott NL, Hamelin EI, Johnson RC. (2017). Quantification of saxitoxin in human blood by ELISA. *Toxicon, 133*, 110–115. https://doi.org/10.1016/j.toxicon.2017.05.009.

Xie, I., Xu, Y., Anderson, M., Wang M, Xue L, Breidinger S, Goykhman D, Woolf EJ, Bateman KP (2018). Extractability-mediated stability bias and hematocrit impact: High extraction recovery is critical to feasibility of volumetric adsorptive microsampling (VAMS) in regulated bioanalysis. *Journal of Pharmaceutical and Biomedical Analysis, 156*, 58–66. https://doi.org/10.1016/j.jpba.2018.04.001.

Youhnovski, N., Mayrand-Provencher, L., Bérubé E-R, Plomley, J., Montpetit H, Furtado M, Keyhani A. (2017). Volumetric absorptive microsampling combined with impact-assisted extraction for hematocrit effect free assays. *Bioanalysis, 9*(22), 1761–1769. https://doi.org/10.4155/bio-2017-0167.

3

Preanalytical Considerations for Implementation of Microsampling Solutions

Bradley B. Collier, Peyton K. Miesse, and Russell P. Grant

Laboratory Corporation of America Holdings (LabCorp), Center for Esoteric Testing, Burlington, NC, USA

3.1 Introduction

In recent years, there has been a growing desire to have a more patient centric approach to the collection of biological samples, to enable a better patient experience. Aspects of patient centric sampling typically include reduced pain to the patient, reduced size or volume of sample, and the ability to perform self-collection of samples away from clinical facilities (James et al., 2020). These advantages may provide a path for additional longitudinal sampling that could potentially lead to an increased understanding of disease states and improve patient treatments and outcomes. The feasibility of utilizing so-called microsamples has been aided with the improvement in performance of analytical instrumentation. However, analytical performance represents only a small aspect of the entire sampling process. With the move toward self-collection, greater emphasis is placed on the patient's ability to obtain, process, and/or transport a sample of sufficient quality, despite receiving little to no training. Even under rigorously defined protocols performed by trained individuals, preanalytical errors account for a large percentage (reports up to 70%) of all mistakes in laboratory diagnostics with analytical and post-analytical errors being the other sources of error (Baskin et al., 2019; Bowen & Adcock, 2016; Giavarina & Lippi, 2017; Plebani & Carraro, 1997; Plebani et al., 2017; Stein & Pum, 2005; Van Vrancken et al., 2012; Zhang et al., 1998). Thus, thorough consideration must be given to the preanalytical processes that are required and how well these processes and the associated technologies are performed by naïve users. As such, efforts to reduce errors must focus on

Patient Centric Blood Sampling and Quantitative Bioanalysis, First Edition.
Edited by Neil Spooner, Emily Ehrenfeld, Joe Siple, and Mike S. Lee.

procedures and processes that have a high likelihood of error generation in order to improve sample integrity (Lewis & Tatsumi, 2006; Plebani et al., 2017).

This chapter will focus on considerations for improving and maintaining sample integrity during the preanalytical phase of sample measurement. This will include a discussion of important preanalytical steps to consider for remote sample collection, including selection of specimen matrix, the means of sample collection, transportation and stability of samples into the laboratory, as well as additional processing required for samples prior to measurement. As the integrity of sample received in the laboratory will be dependent on these and other elements, the quality of analytic results produced will reflect the overall collection process. The preanalytical steps described below are described in the context of clinical diagnostics; however, most of this information can be applied to other types of measurements.

Included in this chapter, to demonstrate the need for preanalytical considerations, is new data obtained from samples collected from donors that provided informed consent under a protocol approved by an Institutional Review Board. In order to collect samples, traditional phlebotomy techniques were utilized to obtain venous and capillary finger-stick samples. Manufacturer recommendations were utilized for novel capillary collection techniques based on vacuum generating devices that adhere to the upper arm. All sample measurements were made on Roche cobas® autoanalyzers using reagents approved by the United States' Food and Drug Administration (FDA). No spiking or contrivance was utilized as the materials utilized can present matrix or recovery effects that are not representative of the matrix being measured (Capiau et al., 2019).

3.2 Sample Matrices

One of the aims of patient centric sampling is the desire to utilize a less invasive means of acquiring samples in order to reduce pain and increase ease of collection for the patient. Some common sample matrices that are considered noninvasive include saliva, urine, tears, sweat, and hair. Although these matrices might prove useful for specific applications, they are susceptible to contamination (exogenous or endogenous) or have unequivocal or unknown reference ranges compared to measurements in standard matrices such as blood, serum, or plasma (Avataneo et al., 2019; Brunet et al., 2009; Corrie et al., 2015; Gallardo & Queiroz, 2008; Malamud, 2011; Strang et al., 1990).

For these reasons, blood (and its components) still remains the most widely utilized bodily matrix for analytical measurements required for clinical and nonclinical purposes (Baskin et al., 2019; Giavarina & Lippi, 2017; Kale et al., 2017; Lei & Prow, 2019; Lewis & Tatsumi, 2006). In order to obtain blood, however, puncturing or lancing of the skin, typically by a trained phlebotomist, is still required. As a result, less invasive and easier to use collection technologies, in which a patient can collect low volumes of blood (i.e. microsamples), are desired. In the following sections, we will be discussing different types of blood sample collection including types of liquid samples (e.g. whole blood or separated blood component), location of sample collection (venous vs. capillary), and type of sample generated (liquid vs. dried).

3.2.1 Venous Sample

Blood is typically collected from a patient by using a hypodermic needle to puncture the skin and withdraw blood directly from a vein using a vacuum evacuated tube or vacutainer. Arterial puncture can also be performed in a similar manner but is not as common. To facilitate additional processing and improve stability, vacutainers often contain additives. For example, anticoagulants such heparin or EDTA are often utilized to prevent blood from clotting. The anticoagulant serves to help preserve blood for measurements that require whole blood samples, such as hematological and pharmaceutical-based measurements. In addition, anticoagulants are utilized to assist with blood fractionation where the blood is separated into its component parts (e.g. plasma and red blood cells) usually through centrifugation. Similar to plasma, serum can be obtained by allowing whole blood to clot prior to separation. In this case, clot activators (e.g. thrombin or silica) can be added to vacutainers to speed the natural clotting process. Of these matrices (whole blood, plasma, and serum), plasma and serum samples are the most prominent specimen types (Londhe & Rajadhyaksha, 2020). Both matrices have relative advantages. Plasma can be acquired more quickly, as there is no time required for clotting prior to centrifugation (Boyanton Jr. & Blick, 2002). In addition, greater volumes of plasma can be obtained relative to serum (Baskin et al., 2019). Despite these advantages, serum is still more frequently utilized due to familiarity with clinical reference ranges in this matrix.

Although similar in constitution for many analytes, plasma and serum samples are not necessarily interchangeable (Baskin et al., 2019). To highlight differences in plasma versus serum measurements, data has been obtained for several analytes routinely used to assess patient health. Samples were collected using traditional venipuncture from donors that provided informed consent under a protocol approved by an Institutional Review Board (ASPIRE Protocol #520100174). All analytes were measured in a certified laboratory on Roche cobas® autoanalyzers with reagents approved by the FDA for the venous sample types collected. Both serum (serum separator vacutainer) and plasma (lithium heparin vacutainer) were collected from multiple donors during the same venipuncture. All samples were centrifuged within 1 hr of collection. Correlation of plasma measurements to serum results was determined as well as mean bias of measurements. Individual bias measurements were also utilized to determine the percentage of results that were within the United States' Clinical Laboratory Improvement Amendments (CLIA) acceptable limits (Verma et al., 2019).

The enzyme aspartate aminotransferase (AST) is frequently measured to assess liver function. The comparison of serum and plasma samples for AST indicated a high degree of correlation ($R^2 = 0.9966$) with a slope of 0.991 and a mean bias of 0.1% with 99.6% of plasma results within CLIA limits of ±15.0% (Figure 3.1). This

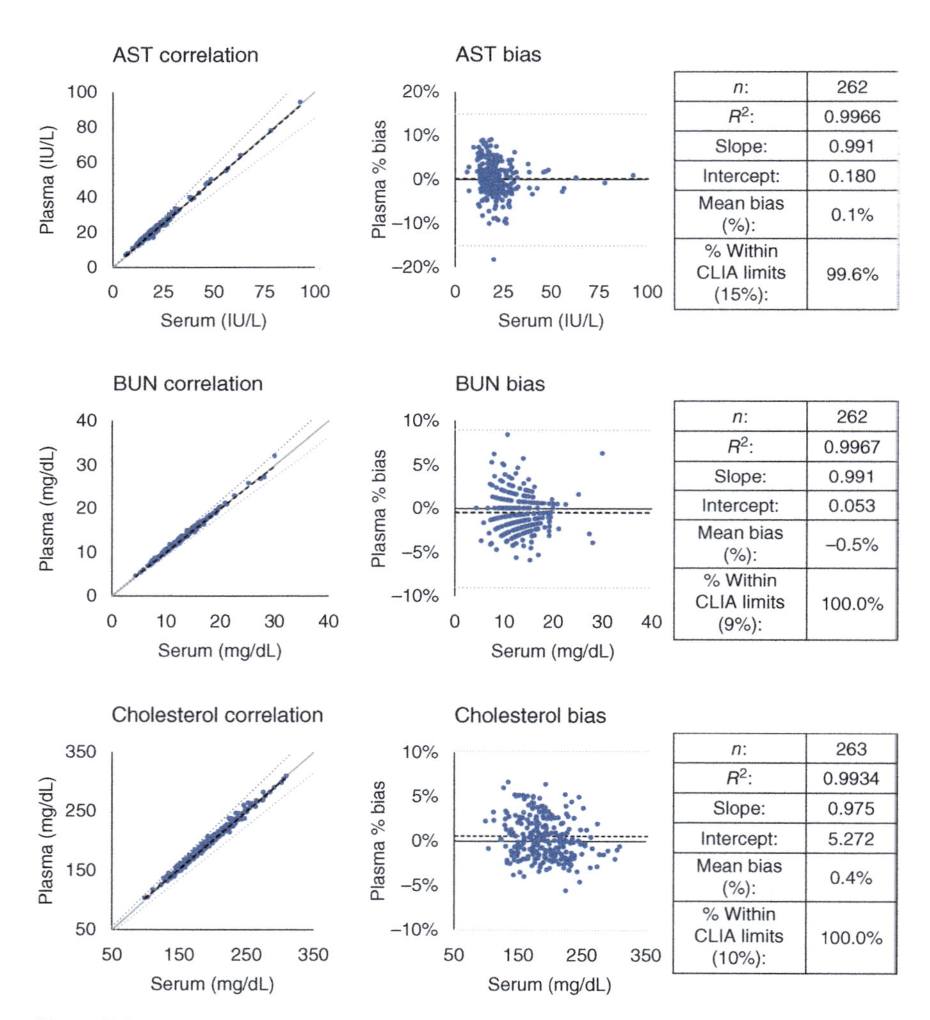

Figure 3.1 Serum versus plasma comparison of AST, BUN, and cholesterol. Black dashed lines represent regression line and mean bias in correlation and percentage bias plots, respectively. Gray dashed lines represent individual CLIA bias limits for each analyte, while the soli gray line represents ideal results.

can be observed in the percentage bias plot that shows the bias of individual plasma measurements relative to the paired serum results. Similar correlative and bias results were observed for blood urea nitrogen (BUN), a common marker of kidney function ($R^2 = 0.9967$, slope = 0.991, mean bias = −0.5%) as well as cholesterol ($R^2 = 0.9934$, slope = 0.975, mean bias = 0.4%). For these three analytes, the mean bias observed in plasma was less than 1%. In addition, 99.6% or more of plasma measurements were within CLIA limits of bias relative to serum measurements.

Other analytes such as creatinine (another marker of kidney function) and glucose (a marker of diabetes) exhibited a high degree of correlation between serum and plasma values ($R^2 > 0.98$, Figure 3.2) but with a slight mean bias observed (2.7 and −2.2%, respectively). These results, although acceptable under CLIA limits, indicate a difference in the measured analyte concentration from matrix-to-matrix. Other markers such as total protein exhibit a larger discrepancy between serum and plasma measurements (Figure 3.2), which is not unexpected (Lum & Gambino, 1974). In this case, the loss of fibrinogen to the clotting cascade leads to elevated total protein measurements in plasma relative to serum (4.2%). On the surface, these results are not immediately obvious as being clinically significant (98.3% of results within CLIA limits). However, total protein is also utilized along with the measurement of albumin to determine the total globulins present in the sample (total globulins = total protein − albumin).

As plasma albumin levels demonstrated a low degree of bias (−0.3%) compared to serum, the calculation of total globulin levels in plasma is greatly influenced by the initial bias of plasma total protein results (Figure 3.3) where a mean bias of 12.6% was observed relative to the calculated total globulin levels in serum. This resulted in only 20.7% of the plasma total globulin levels falling within the same ±8% limits (CLIA limit for both total protein and albumin). In addition, the ratio of albumin to total globulins (A/G in Figure 3.3) was similarly biased in plasma results (−11.2%), and only 27.8% of plasma results fell within 8% of serum results.

These results indicate that although serum and plasma are very similar in constitution, the measurement of analytes is not equivalent, but the difference may not be clinically significant. As such, each individual matrix (and collection device) should be thoroughly investigated prior to utilization (Bowen & Adcock, 2016). To account for differences observed in concentration from matrix-to-matrix, correction factors can be generated with sufficient data (typically ≥120 individuals) to allow measured values to correspond to concentrations expected in standard matrices (Crawford et al., 2020). Alternatively, new reference intervals can be generated for each novel matrix but will also require an adequate number of measurements for each relevant patient subset (e.g. age or gender) prior to implementation (Horowitz et al., 2010).

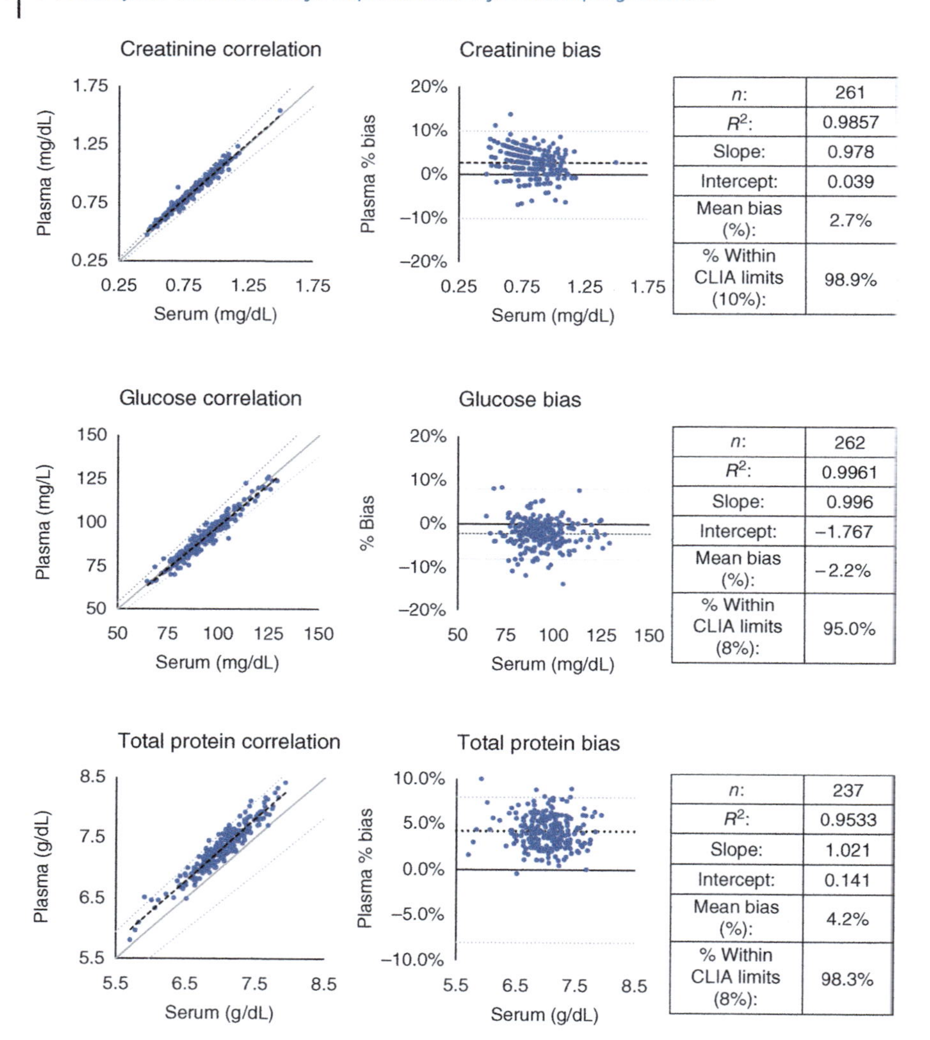

Creatinine	
n:	261
R^2:	0.9857
Slope:	0.978
Intercept:	0.039
Mean bias (%):	2.7%
% Within CLIA limits (10%):	98.9%

Glucose	
n:	262
R^2:	0.9961
Slope:	0.996
Intercept:	−1.767
Mean bias (%):	−2.2%
% Within CLIA limits (8%):	95.0%

Total protein	
n:	237
R^2:	0.9533
Slope:	1.021
Intercept:	0.141
Mean bias (%):	4.2%
% Within CLIA limits (8%):	98.3%

Figure 3.2 Serum versus plasma comparison for creatinine, glucose, and total protein. Glucose results greater than 130 mg/dL were not included as they anchored the regression fit. Black dashed lines represent regression line and mean bias in correlation and percentage bias plots, respectively. Gray dashed lines represent individual CLIA bias limits for each analyte, while the solid gray line represents ideal results.

3.2.2 Capillary Blood

Although ubiquitous to blood collection, venipuncture typically acquires large volumes of blood which is not conducive to patient centric collection. In a professional setting, venipuncture is performed by phlebotomists that have received thorough training and/or certification in order to reduce the possibility of adverse

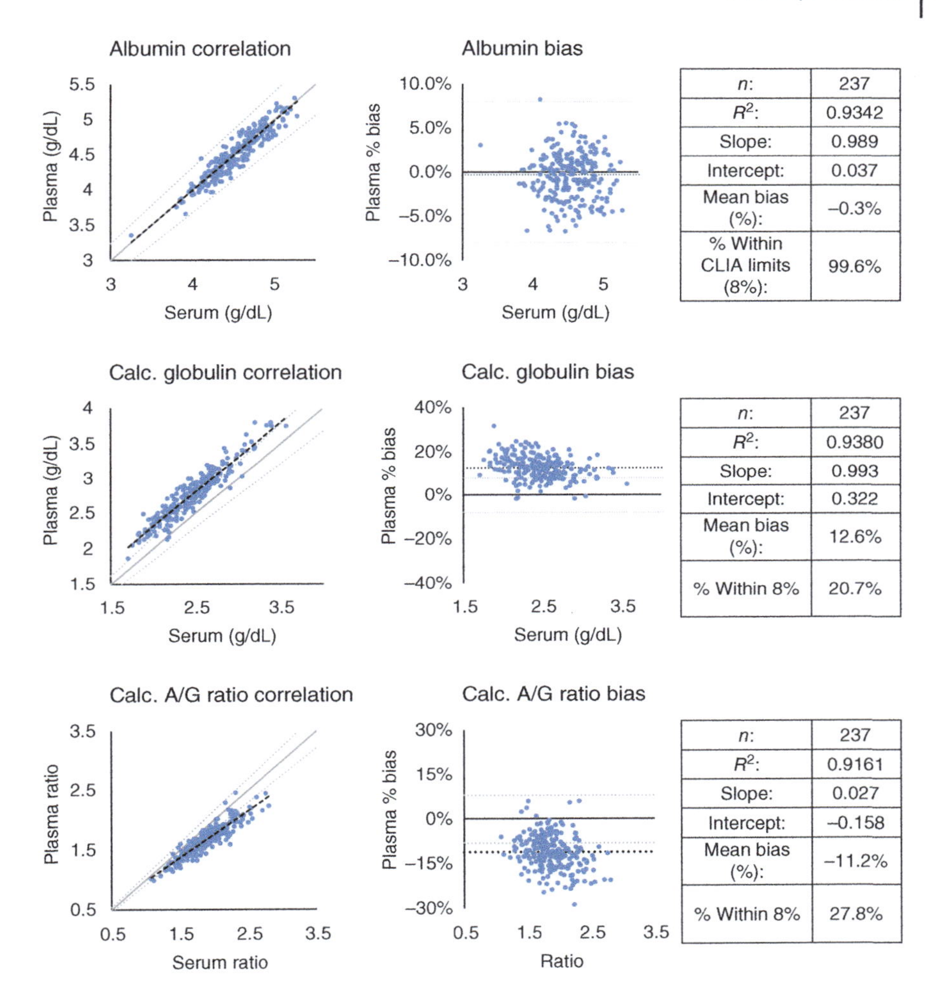

n:	237
R^2:	0.9342
Slope:	0.989
Intercept:	0.037
Mean bias (%):	−0.3%
% Within CLIA limits (8%):	99.6%

n:	237
R^2:	0.9380
Slope:	0.993
Intercept:	0.322
Mean bias (%):	12.6%
% Within 8%	20.7%

n:	237
R^2:	0.9161
Slope:	0.027
Intercept:	−0.158
Mean bias (%):	−11.2%
% Within 8%	27.8%

Figure 3.3 Serum versus plasma comparison data for albumin, calculated globulin, and calculated albumin/globulin ratio. Black dashed lines represent regression line and mean bias in correlation and percentage bias plots, respectively. Gray dashed lines represent 8% bias (CLIA bias limit for albumin and total protein), while the solid gray line represents ideal results.

reactions such as bruising or infection. As an alternate means of sample collection, capillary blood can be obtained by puncturing the skin (e.g. lancing a finger) to obtain a sample from near the surface of the skin. This less invasive approach can be used to acquire small volumes of blood by individuals with minimal training in nonconventional settings.

However, as capillaries help facilitate the exchange of nutrients and waste at the cellular level, blood obtained from capillaries may not have the same composition as blood obtained intravenously. It has been suggested that capillary blood more closely resembles arterial blood than venous blood due to the relative greater arteriolar pressure (Baskin et al., 2019). In addition, capillary blood is also potentially diluted with interstitial and intracellular fluids due to the nature of the puncture wound made. This may lead to increased variability in the composition of the blood (Bond & Richards-Kortum, 2015; Lewis & Tatsumi, 2006; Morris et al., 1999; Stein & Pum, 2005).

Despite these potential sources of error, capillary blood measurements have been investigated for several analytes and found to have similar concentrations as venous blood (Baskin et al., 2019; Crawford et al., 2020; Krleza et al., 2015; Stein & Pum, 2005). Nonetheless, some disagreements still exist on the robustness of acquiring and analyzing capillary blood samples (Boyd et al., 2005; Krleza et al., 2015; Yang et al., 2001). This is in part because studies found in the literature are often small in scale, and there is typically not any standardization in the collection process and measurement between studies (Blicharz et al., 2018; Bond & Richards-Kortum, 2015; Parikh et al., 2009; Yang et al., 2001).

To investigate the potential differences in samples acquired via traditional venipuncture and capillary blood samples, several analytes were investigated by taking paired samples from donors. Venous and capillary samples were collected from donors that provided informed consent under a protocol approved by an Institutional Review Board (ASPIRE Protocol #520100174). Samples were collected within 10 min of each other, and all analytes were measured in a certified laboratory on Roche cobas® autoanalyzers with FDA approved reagents. Venous samples were collected using traditional venipuncture techniques, while capillary samples were collected using two different approaches. For the first approach, capillary samples were collected by lancing a finger using a high blood flow contact-activated lancet (BD 366594). Blood (300 μL) was then collected into lithium heparin-coated glass capillaries (Drummond Scientific 7-000-1000-H) and dispensed into a microcentrifuge tube for centrifugation. The first drop of blood was *not* wiped away, and minor pressure was applied to the finger as needed in order to obtain sufficient volume. In addition, novel collection devices that are temporarily adhered to the upper arm for sample collection were utilized (e.g. Tasso, Inc. Tasso-SST and YourBio Health Tap II collection devices). These devices generate a negative pressure following wound creation to facilitate sample collection. Devices were utilized with either serum or lithium heparin plasma separator tubes containing thixotropic gel to ease sample separation. Results have been anonymized with respect to the device used for sample collection and are referred to as "Device A" and "Device B" in the figures below.

In the data presented later, both serum and plasma samples were acquired from capillary and venous blood. Data on the *x*-axis represents venous results from either serum or plasma, while the *y*-axis represents capillary results from either serum or plasma. The matrices utilized were matched for comparison purposes (i.e. venous plasma to capillary plasma and venous serum to capillary serum). Samples were again measured using clinical autoanalyzers with FDA-approved reagents. Capillary samples were collected from the upper arm using vacuum to facilitate blood flow following lancing. Volumes collected from a single wound site typically ranged from 200 to 500 μL of whole blood.

Investigation of capillary BUN and cholesterol showed a high degree of correlation to venous measurements ($R^2 > 0.98$, Figure 3.4). In addition, the biases observed with respect to venous results for each analyte covered a wider range than was observed when venous serum results were compared to venous plasma results (Figure 3.1). This resulted in a decrease in the number of capillary samples that were within acceptable CLIA limits (90.6 and 98.7% for BUN and cholesterol, respectively). Capillary AST results also showed a greater mean bias (4.3%). As AST is sequestered into red blood cells at a concentration that is approximately 40 times higher than plasma (Caraway, 1962), even a slight amount of hemolysis in capillary samples may lead to elevated AST measurements. Hemolysis can either occur due to the creation of the wound or as a result of applying positive pressure to the wound site to increase blood flow. The latter is typically referred to as "milking" and is frequently indicated as a source of error in capillary blood measurements (refer to Section 3.4.4 for additional details). It should be noted for the AST data, however, that only 27 of the 147 measurements were acquired using capillary blood obtained from lancing a finger where hemolysis due to milking is of greater concern. The mean bias of these 27 measurements was found to be −0.9% indicating that hemolysis due to finger milking was not the cause of elevated AST results. The slight variations in the biases observed from different collection technologies are attributed to day-to-day instrument variation.

Other analytes exhibited correlated results with unacceptable biases (Figure 3.5). For example, capillary creatinine measurements demonstrated a bias of −7.9% relative to venous samples, which resulted in only 71.1% of capillary results within CLIA limits for creatinine (±10%). In addition, capillary glucose measurements displayed a mean bias of 0.1% relative to venous samples but with a low degree of correlation ($R^2 < 0.55$). The increased scatter in capillary glucose results led only 47.2% of capillary results falling within CLIA limits for glucose (±8%).

Analysis of capillary carbon dioxide (CO_2) was also performed, but a large negative bias (−24.0%) was observed relative to venous samples. This is most likely a result of differences in capillary blood concentrations of CO_2 as well as increased outgassing from the relatively low-volume samples (<0.5 mL) collected from capillaries. These results indicate that capillary blood measurements may not suitable for measurement of all analytes when using venous samples as the comparator method for acceptance. A more detailed discussion of analyte acceptance

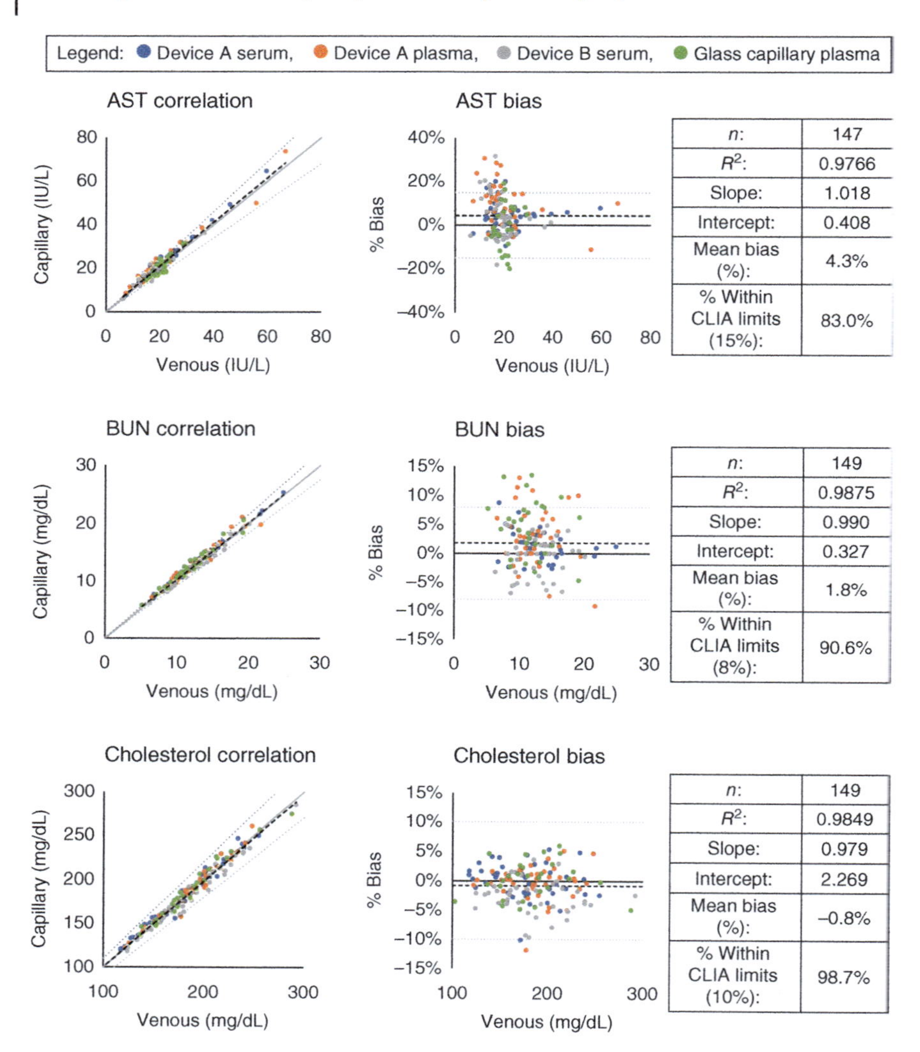

Figure 3.4 Venous serum and plasma versus capillary serum and plasma comparison for AST, BUN, and cholesterol for several collection technologies. Black dashed lines represent regression line and mean bias for all data in correlation and percentage bias plots, respectively. Gray dashed lines represent individual CLIA bias limits for each analyte, while the solid gray line represents ideal results.

can be found in Section 3.3. Due to the greater potential for CO_2 outgassing, the slight variations in the biases observed from different collection technologies may be a result of variation in the time from collection to the time of measurement on the instrument (all samples were centrifuged within 1 hr of collection).

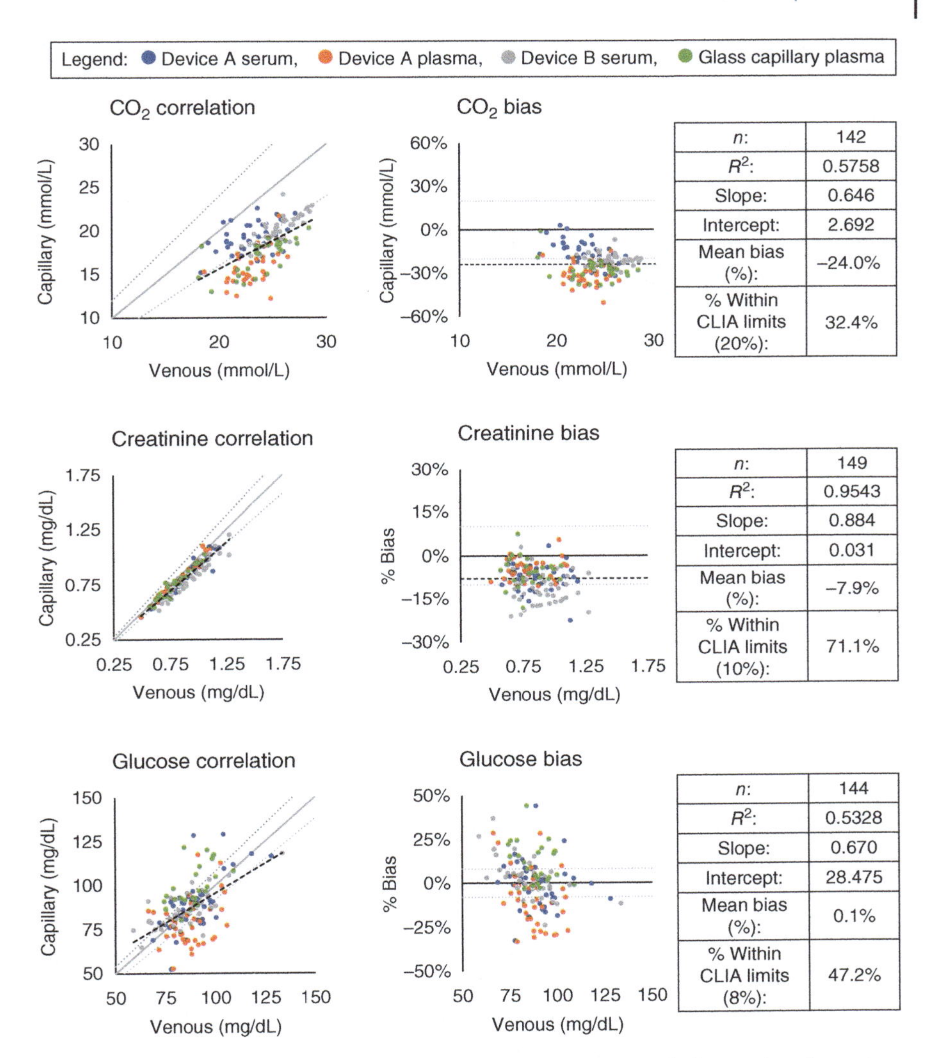

Figure 3.5 Venous serum and plasma versus capillary serum and plasma data for carbon dioxide, creatinine, and glucose for several collection technologies. Black dashed lines represent regression line and mean bias for all data in correlation and percentage bias plots, respectively. Gray dashed lines represent individual CLIA bias limits for each analyte, while the solid gray line represents ideal results.

These results provide evidence that although the same sample matrix is utilized, the collection location (e.g. venous vs. capillary) will have an effect on the quality of results compared to traditional measurements. To investigate the capillary sample measurements further, a serial sampling study was performed to see if results from capillary varied with the volume obtained similar to previous reports (Bond & Richards-Kortum, 2015).

The results in Figure 3.6 below were obtained by lancing a finger and self-collecting blood using lithium heparin-coated capillaries designed to only collect 20 μL of blood. To facilitate blood flow, a high blood flow lancet was used (BD 366594) with a penetration width and depth of 1.5 and 2.0 mm, respectively. Following lancing, the first drop of blood observed was collected and not wiped away as is often recommended. Capillaries were applied to the wound one after another such that no blood was lost during the collection of the samples. All donors were able to collect all 10 samples (200 μL of blood total).

As previously reported, to measure most analytes each whole blood aliquot was diluted 10-fold with phosphate-buffered saline (PBS) and then centrifuged (Crawford et al., 2020). The supernatant was then measured using FDA-approved reagents with an increased aspiration volume (to account for sample dilution). Venous blood was obtained using lithium heparin tubes. To mirror the capillary blood processing, venous blood was also collected into lithium heparin-coated capillaries and diluted 10-fold with PBS. Hemoglobin measurements were made by diluting samples 101-fold with hemolyzing reagent prior to measurement, which is unchanged as per the assay recommendations.

Following measurement of each sample, serial capillary results were normalized to the mean of the three diluted venous replicates (represented by 1.0 in Figure 3.6). The normalized analyte values were then averaged across several donors ($n \geq 3$) and plotted with respect to the total volume collected. Several of the same analytes were investigated as discussed earlier.

Interestingly, AST results were nearly 2 times venous measurements, which again supports the hypothesis that hemolysis is occurring during the collection process. Glucose ($n = 8$) results also showed a slight elevation compared to venous measurements (between 1.1 and 1.2 times) but with a standard deviation that varied from 9 to 17%. The other analytes investigated (BUN, cholesterol, creatinine, and hemoglobin), however, had values that were within 10% of the venous values. Hemoglobin results are important to note as they indicate that capillary blood is *not* being significantly diluted with interstitial or intracellular fluid despite a relatively large wound being generated (with the lancet utilized).

Although the results provided herein are presented to help provide insight into the utility of measurement of capillary blood samples, there are several other areas that have yet to be fully investigated. Clotting differences will not only affect the ability to get a sample from patient-to-patient but could also potentially influence the constituency of the sample and the harvest volume. For instance, anticoagulant may not be immediately mixed into capillary collected blood, and as a result, coagulation can begin at the wound site, which may result in the matrix obtained not being representative of a true plasma sample. This can also result in

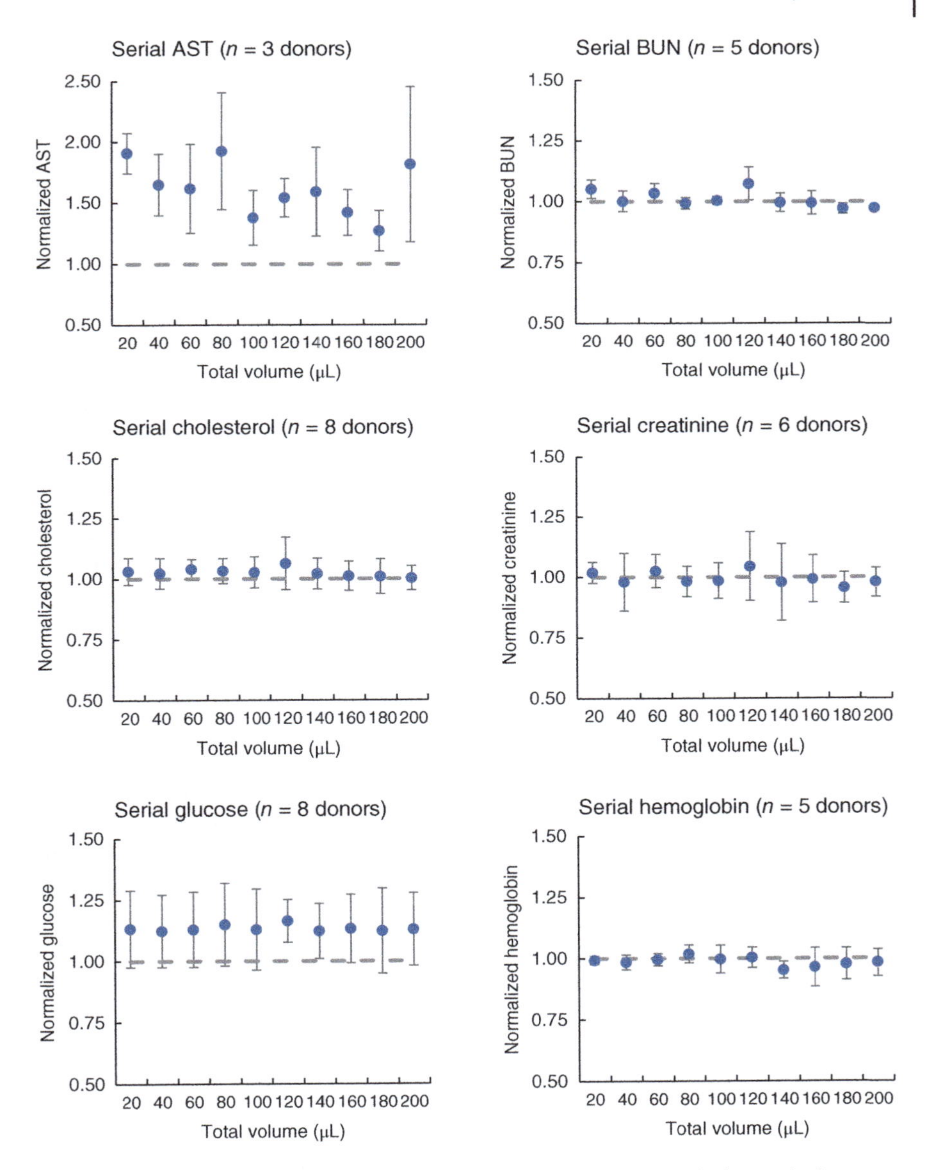

Figure 3.6 Serial capillary sampling data for AST, BUN, creatinine, cholesterol, glucose, and hemoglobin where results on capillary measurements are normalized to triplicate venous measurements processed in the same manner. Error bars represent one standard deviation of normalized values.

limited or no sample being collected. Additionally, the effect of the size and number of lancets or blades have not been examined other than in terms of volume of sample obtained and pain to the patient (Blicharz et al., 2018). As a result, it is unclear if different capillary blood collection techniques produce samples with the same composition. Other common sources of sample deviation are discussed later.

3.2.3 Material Selection

Although we have shown data demonstrating that analyte results can be dependent on the matrix utilized (particularly serum and plasma) and how that matrix is obtained (intravenously and capillary), there are other analytical considerations when selecting how and what type of sample to collect. In particular, the materials that a sample contacts must also be considered (Bowen & Adcock, 2016). Although vacutainers are ubiquitous to the current sample collection process and are seemingly very simple devices, several vacutainer components can affect analyte measurement including the tube itself, the stopper, lubricants, and surfactants, which can all have an effect on analyte measurement through contamination, adsorptive loss, altering biological pathways, or altering stability (Bowen & Adcock, 2016; Bowen & Remaley, 2014). Microsampling technologies have added complexity as they are functionally different than vacutainers because blood is not collected directly into a sterile storage container. Rather, blood flows from an open wound, typically touches the surface of a patient's skin, and may come into contact with various device components other than the storage container. As such, an in depth analysis of the different components that blood comes into contact may be needed in order to avoid any adverse effect on analyte measurement (Bowen & Adcock, 2016). In addition, steps should be taken to reduce the risk of contamination by thoroughly cleansing and sterilizing the draw location prior to sample collection (see Section 3.4 below for additional details).

3.2.4 Dried Samples

The low-volume collection enabled by microsampling technologies also allows for the possibility of desiccating samples following collection. Drying samples has the ability to provide enhanced stability for some analytes but often with the tradeoff of increasing processing time in the laboratory due to the need to reconstitute the sample (Velghe et al., 2016). Many microsampling technologies that utilize drying are limited to the collection of whole blood samples, as the drying process leads to lysis of the blood cells which prevents separation of serum and plasma following sample extraction. In addition, the relative volume of red blood cells (hematocrit), which naturally varies in patients, can affect the measurement of analytes that are traditionally made in plasma or serum (Edge et al., 2016; Londhe &

Rajadhyaksha, 2020; Spooner et al., 2018; Velghe et al., 2019). The lysing of blood cells can lead to dilution of the analyte of interest with increasing hematocrit levels when the analyte is not present in cells. Alternatively, if the analyte is sequestered into blood cells in concentrations greater than the plasma, whole blood measurements of analytes typically measured in serum or plasma will become disproportionately elevated in measurements on whole blood samples. These issues can be potentially be overcome by processing a fixed volume of blood or normalization of the measured analyte concentration to some other factor (Crawford et al., 2020; Velghe et al., 2019). However, the lysing of the cells may also make processing and measurement of the samples difficult or impossible (e.g. hemoglobin can interfere with optical measurements) so separation of plasma or serum may still be necessary even if drying is employed (Edge et al., 2016).

3.2.5 Conclusions

As has been demonstrated, analyte concentrations will be dependent on the blood matrix utilized as well as the location in the vasculature from which it is obtained. Additionally, other aspects of the collection technology may influence analyte measurements. For this reason, validation of the alternative matrix, including bridging studies to demonstrate correlation of measurements to the standard matrix, will need to be performed prior to implementation of any new collection modality. Ultimately, as the specimen type deviates from the preferred specimen type, additional work will be required to allow acceptance of an alternate specimen type for a particular analyte (Beharry, 2010). Correction of (consistent) analyte results to match the reference range to the preferred matrix and collection technology may also be necessary to allow easier interpretation of results and ultimately increase acceptance of the collection technology (Stein & Pum, 2005). Refer to Section 3.3 for additional details.

Despite potential differences to preferred sample matrices, capillary blood remains the most promising approach to achieve patient centric sample collection due to the ease of collection. Further investigation of underlying physiology such as the difference in composition of capillary blood obtained from locations in the body compared to venous blood would be informative and may serve to increase acceptance of measurements based on capillary blood.

3.3 Alternate Sample Acceptance Criteria

Acceptance criteria or analytical performance specifications (APS) have not been discussed in detail to this point. However, the selection of appropriate APS is an important aspect of not only validating an analyte with a specific collection technology but also how the technology and measurement results are utilized. For this reason, APS should be carefully considered for each individual analyte.

Although APS can be defined by a variety of approaches (Klee, 2010), three methods were identified in 2014 by the European Federation of Clinical Chemistry and Laboratory Medicine (EFLM): (a) analytical performance relative to clinical outcome, (b) analyte specific criteria based on biological variation, and (c) technically achievable analytical performance criteria (Aarsand et al., 2021; Ceriotti et al., 2017). For commonly measured analytes, the APS utilized are based on biological variability including both within subject (CV_I) and between subject (CV_G) variability. Analyte-specific biological variation data can be typically found in the EFLM Biological Variation Database (Aarsand et al., 2021) and other databases. Unfortunately, as biological variation data for each analyte is updated, the APS for each analyte is also updated. Ever-changing APS can make it difficult to understand what acceptance criteria are required for each analyte. This is further confounded by the presence of three levels of APS: optimum, desirable, and minimum where the minimum criteria for imprecision, bias, and total error are three times higher than optimum criteria (Table 3.1).

Example APS based on biological variation data obtained from the EFLM database have also been provided in Table 3.2 for the analytes for which data has been given herein. CLIA acceptance limits referenced above are also based on biological variability and are included for reference. These results indicate that expectations for analyte performance specifications can be quite varied for individual analytes. Recommendations from additional regulatory bodies and external assessment programs may further muddle understanding of analyte-specific acceptance criteria. For this reason, thorough scientific and medical review should be performed for all appropriate validation data. This typically includes (but is not limited to) studies investigating precision, accuracy, linearity, sensitivity, and stability of the analyte in alternate sample types (Capiau et al., 2019). Additional device-specific studies such as contamination may also be necessary. Generation of new reference intervals may also be appropriate if the matrix produces analyte results that are different than traditional matrices.

Table 3.1 Optimum, Desirable, and Minimum Analyte Performance Specifications for Imprecision and Bias Based on Within Subject (CV_I) and Between Subject (CV_G) Biological Variability.

Level	Imprecision, *I* (%)	Bias, *B* (%)	Total error, TE (%)
Optimum	$I = 0.25 \times CV_I$	$B = 0.125 \times (CV_I^2 + CV_G^2)^{1/2}$	$TE = 1.65 \times CV_I + CV_G$
Desirable	$I = 0.50 \times CV_I$	$B = 0.250 \times (CV_I^2 + CV_G^2)^{1/2}$	$TE = 1.65 \times CV_I + CV_G$
Minimum	$I = 0.75 \times CV_I$	$B = 0.375 \times (CV_I^2 + CV_G^2)^{1/2}$	$TE = 1.65 \times CV_I + CV_G$

Adapted from Klee (2010).

Table 3.2 Analyte Specific Performance Specifications from CLIA and the EFLM Biological Variation Database.

Analyte	CLIA limit (%)	Biological variability (%)		Optimum criteria (%)			Desirable criteria (%)			Minimum criteria (%)		
		CV_I	CV_G	I	B	TE	I	B	TE	I	B	TE
Albumin	8.0	2.6	5.1	0.7	0.7	1.8	1.3	1.4	3.6	2.0	2.1	5.4
AST	15.0	9.6	20.8	2.4	2.9	6.8	4.8	5.7	13.6	7.2	8.6	20.5
BUN	9.0	13.9	21.0	3.5	3.1	8.9	7.0	6.3	17.8	10.4	9.4	26.6
Carbon dioxide	20.0	4.0	4.8	1.0	0.8	2.4	2.0	1.6	4.9	3.0	2.3	7.3
Cholesterol	10.0	5.3	16.3	1.3	2.1	4.3	2.7	4.3	8.7	4.0	6.4	13.0
Creatinine	10.0	4.5	14.1	1.1	1.9	3.7	2.3	3.7	7.4	3.4	5.6	11.1
Glucose	8.0	8.1	5.0	2.0	1.2	4.5	4.1	2.4	9.1	6.1	3.6	13.6
Total protein	8.0	2.6	4.6	0.7	0.7	1.7	1.3	1.3	3.5	2.0	2.0	5.2

CV_I, within-subject variation; CV_G, between-subject variation; *I*, imprecision; *B*, bias; TE, total error.
Source: Aarsand et al. (2021) and Verma et al. (2019).

In addition to using APS to evaluate an alternate specimen type for an analyte, the utility of the assay may have an impact on the acceptance of an alternate sample type. For example, devices that are only intended for health screening (and not diagnostic level information) may not require as stringent assay criteria, but still be fit for purpose (Capiau et al., 2019; Patel et al., 2021). However, it should be abundantly clear to patients and healthcare providers of any limitations with the information being generated.

Although an alternate sample may be found to be medically and scientifically acceptable for a particular analyte, acceptance criteria describing successful sample collection (e.g. volume and quality) should also be defined prior to implementation of any patient self-collection technology. Low success rates will lead to frustration for the patient and medical and laboratory personnel, which will ultimately lead to poor adoption of new technologies.

3.4 Collection

Although careful selection and analysis of the specimen prior to utilization for the collection of patient samples represents an important aspect of patient centric sampling, thorough investigation of the specimen type (and collection technology) can help to reduce any risk to interpretation of analyte results. The ability of patients to perform self-collection of samples, however, represents perhaps the largest hurdle in the preanalytical process (Giavarina & Lippi, 2017), where quality of sample collection will be directly related to the ability of a person to collect a sample following prescriptive instructions for use or IFUs (Blicharz et al., 2018; Giavarina & Lippi, 2017; Perneger & Etter, 2001). Professionally trained phlebotomists are expected to follow stringent guidelines and techniques in order to properly perform venipuncture collection (Giavarina & Lippi, 2017), but similar training for a patient centric setting is not as straightforward.

For this reason, the entire collection process will need to be optimized for the person performing (self) collection. This includes providing the appropriate materials, training and preparation, sterilizing the wound collection site, creating a wound from which to collect a sample, collecting a sample of adequate volume, and some degree of post-collection processing. As the success of sample collection will influence measurement results (Williamson et al., 2012), each of the aforementioned steps will be investigated herein from the point of view of a patient performing self-collection.

3.4.1 Device and Kit Components

Although not inherently obvious, selection and/or design of every single component of a sample collection kit is crucial to successful sample collection. Each individual component can (and should) be rigorously tested prior to implementation with patients. This will help reduce or eliminate potential sources of error that could lead to improper collection and, more importantly, the possibility of injury during the collection process (U.S. Department of Health and Human Services, Food and Drug Administration, Center for Devices and Radiological Health, Office of Device Evaluation, 2016).

The first step to ensuring patient safety is the sterilization of all applicable components, especially those that come in contact with the patient during the collection process. In addition, appropriate safety mechanisms should be implemented to prevent injury to the patient (e.g. self-retracting lancet blades). Additional human factor engineering considerations may also be necessary to prevent improper or unintentional use of kit components (e.g. accidental activation of a single-use device).

Another aspect of a functional kit is the efficacy of the various individual components. Component function should be easy to understand and easy to activate (if applicable) in order to reduce the training burden needed for each patient. Additional consideration needs to be made in order to reduce the risk of improper collection for devices that are more complex and require manipulation that may not be intuitive. Overall, devices or kit components should work as intended every single time when they are being implemented appropriately. The potential misuse of a single component can result in inadequate sample collection or possible injury to the patient.

3.4.2 Training and Preparation

Prior to sample collection, the person performing the collection should receive adequate training in order to increase the likelihood of performing a successful collection and reduce the possibility of injury. Unfortunately, a limitation of patient centric sampling is the ability to provide consistent and robust training. As such, concise instructions that clearly indicate what is required for a successful sample collection should be provided as part of the kit. In general, the instructions should be adequate to allow collection of the sample without overburdening the patient with excess information that may lead to confusion or even ignoring the instructions completely. The FDA has indicated that all directions provided for at home sample collection should be written no higher than a 7th grade reading level, where most students are 12–13 years old (U.S. Food & Drug Administration, 2020).

Prior to sample collection, additional preparation steps may be recommended for the donor to help facilitate sample collection. These include making sure the patient is hydrated or has warmed the collection site to allow better flow of blood (Freeman et al., 2018). Once these steps are performed, the patient should unpackage each component (as necessary) and place in an easily accessible location. Without proper preparation, patients may perform sample collection inefficiently, which can lead to an overall poor experience. For example, patients who do not easily bleed may not be able to properly collect the small volume of blood that is available. In contrast, patients who bleed very easily may end up bleeding excessively (and getting blood where it is not intended) without the proper gauze or bandages easily accessible.

Following proper component preparation, the patient is typically asked to clean and disinfect the collection site prior to skin puncture. Washing of the collection site with warm, soapy water followed by sterilization with an alcohol pad is generally considered an effective means of preparing the collection site (Dhingra et al., 2010). In addition to protecting the patient, thorough cleaning is also required to prevent the contamination of blood as it is forced through (and onto) the surface of the skin (Arakawa & Ebato, 2012; Dhingra et al., 2010).

Overall, the sample collection process will ideally be consistent from patient-to-patient (Krleza et al., 2015). Additional resources such as video demonstrations and/or remote observation by a trained professional would also be beneficial but should be easily accessible. As with kit components, all resources should receive thorough review and go through usability studies (with complete collection kits) prior to implementation with patients. However, even after significant optimization of the collection kit and process, patients may not always utilize the resources that are provided. Thus, a means of reminding and/or incentivizing the donor to thoroughly review the instructions for use should be investigated to ensure adequate collection.

3.4.3 Wound Generation

Capillary blood is typically collected by activating a sterilized lancet that releases a needle or blade of appropriate size which punctures the skin up to a designed depth (Dhingra et al., 2010). In addition to generating a wound of a specified size, lancets typically function in a single-use manner such that following activation the blade is retracted into a housing to prevent additional, unintended wounds from being made.

In adult patients, capillary blood is typically collected from a finger. However, alternative locations, such as the upper arm, have been investigated (Blicharz et al., 2018) but may result in more difficult capture of the blood depending on the operating principle of the collection device. Unfortunately, there has been limited work published demonstrating correlation of capillary blood collected from different locations on the body (e.g. capillary blood from the finger compared to capillary blood from the shoulder). Locations for collecting capillary blood are instead selected based on the pain to the patient or capillary density, which will be related to ability to collect more blood volume (Blicharz et al., 2018; Li et al., 2013). Ultimately, the rate of bleeding will also vary from person-to-person due to anatomical factors such as skin thickness and blood circulation (Lei & Prow, 2019).

In order to accommodate sample collection from donors that do not bleed very easily, a larger wound is required of all donors to prevent variability in the collection process from donor to donor. In our experience, use of lancets that generate smaller wounds may be less painful to the patient, but multiple lancings during one collection are often required in order to acquire sufficient volume. As an alternative to large single blade lancets, arrays of microneedles have also been demonstrated to reduce pain to the patient but enable sufficient volume to be obtained in most patients (Blicharz et al., 2018). However, the use of smaller microneedles which do not penetrate as deep (1 mm), may not be suitable for all patients, depending on their physiology.

To investigate the effect of the lancet on sample, a short study was performed where four donors were asked to lance three separate fingers with three different types of lancets. A volume of 300 µL blood was acquired from each finger using three 100 µL lithium heparin-coated glass capillaries. Blood from the three glass capillaries was combined and centrifuged to obtain plasma for measurement. Milking of the finger was allowed as needed for each lancet. A description of the lancets used and an image of their respective blades can be found in Figure 3.7. This study was performed under a protocol approved by an Institutional Review Board (ASPIRE Protocol #520100174).

Five analytes were measured in the plasma obtained using the three different lancets. Results for the BD Sentry Safe and Tenderlett Finger Incision Device for each donor were normalized to the results obtained using the BD high blood flow

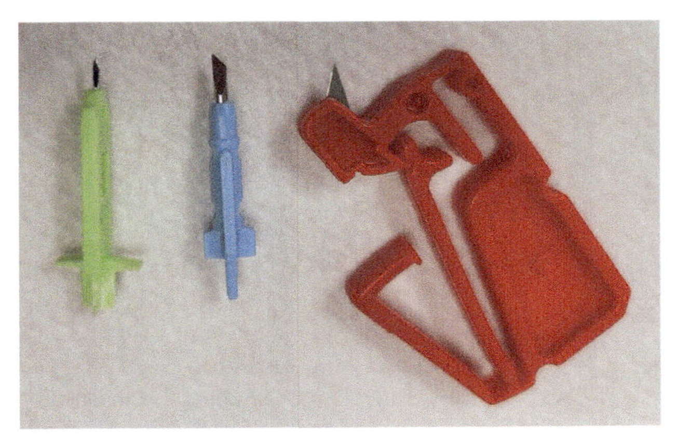

Color above:	Green	Blue	Red
Lancet:	BD Sentry Safe (369523)	BD Microtainer Contact-Activated Lancet, High Blood Flow (366594)	Accriva Diagnostics Tenderlett Finger Incision Device (TL100I)
Specifications:	23 gauge, 1.8 mm depth	1.5 mm wide, 2.0 mm deep	1.75 mm deep

Figure 3.7 Image of the blades and additional information for the lancets utilized for sample comparison.

lancet. The mean and standard deviation of these results suggest that there is not any significant difference in plasma obtained using the three different lancets for AST, BUN, cholesterol, creatinine (enzymatic assay), and glucose (Figure 3.8). However, these results should be considered preliminary as only a small number of donors, analytes, and lancets were utilized. In addition, results were limited to finger lancings and did not include other body locations.

In addition to measurement of the analytes listed above, the degree of sample hemolysis (i.e. red blood cell lysis) was determined for each sample through measurement of the sample's hemolysis index using absorbance-based measurements of the sample on a Roche cobas® autoanalyzer. Although again limited in statistical power, these results indicate that all donors had a higher level of hemolysis using the BD Sentry Safe and Tenderlett lancets (Table 3.3). We hypothesize that this may be a result of increased finger milking used for the BD Sentry Safe lancet due to the smaller puncture wound created. The Tenderlett device functions differently than the other two lancets utilized, as it is designed to create a shallow incision rather than a puncture wound which may impact the level of hemolysis

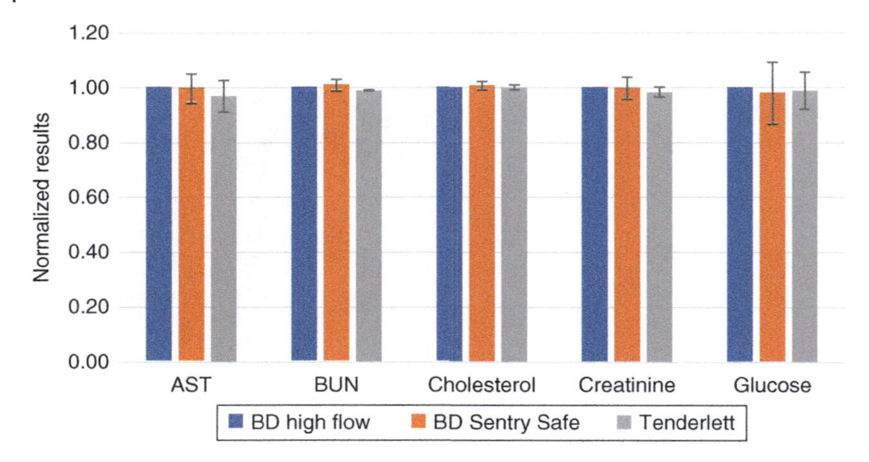

Figure 3.8 Normalized plasma measurements obtained from lancing three different fingers with three different lancets (see Figure 3.7) for four different donors. Results are normalized to BD high blood flow measurements. Error bars represent one standard deviation of normalized results.

Table 3.3 Measured Hemolysis Index for Plasma Samples Obtained Using Three Different Lancets from Four Different Donors. The Hemolysis Index Values are Approximately Equivalent to Hemoglobin Values in mg/dL.

	Hemolysis index			
Lancet	**Donor 1**	**Donor 2**	**Donor 3**	**Donor 4**
BD high flow	20	4	4	7
BD sentry safe	23	22	8	40
Tenderlett	28	12	6	31

observed. It should be noted that all hemolysis levels observed from each lancet utilized will have little to no impact on most clinical analyte measurements.

3.4.4 Collecting Sample

Once the wound is created, as blood exits the body it is either actively added to the device by the person collecting the sample (e.g. adding blood to the membrane of a dried blood spot card) or is passively collected by the device itself (e.g. pulled from the body through vacuum pressure and fed by gravity into the collection tube of devices temporarily adhered to the skin during collection). Although both approaches are relatively straightforward, each process is not without its own concerns.

An issue often debated regarding the collection of capillary blood is whether to wipe away the first drop of blood with gauze (and not collect it). Differences in concentrations of analytes (particularly glucose) have been reported, with the explanation that the sample is being diluted by interstitial and intracellular fluids or being contaminated by alcohol (Bond & Richards-Kortum, 2015; Fruhstorfer & Quarder, 2009; Hortensius et al., 2010). However, the differences in concentration measured have also been reported as not clinically relevant (Crawford et al., 2020; Palese et al., 2016). Unfortunately, there is no general agreement on whether or not wiping away the first drop of blood is necessary for capillary blood collection (Hortensius et al., 2010) and may not even be easily achievable for some collection technologies (e.g. those available from Tasso, Inc. and YourBio Health). For a consensus to be reached regarding wiping the first drop of blood, additional studies will need to be performed investigating the volumetric dependency of sample composition, including hematological measurements and analyte concentrations, as well as sample quality.

Another aspect that is not standardized for the collection of capillary blood is the use of milking where positive pressure is applied to the wound site (i.e. massaged) in a manner that facilitates blood flow through the wound site, especially from a lanced finger. There are concerns that milking may lead to further dilution of the sample with tissue fluids or hemolysis of the sample (Bond & Richards-Kortum, 2015; Fruhstorfer & Quarder, 2009; Palese et al., 2016; Stein & Pum, 2005). Gentle milking has also been reported as acceptable but with little details provided on what amount of force is appropriate (Lewis & Tatsumi, 2006). Despite these concerns, there is limited data to demonstrate that milking can lead to measurement error. Alternatively, it has been demonstrated that blood samples from milked and non-milked fingers produce the same glucose results (Fruhstorfer & Quarder, 2009).

The volume of blood collected has also been suggested to have an impact on the analyte results observed (Bond & Richards-Kortum, 2015). The data provided by Bond and Richards–Kortum indicates that a minimum sample volume of at least 60 μL is required to overcome potential variability in hematological measurements from the initial volume of capillary blood collected, where the variability is again attributed to dilutions from interstitial and intracellular fluids. In addition, if an anticoagulant is being utilized, the volume of blood being collected needs to be within an acceptable range to ensure appropriate sample to anticoagulant ratios. If the concentration of anticoagulant in the collected blood is too low, coagulation may occur and if it is too high, it can adversely affect the sample (e.g. hemolysis) or the measurement of the analyte (Baskin et al., 2019; Heireman et al., 2017). In addition, sufficient sample volume is an absolute requirement in order to perform testing where the volume required will be dependent on aspects further downstream in the sample measurement process (Baskin et al., 2019). Required aspiration volumes as well as instrument dead volumes are examples of these aspects.

In some instances, an "exact" sample volume is desired to facilitate downstream processing and sample measurement. This is of particular interest for dried samples where it is difficult to determine the volume of blood present in a dried sample. However, by limiting the volume of blood that can be absorbed onto a material, measurements of analyte concentration are more precise (Deprez et al., 2019; Protti et al., 2019, 2020; Velghe et al., 2019). This can be achieved through volume metering (*i.e.* limiting collection to a predetermined volume of blood) using capillary tubes or absorbent materials, both of which are designed to allow only a specified volume of sample to be obtained (Crawford et al., 2020; Protti et al., 2019; Velghe et al., 2019). The inability to acquire sufficient volume is a more difficult problem to correct, as such, larger wounds are typically generated.

3.4.5 Post Collection Processing

Following sample collection, the patient will need to perform bandaging of the wound site, as well as proper disposal of used kit components. However, additional steps may be required before the sample can be transported to the laboratory. For example, if anticoagulants are utilized, thorough mixing of blood with the anti-coagulant is required to ensure its utility (Baskin et al., 2019; Stein and Pum, 2005). If the collection technology is based on the drying of specimens, a sufficient drying time and desiccant (in a suitable environment) may be required prior to specimen transportation (Denniff et al., 2013; Freeman et al., 2018). It is also important that samples are properly labeled and/or registered to provide proper identification with the patient. Following any additional processing steps, the sample must be appropriately packaged and shipped into the laboratory for analysis.

3.4.6 Conclusions

Despite many efforts, the collection experience may vary greatly from patient-to-patient, not only due to the collection technology (and additional materials provided for collection) but also on the abilities and expectations of the patient (U.S. Department of Health and Human Services, Food and Drug Administration, Center for Devices and Radiological Health, Office of Device Evaluation, 2016). Without question, patient centric microsampling would benefit from standardization in processes regardless of technology where possible. For example, the skin sterilization procedure could easily be standardized to prevent confusion and promote better understanding (Lewis & Tatsumi, 2006). The posture of the patient during collection has even been shown to have an effect on analyte measurements (Baskin et al., 2019; Lima-Oliveira et al., 2017; Stein & Pum, 2005). However, a recommended posture during collection is never discussed for use with microsampling technologies. Ultimately, increased standardization will likely lead to improved sample fidelity in the hands of unskilled operators (Carter & Card, 2018).

3.5 Transportation and Sample Stability

Following collection, chain of custody of the sample is critical during transportation into the laboratory, in order to ensure sample fidelity and quality of results (Wickremsinhe et al., 2020). Stability of an analyte is typically defined as the ability of the analyte (and specimen matrix) to maintain the concentration that was present at the time of collection for a period of time (Felding et al., 1981; Londhe & Rajadhyaksha, 2020; Stein & Pum, 2005). Although time post-collection is a major factor, analyte stability will ultimately be affected by a variety of factors including the sample type, the time between sample collection and centrifugation (if required), the sample storage conditions, and even the measurement technique utilized (Bautista Balbás et al., 2017; Boyanton Jr & Blick, 2002; Gómez Rioja et al., 2018; Heins et al., 1995; Oddoze et al., 2012; Rühling et al., 1992; Zhang et al., 1998). For this reason, there is inconsistency in the reported stability for individual analytes. Some of the more notable elements of sample stability and how they relate to patient centric microsampling will be discussed below.

3.5.1 Specimen Matrix and Separation

Although selection of specimen matrix will be influenced by the assay being utilized, the matrix chosen plays an important role in the stability of many analytes. For many analytes of clinical interest, separation of plasma or serum from red blood cells is recommended soon after collection to preserve sample stability, whereas other analytes are stable for days in whole blood (Chu and MacLeod, 1986; Clark et al., 2003; Hankinson et al., 1989; Heins et al., 1995; Lewis & Tatsumi, 2006; Stein & Pum, 2005; Oddoze et al., 2012; Ono et al., 1981). Continued biological activity in whole blood is one reason for the decreased stability observed in plasma and serum samples that are not separated within an appropriate time period of collection (Zhang et al., 1998). Metabolism of glucose is a clear example of how the concentration of an analyte can be depleted while exposed to red blood cells even in the presence of additives used to inhibit glycolysis (Roccaforte et al., 2016). Prolonged exposure can also lead to analyte elevation due to diffusion of analytes out of cells or even lysis of the cells. Potassium and AST are notable examples of analytes that become elevated during prolonged exposure to red blood cells (Clark et al., 2003; Rehak & Chiang, 1988).

To assist with separation (and allow easier sample processing), thixotropic separator gels are commonly utilized to serve as a barrier between blood cells and plasma or serum following centrifugation of a sample tube (Oddoze et al., 2012). However, separation gels may also lead to analyte instability due to adsorption, absorption, or even chemical release (Bowen & Remaley, 2014). More recently, mechanical barriers have been demonstrated as an alternative separator between cells and plasma following centrifugation, but are not commonly utilized as this point (Bautista Balbás et al., 2017).

3.5.2 Storage Condition

Once a sample has been collected and any additional processing is performed, the sample is transported into the laboratory for additional processing and measurement. In order for the sample to be appropriately transported into the laboratory, the quality of the sample needs to be maintained through proper storage. For example, although refrigeration of the sample typically helps to maintain analyte stability, it can also lead to hemolysis of the sample (Heins et al., 1995; Koseoglu et al., 2011; Lewis & Tatsumi, 2006). Freezing of whole blood samples is typically avoided where plasma or serum is the matrix required for analyte measurement, as nearly complete hemolysis of the sample will occur (Sloviter, 1962).

In addition, the stability of the analyte of interest needs to be considered in the specimen matrix it will be shipped. This is often investigated by storing samples at predefined temperatures for set periods of time and comparing results to initial (baseline) measurements. Temperatures typically investigated include room temperature (approximately 20–24 °C), refrigerated (approximately 2–8 °C), and frozen (approximately −10 to −25°C). Depending on the application (e.g. shipping samples without temperature control), elevated and freezing conditions may also be investigated. Although this type of study effectively provides stability data in terms of a mean kinetic temperature, shipping conditions will not be as predictable and unless careful packaging and temperature control considerations are made, samples may be exposed to a wide range of temperature conditions (Londhe & Rajadhyaksha, 2020; Nys et al., 2017). For this reason, stability for "ambient" conditions within the context of expected seasonal variability and any associated packaging may also need to be considered.

The data in Figure 3.9 provides an example for the need for temperature control, even under somewhat mild conditions (external temperatures ranging from approximately 7 to 24 °C). A temperature logger (Fisher Scientific 15-059-002) was placed in a polystyrene box (approximately 6×9×11 in. and 1.5 in. thick)

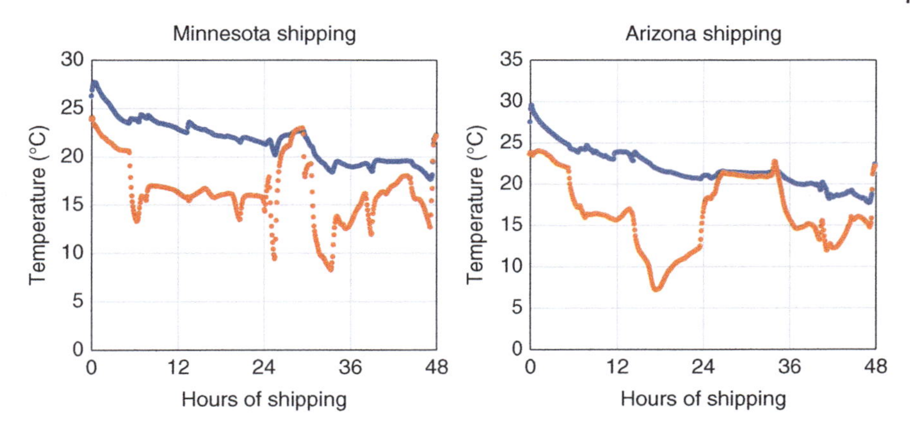

Figure 3.9 Temperature logger data for temperature controlled packages shipped to and from North Carolina in November 2019. Blue dots represent data from the temperature logger that was within the temperature controlled packaging. Orange dots represent data from the temperature logger that was outside of the temperature controlled packaging but within a cardboard box.

containing two warmed temperature packs, and the polystyrene box was then put in a cardboard box for shipping. Another temperature logger was placed inside of the cardboard box but exterior to the polystyrene box to monitor the conditions exterior to the temperature-controlled packaging during shipping. Two separate packages were shipped from North Carolina and returned within the same 48 hr period in early November 2019. One package was sent to Minnesota, while another was mailed to Arizona. As can be seen in the figure, the external temperature of each package was subjected to several spikes and changes in temperature. However, the inside of the package remained at temperatures more consistent with room temperature due to the temperature control mechanisms implemented.

Additional temperature logger data can be found in Figure 3.10 for a package that was shipped from North Carolina to California and back during May of 2019. The data indicates that packages can be exposed to temperatures that are more extreme than the actual outdoor temperatures reported. This is a result of additional external factors that the package is exposed to. In this case, the package appears to have sat outside at the airport in Los Angeles for an extended period of time, possibly in a metal shipping container that does not dissipate heat efficiently. The outdoor temperature ranges observed during the shipping of this package were relatively mild compared to other temperatures that might be expected at different locations throughout the year. As such, even though an ambient temperature may be acceptable for an analyte, appropriate temperature control may be necessary to ensure the sample stays at an appropriate temperature during the

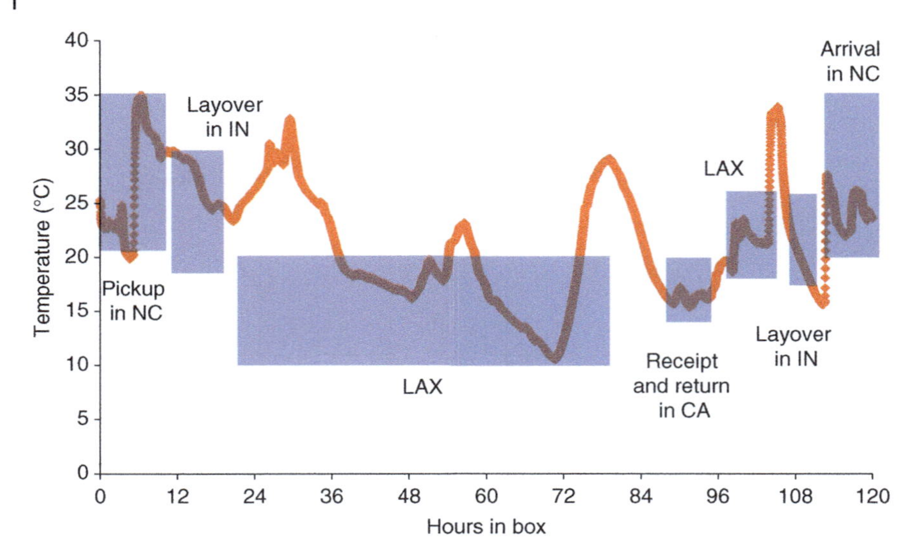

Figure 3.10 Temperature logger data for a package (without temperature control) shipped from North Carolina (NC) through Indiana (IN) and Los Angeles (LAX) airports to a laboratory in California (CA) and then returned. Blue boxes represent the minimum and maximum temperatures reported by a national weather service in that location during the time the package was at that location.

shipping process. In addition, temperature and humidity monitoring may be appropriate to ensure the sample has remained within the validated stability limits of the analyte.

Temperature control, however, may not always be practical or needed for some analytes in certain types of specimens. Alternative testing may be required to demonstrate stability during the shipping process. As an example, the United States Food and Drug Administration has recently proposed excursion studies modeled after the International Safe Transit Association Series 7D test procedure (U.S. Food & Drug Administration, 2020). Both winter and summer profiles (Table 3.4) were proposed for emergency use authorization of dried blood spot samples intended for measurement of SARS-CoV-2 antibodies.

Most of the aspects of stability discussed above are provided in the context of liquid samples. However, one of the advantages of requiring small samples volumes is the ability to dry the sample, which can provide improved analyte stability (Edge et al., 2016; Freeman et al., 2018; Londhe & Rajadhyaksha, 2020; Nys et al., 2017; Spooner et al., 2018; Velghe et al., 2016). Although whole blood is the most common dried matrix, plasma, or serum can also be dried (Crawford et al., 2020; Londhe & Rajadhyaksha, 2020).

Table 3.4 Proposed Summer and Winter Excursions for Testing Analyte Stability.

Cycle period	Winter profile (°C)	Summer profile (°C)	Cycle period (hr)	Total time (hr)
1	−10	40	8	8
2	18	22	4	12
3	−10	40	2	14
4	10	30	36	50
5	−10	40	6	56

3.5.3 Measurement Technique

Although stability is often thought of as the change in concentration of a particular analyte over time, this is not entirely true as the analysis technique itself can be influenced by changes in the overall matrix (Gómez Rioja et al., 2018; Londhe & Rajadhyaksha, 2020). For example, two methods are readily used for measurement of serum and plasma creatinine. The Jaffé method is based on a kinetic colorimetric assay where creatinine forms a complex with picric acid under alkaline conditions to create a yellow-orange complex. This approach, however, is susceptible to interferents from several creatinine-like molecules (Peake & Whiting, 2006), which can become elevated when serum or plasma is not separated from blood cells (Heins et al., 1995). Enzymatic-based approaches for measurement of creatinine, however, are not as susceptible to these interferences, due to the specificity of the enzymes used. Although the methods show a high level of agreement for both plasma and serum samples when centrifuged within 1 hr of collection (Figure 3.11), data in Figure 3.12 demonstrates the elevation of

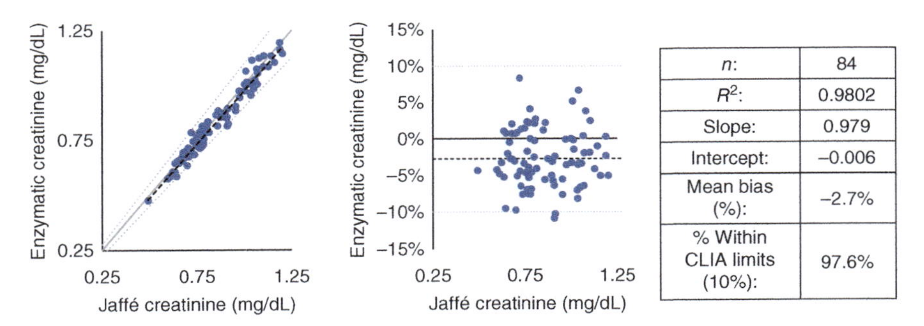

Figure 3.11 Jaffé versus enzymatic creatinine assay comparison data. Black dashed lines represent regression line and mean bias in correlation and percentage bias plots, respectively. Gray dashed lines represent individual CLIA bias limits for each analyte, while the solid gray line represents ideal results.

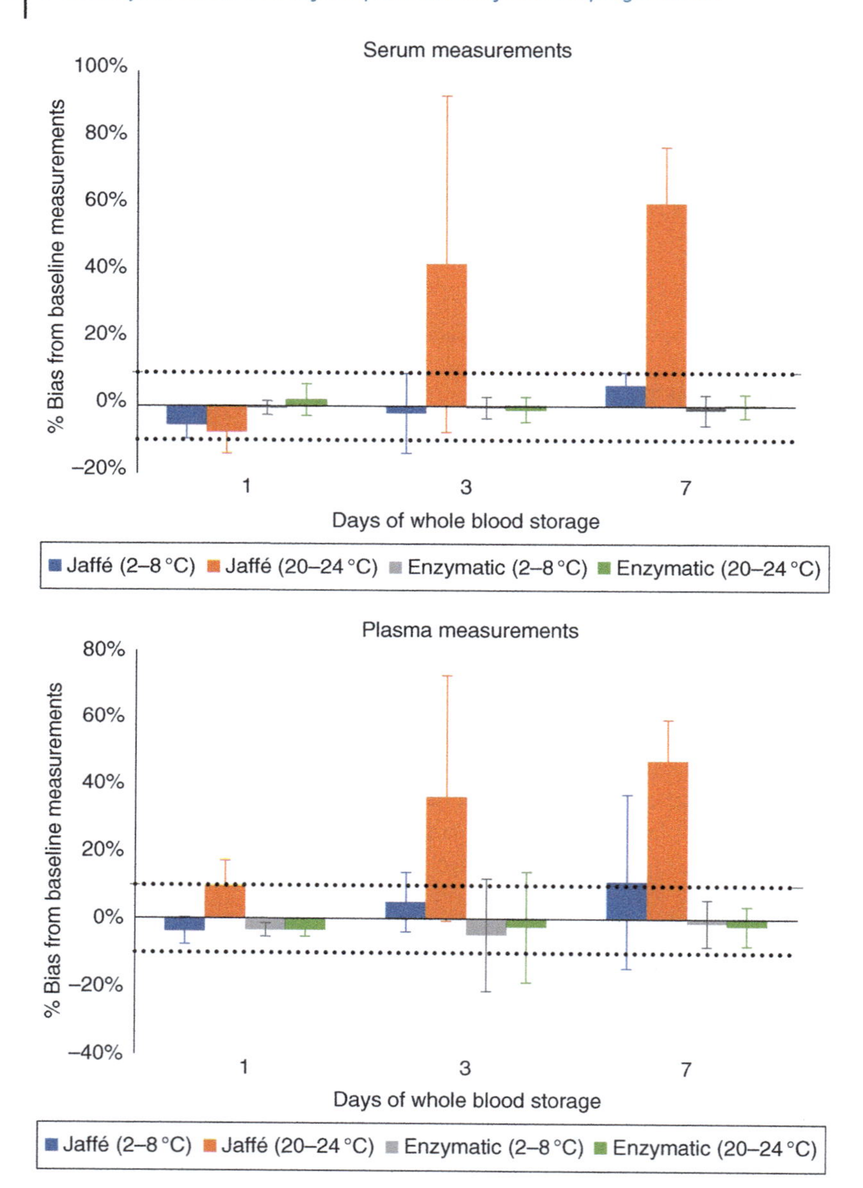

Figure 3.12 Creatinine whole blood stability for two different assays for both serum- and plasma-based measurements. Samples were centrifuged prior to measurement. Error bars represent one standard deviation, while the dashed lines represent the recommended CLIA limits of ±10%. Eight (8) to 14 measurements were used for each data point.

creatinine measurements over time when using the Jaffé assay with samples stored as whole blood prior to centrifugation and measurement. Refrigeration of the samples helps to reduce the elevation, but both refrigerated and room temperature samples show an elevation of measured "creatinine" over time for samples measured as serum and plasma. Enzymatic measurement of creatinine demonstrates less influence from interfering molecules due to the specificity of the enzyme utilized (Figure 3.12).

3.5.4 Hematocrit Effects

Another consideration to make for whole blood samples, especially those that are dried, is the effect that varying hematocrit levels can ultimately have on test results. Increased levels of hematocrit have been shown to have smaller dried blood spot sizes due to the increased viscosity limiting the spread of blood and ultimately leading to a smaller volume of blood being sampled when a fixed punch size is utilized for sample processing (Denniff & Spooner, 2010; de Vries et al., 2013; Fan & Lee, 2012; O'Mara et al., 2011). However, hematocrit correction can be performed to some degree by measuring hematocrit levels directly or indirectly (Ackermans et al., 2021; Capiau et al., 2016, 2018; Oostendorp et al., 2016) or by fixing the volume of dried sample (Deprez et al., 2019; Londhe & Rajadhyaksha, 2020; Meesters et al., 2012; Protti et al., 2019; Velghe & Stove, 2018; Youhnovski et al., 2011), recovery of the analyte during extraction can also be affected by hematocrit levels as well as the age of the sample due to molecular changes in hemoglobin which (de Vries et al., 2013; Li et al., 2011; Lomeo et al., 2008; Xie et al., 2018). Elevation of analyte concentration may also play a role in analyte recovery (Koster et al., 2013, 2015).

3.5.5 Conclusions

A broad overview of factors that affect analyte stability has been provided in this section, but these details should not be considered complete. Other environmental considerations (e.g. light exposure, humidity), device specific considerations, even the acceptance criteria utilized for studies will influence reported analyte stability or even the intended use of the analyte results. Assessment of sample stability may ultimately require testing for each collection device, analyte of interest, and assay utilized. The individual analyte stability along with its proposed utility should determine the appropriate transportation requirements for the sample as well as an attainable chain of custody (Stein & Pum, 2005).

It should also be noted that overly strict sample requirements may be difficult to achieve without the appropriate infrastructure in place and can lead to unnecessary rejection of samples (Zhang et al., 1998). Conversely, requirements that are too lenient may produce poor-quality results.

3.6 Preanalytical Processing

Once received and accessioned into the laboratory's specimen management system, additional sample processing is often required prior to analyte measurement. The type of processing required will be dependent on the specimen matrix received and may be as straightforward as transferring the matrix to the appropriate vessel for measurement but can also be more complex. Dried samples will require reconstitution or extraction, which can be time consuming and require extensive training of lab technicians, and thus may be considered a barrier for large-scale adoption (Déglon et al., 2015; Velghe et al., 2019). In addition, low-volume samples require processing steps to ensure that volume is not lost, limit preanalytic variation, and maximize the utility of the sample (Poitout-Belissent et al., 2016). Some common processing techniques will be discussed herein.

3.6.1 Separation of Plasma and Serum

One of the most common processing techniques is the separation (fractionation) of blood components in order to obtain the preferred matrix for measurement. Serum and plasma are routinely separated using a centrifuge. In this process, the blood is spun at a high speed to force denser cells and platelets to the bottom of the tube leaving the separated plasma or serum on top of the tube, which can easily be aspirated away from the cells. As mentioned before, thixotropic separation gels or mechanical barriers can also be utilized to form a rigid barrier between the cells and the plasma or serum, which allows the sample to be poured off more easily and greater volumes to be obtained for sample measurement. Unfortunately, centrifugation (at the time of collection) is limited to locations that can accommodate a centrifuge, which are often bulky and require external power. Single use or portable centrifuges have been proposed for remote use (e.g. in the home); however, they have not yet reached full viability (Strohmeier et al., 2015). In addition, although centrifugation is the traditional method for separation, it is not as easily implemented with low volume (<200 μL) samples, as they do not easily "pour off" and are not readily amenable to use with available automated liquid handlers. Currently, liquid microsamples are typically processed manually for this reason.

As an alternative, plasma separation can be achieved through filtration of cells (Velghe et al., 2019). This is functionally achieved by limiting cell transport either through or along a filter membrane (Crawford et al., 2020; Liu et al., 2013). For through- or trans-membrane filtration, whole blood is applied to one side of the membrane which restricts cells and platelets from passing through it. Liquid plasma is allowed to pass through the membrane and is collected either as a liquid or onto an additional membrane for drying. Although this approach has been proposed for a variety of applications, it is limited in the volume of plasma that can be obtained due to fouling (clogging) of the membrane with larger volumes of blood (Ďurč et al., 2018; Liu et al., 2013). In addition, hemolysis of blood can occur if too much pressure is applied to achieve separation (Philp et al. 1994).

Lateral flow filtration is achieved by applying blood to a membrane and then, due to reduced diffusion along the membrane relative to plasma, cell migration is restricted to near the dosing site while plasma (or serum) can migrate a greater distance within the membrane. For this type of separation, plasma samples are typically retained in the membrane and dried with a desiccant (Crawford et al., 2020; Kaiser et al., 2020). Utilization of these two different membrane filtration approaches is more commonly performed with dried samples to reduce the processing requirements at the time of collection. Although there are advantages to utilizing filtered plasma, the resulting matrix on which measurements are ultimately made will need to be thoroughly compared to plasma obtained through standard approaches (i.e. centrifugation of blood obtained intravenously) for each analyte of interest due to the potential for reduced recovery (Crawford et al., 2020; Homsy et al., 2012; Liu et al., 2013, 2016; Velghe et al., 2019).

3.6.2 Sample Dilution

As already mentioned above, current automated systems are not able to reproducibly process low-volume samples (<100 µL). This is prohibitive for analytical processes where exact volumes are typically required (Londhe & Rajadhyaksha, 2020). For this reason, liquid microsamples may be diluted to allow easier processing and/or reducing analyte concentration into the analytical measurement range of the assay. Dried samples are typically diluted through the extraction process, as reconstituting to the exact same volume of dried sample does not allow efficient volume recovery. In all cases, the diluent utilized should be carefully selected to reduce matrix effects that may influence the measurement and in the case of dried samples improve analyte recovery (Londhe & Rajadhyaksha, 2020; Poitout-Belissent et al., 2016). Optimization of the extraction procedure is required for the analytical process and optimization for each individual analyte may be required (Protti et al., 2019; Velghe et al., 2019). To improve analytical performance and streamline processing, collection materials have also been developed that are designed to absorb only a specified volume of blood provided that adequate volume is provided (Protti et al., 2019; Velghe et al., 2019). This approach is still limited by patients that are unable to provide sufficient volume. Alternatively, volume correction can be performed analytically by measuring an analyte with a narrow clinical range in parallel for volume correction (Crawford et al., 2020).

3.6.3 Conclusions

Preanalytical processing of samples can be a major source of errors especially in the case that manual steps are required (Poitout-Belissent et al., 2016). Easily repeatable processes (and ideally universal) should be implemented wherever possible. As such, automated solutions are desirable but not always feasible depending on the microsampling technology utilized. Reconstitution and extraction of dried samples have been able to take advantage of this type of instrumentation and become more prevalent in recent years due to the use of larger extraction volumes (Edge et al., 2016).

3.7 Overall Conclusions

Overall, there are a variety of areas that need to be addressed prior to implementation of patient centric sampling approaches. This includes appropriate selection of a sample matrix, optimization of the collection process, development of a robust transportation solution, and refinement of any sample processing steps. As each step is vital to the quality of analytical results, each should be carefully reviewed prior to implementation (Giavarina & Lippi, 2017).

During initial investigation of alternate collection technologies, extensive studies will need to be performed in order to ensure the quality of analytical results before adoption of any proposed sampling technology (Capiau et al., 2019; Viswanathan, 2012). This includes bridging studies to compare the preferred matrix from the standard collection technology with the new matrix (Déglon et al., 2015; Freeman et al., 2018; Wickremsinhe et al., 2020) where any observed differences in the new matrix may be detrimental to its acceptance as an alternate sample (Viswanathan, 2012). However, acceptance criteria utilized may be dependent on the intended application (e.g. clinical diagnosis, health screening, and nutritional).

Once the new technology has been demonstrated to be analytically acceptable, the self-collection process must be thoroughly reviewed and refined to optimize the rate of collection success. In general, the self-collection process (regardless of technology) would benefit from standardization in processes to allow better understanding from patients about what is required (Lei & Prow, 2019). Additionally, any post-collection processing required by the patient (including packaging and shipping) should be clear and intuitive in order to ensure sample viability (Baskin et al., 2019; Velghe et al., 2016).

After receipt in the laboratory, sample fidelity can be more easily maintained through adoption and utilization of standard operating procedures for any additional preanalytical processing steps. However, as the demand for patient centric microsamples continues to grow, there will be an increased need for laboratories (and analytical instrumentation) to do more and/or improved testing with reduced sample volumes.

Although utilization of microsample blood collection technologies seems to be preferred by patients to venous sample collection (Mbughuni et al., 2020), implementation of any patient centric sampling technology will ultimately require thorough review of the *entire* sampling process. Some common preanalytical considerations and representative data have been provided herein, but these topics are not exhaustive and additional considerations will be dependent on the type of technology being utilized.

References

Aarsand, A. K., Fernandez-Call, P., Webster, C., Coskun, A., Gonzales-Lao, E., Diaz-Garzon, J., Jonker, N., Minchinela, J., Simon, M., Braga, F., Perich, C., Bonde, B., Roraas, T., Marques-Garcia, F., Carobene, A., Aslan, B., Tejedor, X., Barlett, W. A., & Sandberg, S. (2021). *The EFLM Biological Variation Database*. Retrieved May 17, 2021, from https://biologicalvariation.eu.

Ackermans, M. T., de Kleijne, V., Martens, F., & Heijboer, A. C. (2021). Hematocrit and standardization in DBS analysis: A practical approach for hormones mainly present in the plasma fraction. *Clinica Chimica Acta*, *520*, 179–185. https://doi.org/10.1016/j.cca.2021.06.014.

Arakawa, M., & Ebato, C. (2012). Influence of fruit juice on fingertips and patient behavior on self-monitoring of blood glucose. *Diabetes Research and Clinical Practice*, *96*(2), 9–11. https://doi.org/10.1016/j.diabres.2012.01.028.

Avataneo, V., D'Avolio, A., Cusato, J., Cantù, M., & de Nicolò, A. (2019). LC-MS application for therapeutic drug monitoring in alternative matrices. *Journal of Pharmaceutical and Biomedical Analysis*, *166*, 40–51. https://doi.org/10.1016/j.jpba.2018.12.040.

Baskin, L., Chin, A., Abdullah, A., & Naugler, C. (2019). Errors in patient preparation, specimen collection, anticoagulant and preservative use: How to avoid such pre-analytical errors. In *Accurate results in the clinical laboratory* (2nd ed., Elsevier pp. 11–26).

Bautista Balbás, L. A., Amaro, M. S., Rioja, R.G., Alcaide Martin, M. J., Soto, A. B. (2017). Stability of plasma electrolytes in barricor and PST II tubes under different storage conditions. *Biochemia Medica*, *27*(1), 225–230. https://doi.org/10.11613/BM.2017.024.

Beharry, M. (2010). DBS: A UK (MHRA) regulatory perspective. *Bioanalysis*, *2*(8), 1363–1364. https://doi.org/10.4155/bio.10.109.

Blicharz, T. M., Gong, P., Bunner, B. M., Chu, L. L., Leonard, K. M., Wakefield, J. A., Williams, R. E., Dadgar, M., Tagliabue, C. A., el Khaja, R., Marlin, S. L., Haghgooie, R., Davis, S. P., Chickering, D. E., & Bernstein, H. (2018). Microneedle-based device for the one-step painless collection of capillary blood samples. *Nature Biomedical Engineering*, *2*(3), 151–157. https://doi.org/10.1038/s41551-018-0194-1.

Bond, M. M., & Richards-Kortum, R. R. (2015). Drop-to-drop variation in the cellular components of fingerprick blood: Implications for point-of-care diagnostic development. *American Journal of Clinical Pathology*, *144*(6), 885–894. https://doi.org/10.1309/AJCP1L7DKMPCHPEH.

Bowen, R. A. R., & Adcock, D. M. (2016). Blood collection tubes as medical devices: The potential to affect assays and proposed verification and validation processes for the clinical laboratory. *Clinical Biochemistry*, *49*(18), 1321–1330. https://doi.org/10.1016/j.clinbiochem.2016.10.004.

Bowen, R. A. R., & Remaley, A. T. (2014). Interferences from blood collection tube components on clinical chemistry assays. *Biochemia Medica*, *24*(1), 31–44. https://doi.org/10.11613/BM.2014.006.

Boyanton, Jr., B. L., & Blick, K. E. (2002). Stability studies of twenty-four analytes in human plasma and serum. *Clinical Chemistry*, *48*(12), 2242–2247. https://doi.org/10.1093/clinchem/18.12.1498.

Boyd, R., Leigh, B., & Stuart, P. (2005). Capillary versus venous bedside blood glucose estimations. *Emergency Medicine Journal*, *22*(3), 177–179. https://doi.org/10.1136/emj.2003.011619.

Brunet, B. R., Barnes, A. J., Scheidweiler, K. B., Mura, P., & Marilyn, A. (2009). Development and validation of a solid-phase extraction gas chromatography–mass spectrometry method for the simultaneous quantification of methadone, heroin, cocaine and metabolites in sweat. *Sweat for DOAs 392*, 115–127. https://doi.org/10.1007/s00216-008-2228-0.Development.

Capiau, S., Wilk, L. S., Aalders, M. C. G., & Stove, C. P. (2016). A novel, nondestructive, dried blood spot-based hematocrit prediction method using noncontact diffuse reflectance spectroscopy. *Analytical Chemistry*, *88*(12), 6538–6546. https://doi.org/10.1021/acs.analchem.6b01321.

Capiau, S., Wilk, L. S., de Kesel, P. M. M., Aalders, M. C. G., & Stove, C. P. (2018). Correction for the hematocrit bias in dried blood spot analysis using a nondestructive, single-wavelength reflectance-based hematocrit prediction method. *Analytical Chemistry*, *90*(3), 1795–1804. https://doi.org/10.1021/acs.analchem.7b03784.

Capiau, S., Veenhof, H., Koster, R. A., Bergqvist, Y., Boettcher, M., Halmingh, O., Keevil, B. G., Koch, B. C. P., Linden, R., Pistos, C., Stolk, L. M., Touw, D. J., Stove, C. P., & Alffenaar, J.-W. C. (2019). Official international association for therapeutic drug monitoring and clinical toxicology guideline: Development and validation of dried blood spot-based methods for therapeutic drug monitoring. *Therapeutic Drug*

Monitoring, *41*(4). https://journals.lww.com/drug-monitoring/Fulltext/2019/08000/Official_International_Association_for_Therapeutic.1.aspx.

Caraway, W. T. (1962). Chemical and diagnostic specificity of laboratory tests. *The American Journal of Clinical Pathology*, *37*(5), 445–464. http://ajcp.oxfordjournals.org.ezproxy.its.uu.se/content/ajcpath/37/5/445.full.pdf.

Carter, G., & Card, D. J. (2018). *Methods for assessment of vitamin D, Laboratory Assessment of Vitamin Status*. Elsevier Inc. https://doi.org/10.1016/B978-0-12-813050-6.00003-6.

Ceriotti, F., Fernandez-Calle, P., Klee, G. G., Nordin, G., Sandberg, S., Streichert, T., Vives-Corrons, J. L., & Panteghini, M. (2017). Criteria for assigning laboratory measurands to models for analytical performance specifications defined in the 1st EFLM Strategic Conference. *Clinical Chemistry and Laboratory Medicine*, *55*(2), 189–194. https://doi.org/10.1515/cclm-2016-0091.

Chu, S. Y., & MacLeod, J. (1986). Effect of three-day clot contact on results of common biochemical tests with serum. *Clinical Chemistry*, 2100. https://doi.org/10.1093/clinchem/32.11.2100a.

Clark, S., Youngman, L. D., Palmer, A., Parish, S., Peto, R., & Collins, R. (2003). Stability of plasma analytes after delayed separation of whole blood: Implications for epidemiological studies. *International Journal of Epidemiology*, *32*(1), 125–130. https://doi.org/10.1093/ije/dyg023.

Corrie, S. R., Coffey, J. W., Islam, J., Markey, K. A., & Kendall, M. A. F. (2015). Blood, sweat, and tears: Developing clinically relevant protein biosensors for integrated body fluid analysis. *Analyst*, *140*(13), 4350–4364. https://doi.org/10.1039/c5an00464k.

Crawford, M. L., Collier, B. B. B. B., Bradley, M. N., Holland, P. L., Shuford, C. M., & Grant, R. P. (2020). Empiricism in microsampling: Utilizing a novel lateral flow device and intrinsic normalization to provide accurate and precise clinical analysis from a finger stick. *Clinical Chemistry*, *66*(6), 821–831. https://doi.org/10.1093/clinchem/hvaa082.

de Vries, R., Barfield, M., van de Merbel, N., Schmid, B., Siethoff, C., Ortiz, J., Verheij, E., van Baar, B., Cobb, Z., White, S., & Timmerman, P. (2013). The effect of hematocrit on bioanalysis of DBS: Results from the EBF DBS-microsampling consortium. *Bioanalysis*, *5*(17), 2147–2160. https://doi.org/10.4155/bio.13.170.

Déglon, J., Leuthold, L. A., & Thomas, A. (2015). Potential missing steps for a wide use of dried matrix spots in biomedical analysis. *Bioanalysis*, *7*(18), 2375–2385. https://doi.org/10.4155/bio.15.166.

Denniff, P., & Spooner, N. (2010). The effect of hematocrit on assay bias when using DBS samples for the quantitative bioanalysis of drugs. *Bioanalysis*, *2*(8), 1385–1395. https://doi.org/10.4155/bio.10.103.

Denniff, P., Woodford, L., & Spooner, N. (2013). Effect of ambient humidity on the rate at which blood spots dry and the size of the spot produced. *Bioanalysis*, *5*(15), 1863–1871. https://doi.org/10.4155/bio.13.137.

Deprez, S., Paniagua-González, L., Velghe, S., & Stove, C. P. (2019). Evaluation of the performance and hematocrit independence of the HemaPEN as a volumetric dried blood spot collection device. *Analytical Chemistry*, *91*(22). https://doi.org/10.1021/acs.analchem.9b03179.

Dhingra, N., M. Diepart, G. Dziekan, S. Khamassi, F. Otazia, S. Wilburn, (2010). *WHO Guidelines on drawing blood: Best practices in phlebotomy* (pp. 1–105). World Health Organization. https://apps.who.int/iris/handle/10665/44294.

Ďurč, P., Foret, F., & Kubáň, P. (2018). Fast blood plasma separation device for point-of-care applications. *Talanta*, *183*(January), 55–60. https://doi.org/10.1016/j.talanta.2018.02.004.

Edge, T., Smith, C., Elanco, B., Edge, T., Smith, C., Elanco, B., Edge, T., & Smith, C. (2016). Review of microsampling techniques in bioanalysis. *Chromatography Today*, 34–37. https://www.chromatographytoday.com/article/bioanalytical/40/university-of-liverpool/review-of-microsampling-techniques-in-bioanalysis/2010.

Fan, L., & Lee, J. A. (2012). Managing the effect of hematocrit on DBS analysis in a regulated environment. *Bioanalysis*, *4*(4), 345–347. https://doi.org/10.4155/bio.11.337.

Felding, P., Petersen, P. H., & Hørder, M. (1981). The stability of blood, plasma and serum constituents during simulated transport. *Scandinavian Journal of Clinical and Laboratory Investigation*, *41*(1), 35–40. https://doi.org/10.3109/00365518109092012.

Freeman, J. D., Rosman, L. M., Ratcliff, J. D., Strickland, P. T., Graham, D. R., & Silbergeld, E. K. (2018). State of the science in dried blood spots. *Clinical Chemistry*, *64*(4), 656–679. https://doi.org/10.1373/clinchem.2017.275966.

Fruhstorfer, H., & Quarder, O. (2009). Blood glucose monitoring: Milking the finger and using the first drop of blood give correct glucose values. *Diabetes Research and Clinical Practice*, *85*(1), 184–185. https://doi.org/10.1016/j.diabres.2009.04.019.

Gallardo, E., & Queiroz, J. A. (2008). The role of alternative specimens in toxicological analysis. *Biomedical Chromatography*, *22*(8), 795–821. https://doi.org/10.1002/bmc.1009.

Giavarina, D., & Lippi, G. (2017). Blood venous sample collection: Recommendations overview and a checklist to improve quality. *Clinical Biochemistry*, *50*(10), 68–573. https://doi.org/10.1016/j.clinbiochem.2017.02.021.

Gómez Rioja, R., Segovia Amaro, M., Diaz-Garzón, J., Bauçà, J. M., Martínez Espartosa, D., & Fernández-Calle, P. (2018). Laboratory sample stability. Is it possible to define a consensus stability function? An example of five blood magnitudes. *Clinical Chemistry and Laboratory Medicine*, *56*(11), 1806–1818. https://doi.org/10.1515/cclm-2017-1189.

Hankinson, S. E., London, S. J., Chute, C. G., Barbieri, R. L., Jones, L., Kaplan, L. A., Sacks, F. M., & Stampfer, M. J. (1989). Effect of transport conditions on the stability of biochemical markers in blood. *Clinical Chemistry*, *35*(12), 2313–2316. https://doi.org/10.1093/clinchem/35.12.2313.

Heins, M., Heil, W., & Withold, W. (1995). Storage of serum or whole blood samples? Effects of time and temperature on 22 serum analytes. *Clinical Chemistry and Laboratory Medicine*, *33*(4), 231–238. https://doi.org/10.1515/cclm.1995.33.4.231.

Heireman, L., van Geel, P., Musger, L., Heylen, E., Uyttenbroeck, W., & Mahieu, B. (2017). Causes, consequences and management of sample hemolysis in the clinical laboratory. *Clinical Biochemistry 50*(18), 1317–1322. https://doi.org/10.1016/j.clinbiochem.2017.02.021.

Homsy, A., van der Wal, P. D., Doll, W., Schaller, R., Korsatko, S., Ratzer, M., Ellmerer, M., Pieber, T. R., Nicol, A., & de Rooij, N. F. (2012). Development and validation of a low cost blood filtration element separating plasma from undiluted whole blood. *Biomicrofluidics*, *6*(1), 1–9. https://doi.org/10.1063/1.3672188.

Horowitz, G. L., Altaie, S., Boyd, J. C., Ceriotti, F., Garg, U., Horn, P., Pesce, A., Sine, H. E., & Zakowski, J. (2010). *EP28-A3C: Defining, establishing, and verifying reference intervals in the clinical laboratory; approved guideline* (Vol. 28, 3rd ed). CLSI (Issue 30).

Hortensius, J., Kleefstra, N., Slingerland, R. J., Fokkert, M. J., Groenier, K. H., Houweling, S. T., & Bilo, H. J. G. (2010) The influence of a soiled finger in capillary blood glucose monitoring. *Netherlands Journal of Medicine*, *68*(7-8), 330–331.

James, C. A., Barfield, M. D., Maass, K. F., Patel, S. R., & Anderson, M. D. (2020). Will patient-centric sampling become the norm for clinical trials after COVID-19? *Nature Medicine*, *26*(12), 1810. https://doi.org/10.1038/s41591-020-01144-1.

Kaiser, N. K., Steers, M., Nichols, C. M., Mellert, H., & Pestano, G. A. (2020). Design and characterization of a novel blood collection and transportation device for proteomic applications. *Diagnostics*, *10*(12), 1032. https://doi.org/10.3390/diagnostics10121032.

Kale, K., Iyengar, A., Kapila, R., & Chhabra, V. (2017). Saliva—A diagnostic tool in assessment of lipid profile. *Scholars Academic Journal of Biosciences*, *5*(8), 574–584.

Klee, G. G. (2010). Establishment of outcome-related analytic performance goals. *Clinical Chemistry*, *56*(5), 714–722. https://doi.org/10.1373/clinchem.2009.133660.

Koseoglu, M., Hur, A., Atay, A., Cuhadar, S., Koseoglu, M., Hur, A., Atay, A., & Cuhadar, S. (2011). Effects of hemolysis interferences on routine biochemistry parameters. *Biochemia Medica*, *21*(1), 79–85.

Koster, R. A., Alffenaar, J.-W. C., Greijdanus, B., & Uges, D. R. A. (2013). Fast LC-MS/MS analysis of tacrolimus, sirolimus, everolimus and cyclosporin A in dried blood spots and the influence of the hematocrit and immunosuppressant concentration on recovery. *Talanta*, *115*, 47–54. https://doi.org/10.1016/j.talanta.2013.04.027.

Koster, R. A., Alffenaar, J.-W. C., Botma, R., Greijdanus, B., Uges, D. R. A., Kosterink, J. G. W., & Touw, D. J. (2015). The relation of the number of hydrogen-bond acceptors with recoveries of immunosuppressants in DBS analysis. *Bioanalysis*, *7*(14), 1717–1722. https://doi.org/10.4155/bio.15.94.

Krleza, J. L., Dorotic, A., Grzunov, A., Maradin, M. (2015). Capillary blood sampling: National recommendations on behalf of the Croatian Society of Medical

Biochemistry and Laboratory Medicine. *Biochemia Medica, 25*(3), 335–358. https://doi.org/10.11613/BM.2015.034.

Lei, B. U. W., & Prow, T. W. (2019). A review of microsampling techniques and their social impact. *Biomedical Microdevices, 21*(4). https://doi.org/10.1007/s10544-019-0412-y.

Lewis, S. M., & Tatsumi, N. (2006). Collection and handling of blood. In *Dacie and Lewis practical haematology* (2nd ed.). Elsevier Ltd. https://doi.org/10.1016/B0-44-306660-4/50005-2.

Li, C. G., Lee, C. Y., Lee, K., & Jung, H. (2013). An optimized hollow microneedle for minimally invasive blood extraction. *Biomedical Microdevices, 15*(1), 17–25. https://doi.org/10.1007/s10544-012-9683-2.

Li, F., Zulkoski, J., Fast, D., & Michael, S. (2011). Perforated dried blood spots: A novel format for accurate microsampling. *Bioanalysis, 3*(20), 2321–2333. https://doi.org/10.4155/bio.11.219.

Lima-Oliveira, G., Guidi, G. C., Salvagno, G. L., Danese, E., Montagnana, M., & Lippi, G. (2017). Patient posture for blood collection by venipuncture: Recall for standardization after 28 years. *Revista Brasileira de Hematologia e Hemoterapia. 2017/02/22, 39*(2), 127–132. https://doi.org/10.1016/j.bjhh.2017.01.004.

Liu, C., Mauk, M., Gross, R., Bushman, F. D., Edelstein, P. H., Collman, R. G., & Bau, H. M. (2013). Membrane-based, sedimentation-assisted plasma separator for point-of-care applications. *Analytical Chemistry, 85*(21), 10463–10470.

Liu, C., Liao, S.-C., Song, J., Mauk, M. G., Li, X., Wu, G., Ge, D., Greenberg, R. M., Yang, S., & Bau, H. H. (2016). A high-efficiency superhydrophobic plasma separator. *Lab on a Chip, 16*(3), 553–560. https://doi.org/10.1039/c5lc01235j.A.

Lomeo, A., Bolner, A., Scattolo, N., Guzzo, P., Amadori, F., Sartori, S., & Lomeo, L. (2008). HPLC analysis of HbA1c in dried blood spot samples (DBS): A reliable future for diabetes monitoring. *Clinical Laboratory, 54*, 161–167.

Londhe, V., & Rajadhyaksha, M. (2020). Opportunities and obstacles for microsampling techniques in bioanalysis: Special focus on DBS and VAMS. *Journal of Pharmaceutical and Biomedical Analysis, 182*, 113102. https://doi.org/10.1016/j.jpba.2020.113102.

Lum, G., & Gambino, S. R. (1974). A comparison of serum versus heparinized plasma for routine chemistry tests. *American Journal of Clinical Pathology, 61*(1), 108–113. https://doi.org/10.1093/ajcp/61.1.108.

Malamud, D. (2011). Saliva as a diagnostic fluid. *Dental Clinics of North America, 55*(1), 159–178. https://doi.org/10.1016/j.cden.2010.08.004.

Mbughuni, M. M., Stevens, M. A., Langman, L. J., Kudva, Y. C., Sanchez, W., Dean, P. G., & Jannetto, P. J. (2020). Volumetric microsampling of capillary blood spot vs whole blood sampling for therapeutic drug monitoring of tacrolimus and cyclosporin A: Accuracy and patient satisfaction. *The Journal of Applied Laboratory Medicine*, 1–15. https://doi.org/10.1093/jalm/jfaa005.

Meesters, R. J. W., Zhang, J., van Huizen, N. A., Hooff, G. P., Gruters, R. A., & Luider, T. M. (2012). Dried matrix on paper disks: The next generation DBS microsampling technique for managing the hematocrit effect in DBS analysis. *Bioanalysis*, *4*(16), 2027–2035. https://doi.org/10.4155/bio.12.175.

Morris, S. S., Ruel, M. T., Cohen, R. J., Dewey, K. G., de la Brière, B., & Hassan, M. N. (1999). Precision, accuracy, and reliability of hemoglobin assessment with use of capillary blood. *The American Journal of Clinical Nutrition*, *69*(6), 1243–1248. https://doi.org/10.1093/ajcn/69.6.1243.

Nys, G., Kok, M. G. M., Servais, A. C., & Fillet, M. (2017). Beyond dried blood spot: Current microsampling techniques in the context of biomedical applications. *TrAC—Trends in Analytical Chemistry*, *97*, 326–332. https://doi.org/10.1016/j.trac.2017.10.002.

O'Mara, M., Hudson-Curtis, B., Olson, K., Yueh, Y., Dunn, J., Spooner, N. (2011). The effect of hematocrit and punch location on assay bias during quantitative bioanalysis of dried blood spot samples. *Bioanalysis*, *3*(20), 2335–2347. https://doi.org/10.4155/bio.11.220.

Oddoze, C., Lombard, E., & Portugal, H. (2012). Stability study of 81 analytes in human whole blood, in serum and in plasma. *Clinical Biochemistry*, *45*(6), 464–469. https://doi.org/10.1016/j.clinbiochem.2012.01.012.

Ono, T., Kitaguchi, K., Takehara, M., Shiiba, M., & Hayami, K. (1981). Serum-constituents analyses: Effect of duration and temperature of storage of clotted blood. *Clinical Chemistry*, *27*(1), 35–38. https://doi.org/10.1093/clinchem/27.1.35.

Oostendorp, M., el Amrani, M., Hekman, D., & van Maarseveen, E. M. (2016). Measurement of hematocrit in dried blood spots using near-infrared spectroscopy: Robust, fast, and nondestructive. *Clinical Chemistry*, *62*(11), 1534–1536.

Palese, A., Fabbro, E., Casetta, A., & Mansutti, I. (2016). First or second drop of blood in capillary glucose monitoring: Findings from a quantitative study. *Journal of Emergency Nursing*, *42*(5), 420–426. https://doi.org/10.1016/j.jen.2016.03.027.

Parikh, P., Mochari, H., & Mosca, L. (2009). Measurement issues: Clinical utility of a fingerstick technology to identify individuals with abnormal blood lipids and high-sensitivity C-reactive protein levels. *American Journal of Health Promotion*, *23*(4), 279–282. https://doi.org/10.4278/ajhp.071221140.

Patel, K., El-khoury, J. M., Aarsand, A. K., Badrick, T., Jones, G. R. D., Sikaris, K., & Parnas, M. L. (2021). Current utility and reliability of biological variability. *Clinical Chemistry*, 1–6. https://doi.org/10.1093/clinchem/hvab055.

Peake, M., & Whiting, M. (2006). Measurement of serum creatinine—Current status and future goals. *The Clinical Biochemist Reviews*, *27*(4), 173–184. https://pubmed.ncbi.nlm.nih.gov/17581641.

Perneger, T. V, & Etter, J.-F. (2001). Commentary: Extending the boundaries of data collection by mail. *International Journal of Epidemiology*, *30*(2), 301–302. https://doi.org/10.1093/ije/30.2.301.

Philp, J. L., Jaffrin, M. Y., & Ding, L. (1994). Hemolysis during membrane plasma separation with pulsed flow filtration enhancement. *Journal of Biomechanical Engineering*, *116*(4), 514–520. https://doi.org/10.1115/1.2895803.

Plebani, M., & Carraro, P. (1997). Mistakes in a stat laboratory: Types and frequency. *Clinical Chemistry*, *43*(8), 1348–1351. https://doi.org/10.1093/clinchem/43.8.1348.

Plebani, M., Sciacovelli, L., & Aita, A. (2017). Quality indicators for the total testing process. *Clinics in Laboratory Medicine*, *37*(1), 187–205. https://doi.org/10.1016/j.cll.2016.09.015.

Poitout-Belissent, F., Aulbach, A., Tripathi, N., & Ramaiah, L. (2016). Reducing blood volume requirements for clinical pathology testing in toxicologic studies—Points to consider. *Veterinary Clinical Pathology*, *45*(4), 534–551. https://doi.org/10.1111/vcp.12429.

Protti, M., Mandrioli, R., & Mercolini, L. (2019). Tutorial: Volumetric absorptive microsampling (VAMS). *Analytica Chimica Acta*, *1046*, 32–47. https://doi.org/10.1016/j.aca.2018.09.004.

Protti, M., Marasca, C., Cirrincione, M., Cavalli, A., Mandrioli, R., & Mercolini, L. (2020). Assessment of capillary volumetric blood microsampling for the analysis of central nervous system drugs and metabolites. *Analyst*. https://doi.org/10.1039/d0an01039a.

Rehak, N. N., & Chiang, B. T. (1988). Storage of whole blood: Effect of temperature on the measured concentration of analytes in serum. *Clinical Chemistry*, *34*(10), 2111–2114. https://doi.org/10.1093/clinchem/34.10.2111.

Roccaforte, V., Daves, M., Platzgummer, S., & Lippi, G. (2016). The impact of different sample matrices in delayed measurement of glucose. *Clinical Biochemistry*, *49*(18), 1412–1415. https://doi.org/10.1016/j.clinbiochem.2016.08.015.

Rühling, K., Lang, A., Holtz, H., Winkler, L., Schlag, B., & Till, U. (1992). Increase in plasma total and lipoprotein cholesterol during incubation of whole blood samples at 37°C—Influence of LCAT inhibitors. *Clinica Chimica Acta*, *205*(3), 205–212. https://doi.org/10.1016/0009-8981(92)90061-T.

Sloviter, H. A. (1962). Mechanism of hæmolysis caused by freezing and its prevention. *Nature*, *193*(4818), 884–885. https://doi.org/10.1038/193884a0.

Spooner, N., Olatunji, A., & Webbley, K. (2018). Investigation of the effect of blood hematocrit and lipid content on the blood volume deposited by a disposable dried blood spot collection device. *Journal of Pharmaceutical and Biomedical Analysis*, *149*, 419–424. https://doi.org/10.1016/j.jpba.2017.11.036.

Stein, E. A., & Pum, J. (2005). Blood and plasma. In P. Worsfold, A. Townshend, and C. F. Poole (Eds.), *Encyclopedia of analytical science* (pp. 294–300). Oxford: Elsevier. http://dx.doi.org/10.1016/B0-12-369397-7/00047-9.

Strang, J., Marsh, A., & Desouza, N. (1990). Hair analysis for drugs of abuse. *The Lancet*, *335*(8691), 740. https://doi.org/10.1016/0140-6736(90)90865-3.

Strohmeier, O., Keller, M., Schwemmer, F., Zehnle, S., Mark, D., von Stetten, F., Zengerle, R., & Paust, N. (2015). Centrifugal microfluidic platforms: Advanced unit operations and applications. *Chemical Society Reviews*, *44*(17), 6187–6229. https://doi.org/10.1039/c4cs00371c.

U.S. Department of Health and Human Services, Food and Drug Administration, Center for Devices and Radiological Health, Office of Device Evaluation. (2016). *Applying human factors and usability engineering to medical devices—Guidance for industry and food and drug administration staff.*

U.S. Food & Drug Administration. (2020). *Home specimen collection serology template for fingerstick dried blood spot.* Retrieved September 30, 2021, from https://www.fda.gov/medical-devices/coronavirus-disease-2019-covid-19-emergency-use-authorizations-medical-devices/vitro-diagnostics-euas.

Van Vrancken, M. J., Briscoe, D., Anderson, K. M., & Wians, F. H. (2012). Time-dependent stability of 22 analytes in lithium-plasma specimens stored at refrigerator temperature for up to 4 days. *Laboratory Medicine*, *43*(6), 268–275. https://doi.org/10.1309/lm53sbz4vqgwvqom.

Velghe, S., Capiau, S., & Stove, C. P. (2016). Opening the toolbox of alternative sampling strategies in clinical routine: A key-role for (LC-)MS/MS*TrAC—Trends in Analytical Chemistry*, *84*, 61–73. https://doi.org/10.1016/j.trac.2016.01.030.

Velghe, S., Delahaye, L., & Stove, C. P. (2019). Is the hematocrit still an issue in quantitative dried blood spot analysis? *Journal of Pharmaceutical and Biomedical Analysis*, *163*, 188–196. https://doi.org/10.1016/j.jpba.2018.10.010.

Velghe, S., & Stove, C. P. (2018). Evaluation of the Capitainer-B microfluidic device as a new hematocrit-independent alternative for dried blood spot collection. *Analytical Chemistry*, *90*(21), 12893–12899. https://doi.org/10.1021/acs.analchem.8b03512.

Verma, S., Redfield, R., Azar II, A. M. (2019). Clinical Laboratory Improvement Amendments of 1988 (CLIA) proficiency testing regulations related to analytes and acceptable performance. *Federal Register*, *84*(23), 1536–1567. https://www.federalregister.gov/documents/2019/02/04/2018-28363/clinical-laboratory-improvement-amendments-of-1988-clia-proficiency-testing-regulations-related-to.

Viswanathan, C. T. (2012). Perspectives on microsampling: DBS. *Bioanalysis*, *4*(12), 1417–1419. https://doi.org/10.4155/bio.12.123.

Wickremsinhe, E. R., Ji, Q. C., Gleason, C. R., Anderson, M., & Booth, B. P. (2020). Land O'Lakes workshop on microsampling: Enabling broader adoption. *AAPS Journal*, *22*(6), 1–7. https://doi.org/10.1208/s12248-020-00524-2.

Williamson, S., Munro, C., Pickler, R., Grap, M. J., & Elswick, R. K. (2012). Comparison of biomarkers in blood and saliva in healthy adults. *Nursing Research and Practice*, *2012*, 1–4. https://doi.org/10.1155/2012/246178.

Xie, I., Xu, Y., Anderson, M., Wang, M., Xue, L., Breidinger, S., Goykhman, D., Woolf, E. J., & Bateman, K. P. (2018). Extractability-mediated stability bias and hematocrit impact: High extraction recovery is critical to feasibility of volumetric adsorptive microsampling (VAMS) in regulated bioanalysis. *Journal of Pharmaceutical and Biomedical Analysis, 156*, 58–66. https://doi.org/10.1016/j.jpba.2018.04.001.

Yang, Z. W., Yang, S. H., Chen, L., Qu, J., Zhu, J., & Tang, Z. (2001). Comparison of blood counts in venous, fingertip and arterial blood and their measurement variation. *Clinical and Laboratory Haematology, 23*(3), 155–159. https://doi.org/10.1046/j.1365-2257.2001.00388.x.

Youhnovski, N., Bergeron, A., Furtado, M., & Garofolo, F. (2011). Pre-cut dried blood spot (PCDBS): An alternative to dried blood spot (DBS) technique to overcome hematocrit impact. *Rapid Communications in Mass Spectrometry, 25*(19), 2951–2958. https://doi.org/10.1002/rcm.5182.

Zhang, D. J., Elswick, R. K., Miller, W. G., & Bailey, J. L. (1998). Effect of serum-clot contact time on clinical chemistry laboratory results. *Clinical Chemistry, 44*(6), 1325–1333. https://doi.org/10.1093/clinchem/44.6.1325.

4

Collection and Bioanalysis of Quantitative Microsamples

Technological Innovations and Practical Implications

Regina V. Oliveira[1], *Marc Yves Chalom*[2], *and Carlos Roberto V. Kiffer*[3]

[1] *Núcleo de Pesquisa em Cromatografia (Separare), Departamento de Química, Universidade Federal de São Carlos, São Carlos, São Paulo, Brazil*
[2] *Sens Representações Comerciais, São Paulo, São Paulo, Brazil*
[3] *Laboratório Especialista de Microbiologia Clínica, Disciplina de Infectologia, Escola Paulista de Medicina, Universidade Federal de São Paulo (UNIFESP), São Paulo, São Paulo, Brazil*

4.1 Introduction

Numerous healthcare institutions and life sciences corporations have active initiatives to promote patient centric approaches as one of their most urgent priorities (Eklund et al., 2019; Rossiter et al., 2020). Different organizations in the pharmaceutical industry and across health and emergency care facilities throughout the world are searching for ways to promote a business transformation by prioritizing patient needs, providing services more responsive to patients and supporting consumer expectations for on-demand care delivered on their terms (World Health Organization, OECD, International Bank for Reconstruction and Development 2018). A personalized patient approach can substantially improve patients and their families' experience, as well as being clinically and economically effective (Dorofaeff et al., 2016).

The term, patient-centered healthcare or patient centricity is defined as the actions that healthcare systems can establish, aligned with patients by delivering a service or solution focusing on the patient needs and preferences (Yeoman et al., 2017), allowing the best experience and outcome for patients and their families.

Patient Centric Blood Sampling and Quantitative Bioanalysis, First Edition.
Edited by Neil Spooner, Emily Ehrenfeld, Joe Siple, and Mike S. Lee.

In clinical research, patient centricity may be employed at three levels: (a) patient engagement at the strategic level, providing insights into the study design; (b) incorporation of patient-reported outcomes measures (PROMs—rate the quality of care delivered to a patient) in a study and/or contributions in the development and validation of a PROM; (c) patient engagement in study recruitment and retention of patients in the clinical studies (Yeoman et al., 2017). Therefore, patient centricity is an essential aspect in drug research, development, and disease management. Moreover, to ensure adequate treatment and to improve patients' quality of life, a thorough understanding of the challenges in the patient's daily lives is required. This includes medical conditions, experiences with other treatments, the experience of symptoms, cost of the interventions, treatment options, and objectives for their healthcare (Lavallee et al., 2016).

Blood sampling and analysis are important parts of the healthcare system, and currently many aspects of both do not bring the requirements of the patients to the fore. However, several technologies are emerging that will change this. For instance, in drug discovery and development programs, nonclinical and clinical studies have explored the potential of using microsampling to monitor drug exposure in biological samples (Nys et al., 2017; Spooner et al., 2019).

Microsampling refers to the collection of small amounts of sample (typically <50 μL) to evaluate biochemistry parameters of a drug or chemical exposure in whole blood, plasma, serum, and/or other biological samples.

In this chapter, we outline the benefits of the utilization of patient centric sampling and microsampling for the quantitative bioanalysis of biological samples on clinical drug development studies and clinical applications, such as therapeutic drug monitoring (TDM) and disease diagnostics. We also discuss the new technologies currently becoming available to promote an improved practice for the collection and analysis of biofluids, which could also contribute to patient education, engagement, and retention in healthcare systems and drug development programs.

4.2 Practical Implications in Clinical Settings

Significant advances in the routine application of microsampling methods have been achieved, particularly due to their utilization in nonclinical and clinical phases of drug development. Furthermore, the benefits of this approach have been demonstrated over conventional sampling in clinical settings due to the facilitation of ease of sampling, storage, and shipment (Londhe & Rajadhyaksha, 2020). In addition to discussing the technological innovations that have been made with microsampling techniques, we will also explore the practical implications of their use in clinical settings and look at the range of the technique's benefits on clinical development and laboratories.

4.2.1 Clinical Development

Due to its applications in nonclinical and clinical phases of drug development, microsampling has gained the attention of the International Conference on Harmonization (ICH) (ICH/EMA, 2019; ICH S11, 2020; ICH S3A-S3B, 1995a, 1995b). Although the benefits of ease of sampling, storage, and shipment for microsampling approaches over conventional phlebotomy are clear and highly significant, their impacts may be slightly different depending on the phase of clinical development.

Clinical pharmacokinetic (PK) studies often require the collection of large blood volumes from patients (i.e., several mL's). However, various PK study scenarios may be impacted differently by the implementation of patient centric sampling and/or microsampling (<50 μL). Adult healthy volunteer PK studies may benefit from the implementation of microsampling mainly due to the emotional burden on the patient of large volume blood draws. Furthermore, benefits may be derived regarding the storage and transport of samples, which are time- and resource-consuming for conventional approaches. It is notable that conventional sampling methods for PK determination are physiologically troublesome for special populations, particularly neonates and pediatric patients or nonhealthy patients in phase I/II studies. In the latter situation, there are evident ethical and technical advantages for the collection of smaller amounts of samples over conventional approaches (i.e. mL vs. few μL) (Spooner et al., 2019).

In later phases of clinical trials (II, III, and IV or observational studies), the amount of blood drawn may not be a significant issue for adult or noncritically ill patients. Nevertheless, it is still significant for the critically ill and pediatric populations. Also, storage and transportation of conventional samples become much more significant in this scenario. Late-phase clinical trials commonly use centralized laboratories to standardize methodologies and procedures. This, allied to conventional blood sample collection methods, leads to the need for storage and transport over considerable distances and from multiple locations. Safe and efficient storage and transport of conventional samples are practically unfeasible without the use of ice or refrigeration, which significantly limits the implementation of clinical trials to certain geographical areas, while significantly increasing its costs. It has been previously estimated that a modestly regular multicenter clinical trial would involve around 2,000 shipments of samples at ambient conditions and 1,000 on dry ice for biomarkers and/or PK analysis, even after considering efficiency gain by combining shipments and temporary storage (van Amsterdam and Waldrop, 2010). In the authors' experience, biological sample storage and transport may consume up to 30–40% of the centralized laboratory costs in a late-phase clinical trial. Of note, transport of PK and biomarker samples are commonly performed on dry ice, which is highly regulated (International Air

Transport Association (IATA), the class 9 of dangerous goods), requires personnel training, is highly expensive, and is unfeasible in some regions due to dry ice unavailability. In this regard, samples in dried form can be shipped at ambient temperatures if analyte stability is demonstrated, with significant cost savings and process convenience advantages. Other aspects related to conventional sampling, such as training, kit supply, sample processing, interim storage, sample handling, and destruction are relevant and resource-consuming (van Amsterdam and Waldrop, 2010). On the other hand, microsampling in general, and dried blood spot (DBS) sampling in particular, may not require special biohazard arrangements (Knudsen et al., 1993; Sharma et al., 2014). This leads to simpler processes and, hence, less resource-consuming work, with potential cost savings. In addition, patient centric sampling provides convenience for the patients, which has been demonstrated by the success of home sampling during the COVID-19 pandemic (Anderson, 2021; Bateman, 2020).

It is therefore apparent that the implementation of microsampling techniques for drug development, apart from facilitating clinical trial execution in previously intractable locations with potential cost savings, also delivers emotional and ethical advantages over conventional larger blood volumes. This is particularly true for studies in special populations, such as children and the critically ill, who could certainly benefit from less invasive procedures with smaller amounts of blood (Spooner et al., 2019). There are also numerous benefits for the implementation of microsampling for studies with regular adult populations, including the possibility of having additional sample collections, such as more time points in adaptive clinical designs, newly available biomarkers in special situations, sampling during an adverse event, and others. Thus, potential gains of microsampling in different clinical trial phases or populations may vary but are evident over conventional blood collection methods with larger volumes, since more convenient sampling to the patient will most likely impact in the recruitment and retention of subjects for clinical trials.

4.2.2 Clinical Analyses

Microsampling may also add potential benefits for routine clinical chemistry testing of challenging populations (particularly neonates, elderlies, and the critical ills), for patient convenience, and for remote settings. In a recent review, Freeman (Freeman et al., 2018) identified more than 2,000 analytes measured in DBSs, including genes, transcripts, proteins, and metabolites. The authors also verified that analytical methods applied to traditional and DBS samples demonstrated that sampling methodology was not a limiting factor in most clinical laboratories for the implementation of microsampling techniques. However, microsampling in general, and DBS in particular, may often require the conversion of regular reference values.

One area of particular interest for microsampling in clinical laboratories is TDM, especially for controlling antimicrobial serum levels, both for traditional TDM such as vancomycin and aminoglycosides with a narrow therapeutic index, as well as for nontraditionally monitored antimicrobials, such as beta-lactams that have a wide therapeutic index. With the increasing emergence of antimicrobial resistance, more precise control of antimicrobial serum levels is required to better achieve pharmacodynamic (PD) targets and maximize drug activity, particularly for antimicrobials with less predictable PK/PD models that characterizes the full-time course-kill curve experiments and safety profiles (Huttner et al., 2015; Roberts et al., 2012; Scaglione et al., 2009; Touw et al., 2005). Besides, considerable knowledge gaps remain, regarding the PK properties of antibiotics in critically ill pediatric patients and pediatric populations. In this scenario, microsampling may make a significant impact for TDM in such vulnerable patient populations, where smaller sample volumes can be collected (Guerra Valero et al., 2018). In a recent review, Dorofaeff et al. (2016) showed that most of the 32 antibiotic PK studies performed on critically ill neonate or pediatric patients identified changes to the PK properties in the respective patient group, compared to usually expected parameters. Thus, the implementation of TDM using microsamples to ensure antibiotic doses are suitable may have a vast field of clinical application, especially for vulnerable populations.

For clinical pathology/chemistry, laboratories routinely perform assays to detect the presence and concentrations of a broad range of exogenous (dosed drugs) and endogenous biomarkers, including metabolites, peptides, proteins, and lipids. Routinely, quantitative and/or qualitative determinations are often performed on an array of biospecimens: plasma, serum, or urine, and, occasionally, on other fluids such as cerebrospinal fluid, saliva, bronchoalveolar lavage fluids, and oro-nasopharyngeal fluids. While some clinical alterations relate directly to a pathology, biological markers (biomarkers) offer the means for classification of a disease and risk factors, indicating normal biological processes, pathogenic processes, or pharmacological responses to a therapeutic intervention (Naylor, 2003; Watson, 2005). Therefore, the use of microsampling techniques for exogenous drugs and biomarkers may promote convenience for patients and broader healthcare since it requires fewer blood draws, allows its application in remote settings, adds flexibility in the collection of PK data, improves logistical feasibility for sample storage and shipping, and opens new perspectives for new approaches to clinical trials.

Traditionally, enzyme-linked immunosorbent assay (ELISA) and other enzyme immunoassays (EIA) have been the analytical technologies of choice for the detection and quantitation of endogenous and exogenous molecules in clinical laboratories (Darwish, 2006). Rapid immunoassays play an important role in patient-centered sampling as they can impact the quality of care in point-of-care

(POC) setting such as emergency rooms, walk-in clinics, pediatric and geriatric clinics, and healthcare centers. An example of an immunoassay for the measurement of dosed drugs includes the determination of sirolimus, an immunosuppressive agent administered as prophylaxis of acute rejection to patients after kidney transplantation, using an EIA composed of microparticles coated with anti-sirolimus antibodies (Zochowska et al., 2006). Cortisol awakening response (CAR), an endogenous biomarker related to psychosocial and mental health characteristics, including anticipated stress, can also be measured in saliva using a commercially available ELISA kit, with intra-assay coefficients of variance less than 10% (Losiak and Losiak-Pilch, 2020). Neurosteroids (Teubel and Parr, 2020), disease biomarkers (Zhang et al., 2020), hormones (Lacombe et al., 2020), 25-hydroxyvitamin D (Alghamdi et al., 2020), and, recently, SARS-CoV-2 (Li et al., 2020) are also examples of biomolecules that are efficiently measured by immunoassays.

The extensive use of immunoassay methods in the clinical analysis is attributed to their straightforward protocols, acceptance in regulated environments, inexpensive instrumentation, rapid data turnaround, and sensitivity. However, immunoassays also have some limitations, including lack of specificity, that may be observed due to nonspecific binding of the ligand antibody to matrix components, selectivity issues, required samples volumes when using rare matrices (ocular fluids, fluids/tissues biopsies), cost-per-sample due to reagent prices, reproducibility, multiplex EIA, and long development times for a new immunoassay method, mainly associated with the generation of the desired antibody (Cross and Hornshaw, 2016; Darwish, 2006).

Recent decades have witnessed large technological advances in analytical technologies, in particular, liquid chromatography-mass spectrometry (LC-MS), the use of which is on the rise for accurate bioanalytical determination in clinical laboratories (Jannetto and Fitzgerald, 2016; Tanna et al., 2020; Teclemariam et al., 2020). LC-MS technologies are highly sensitive, therefore amenable to low sample volumes ($<50\,\mu L$), promoting the easy implementation of less invasive alternatives to blood collection, such as patient-centered sampling technologies. With the upsurge of these technologies, and a growing concern for the needs of the patient, the path was made to innovate and create patient-centered microdevices for biosample collection (Bateman, 2020).

Despite the ever-increasing demand for LC-MS methods for clinical applications, there are specific analytical considerations that can impact the ability to accurately quantitate low levels of analytes from biological microsamples and, therefore, impair reliable disease diagnosis and prognosis (Klak et al., 2019; Londhe and Rajadhyaksha, 2020; Volani et al., 2017). The following sections of this chapter will focus on important considerations for the quantitative bioanalysis of microsamples in the sample collection (pre-analytical) and analytical phases.

4.3 Microsampling Devices—A Patient-Centered Approach

In contrast to microsampling (typically <50 μL), conventional samples drawn by venipuncture from normal adults for current clinical diagnostics tests in a clinical setting can be up to 30–50 mL of whole blood in vacutainer tubes containing anticoagulants, which additionally require appropriate sample handling and suitable storage temperatures (Henion et al., 2013). Besides sample volume, there are many other advantages associated with microsampling techniques, which includes benefits to animals during toxicology and nonclinical studies and to humans during clinical trials involving TDM and in the field of diagnostic tests, such as POC devices. Microsampling has the potential for implementation in self- and home-sampling procedures, allows for a quicker, cost-effective, and less invasive diagnostic procedure, and is also less painful and stressful to patients (adults or pediatrics), and often the microsampling approaches demand few requirements on handling, storage, and shipping (Lee and Lee, 2013; Lei and Prow, 2019). Microsampling applied to animal studies is beyond the scope of this chapter, but there is relevant literature addressing this topic (Chapman et al., 2014; Harstad et al., 2016; Prior et al., 2017; Spooner et al., 2019).

The utilization of microsampling provides a revolution in medicine, healthcare, and research, by improving the comfort and convenience for patients of any age, especially for those unable to visit a clinic due to a vulnerable medical condition or for those living in remote regions. Microsampling allows an effective treatment by performing POC TDM for precision monitoring of drug doses, improving efficacy, and patient adherence to therapy, while reducing side effects (Taddeo et al., 2020). Moreover, microsampling may facilitate volunteer's and patient's participation in clinical trials. Figure 4.1 illustrates the main advantages of using microsampling devices for biological fluids collection and the main benefits for patients.

4.3.1 Collection Devices for Microsampling Analysis

Over the last decade, different analytical methods, devices, and techniques utilizing microsampling and microfluidics have been described in the literature. The microsampling devices are versatile, and their use as a sample collector in the pre-analytical step is applicable to any analytical technique as well as for LC-MS systems, which will be further discussed.

Intense research efforts have led to the approval and commercialization of microsampling devices for clinical use. Some examples include the use of cellulose-based filter paper for DBS. For nearly 60 years, DBS sampling has been widely used for newborn screening, and other biological fluids such as plasma,

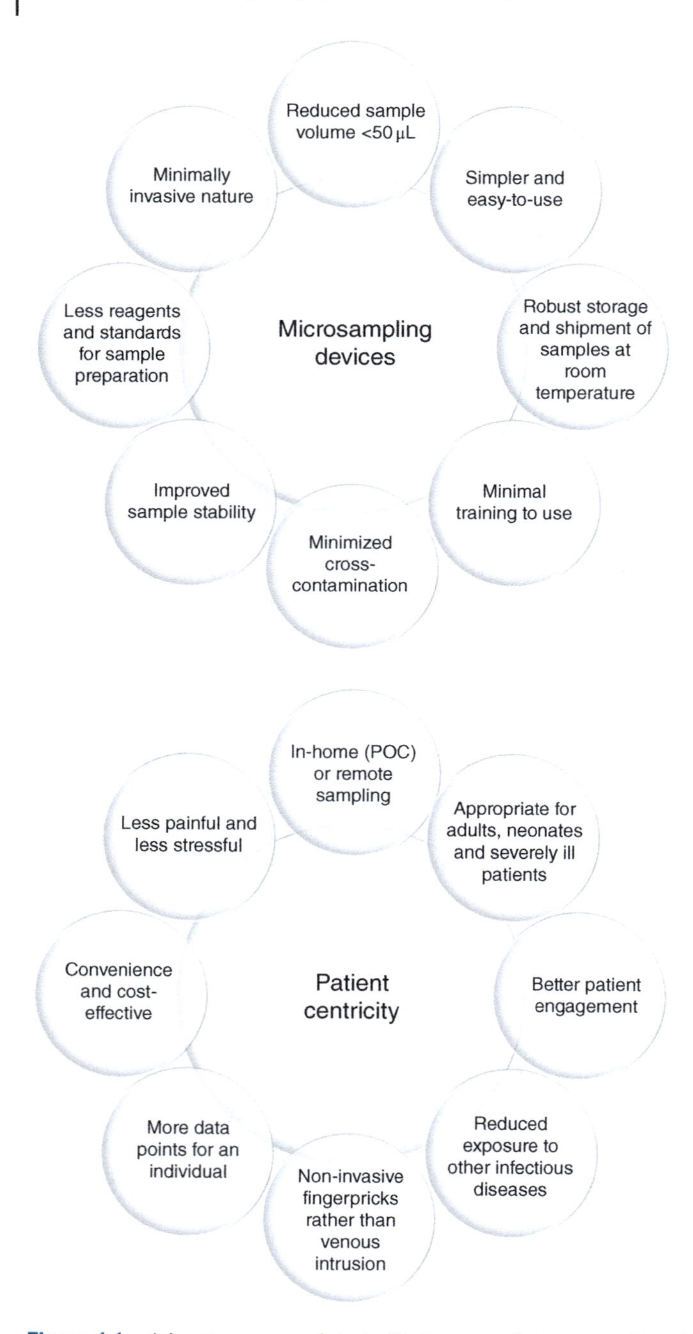

Figure 4.1 Advantages associated with the use of microsampling devices and the main benefits to patients. POC, point-of-care.

saliva, urine, and breast milk have also been spotted on paper substrates for research purposes (Tey and See, 2021). Volumetric absorptive microsampling devices (VAMS®) (Denniff and Spooner, 2014; Harahap et al., 2020) are commercialized for remote collection of capillary blood and other biofluids. Microneedles for cosmetics, therapeutics, and diagnostics applications including fluid sampling, allergy testing, vaccination, and photodynamic therapy are already commercial devices with a considerable number of them being tested in clinical trials (Bhatnagar et al., 2017; Blicharz et al., 2018). Capillary microsampling (CMS) (HemaXis DB10) (Verhaeghe et al., 2017) and plasma separation membranes (PSMs) (Vivid™ and Cobetter OneStep PSM) for the generation of dried plasma spot (DPS) (Li et al., 2012b) are also commercially available. Some other very interesting technologies with great potential are under development, such examples include microbiopsy devices for skin sampling (Churiso et al., 2020; Lin et al., 2013) that are of great interest for diagnostic dermatology and skin research, while volumetric absorptive paper disc (VAPD) and mini-disc (VAPD-mini) (Nakahara et al. 2018), perforated DBS (PDBS) (Li et al., 2011, 2012a), or precut DBS (PCDBS) (Youhnovski et al., 2011) have been investigated as alternatives for blood collection.

Microsampling and miniaturized POC devices are becoming more popular and convenient than conventional sampling practices because they allow remote patient monitoring, only a very small amount of blood is needed, are less invasive, and collection of specimens is simple and minimizes subject pain, accompanied by the increasing interest in the patient-centered approach. Traditional blood collection methods include phlebotomy, where venous blood is collected by penetrating a vein with a needle and a collection apparatus or syringe, while capillary blood sampling refers to sampling blood from a puncture on the finger, earlobe, heel base, big toe, or from a location on the body such as the upper arm. Although capillary and venous blood are not necessarily equivalent, the differences between them in most cases are minimal. Capillary blood is a combination of arterial (oxygenated blood) and venous blood (deoxygenated blood), and it contains some interstitial and intracellular fluid. The hemoglobin, hematocrit, and platelet count of capillary blood are slightly higher than those of venous blood. Small differences in glucose (generally higher than in capillary blood), potassium, total protein, and calcium are also described (Becker et al., 2022). However, there are different reports showing that capillary blood promotes reliable results that correlate well to traditional phlebotomy reference values (Keevil et al., 2009; Mbughuni et al., 2020; Simmonds et al., 2011; Spooner et al., 2010) demonstrating the potential o microsampling for clinical applications. The main reasons to adopt capillary blood collection should be considered when small volumes of blood are available, and the analysis of whole blood is possible. Additionally, this approach is recommended when veins are not accessible or need to be spare for intravenous therapy, in older patients, patients with obesity, burn victims, newborns, and babies' patients to avoid anemia or cardiac arrest when large amounts of blood are drawn and to minimize risks in patients with thrombophilia or hemophilia (Krleza et al., 2015).

4.3.1.1 Blood Sampling Techniques

4.3.1.1.1 Cellulose-Based Devices for Samples Collection

Dried Blood Spot DBS sampling is a well-established, simple, and noninvasive method for whole blood collection, storage, and transport using cellulose-based paper cards. DBS was introduced for the screening of phenylketonuria (PKU), a metabolic disease diagnosed in neonates in 1963 (Guthrie and Susi, 1963). This approach has since been widely used to monitor organic acid conditions, metabolic, endocrine, amino acid, and hemoglobin disorders. Other disorders such as hearing loss, severe combined immunodeficiencies, spinal muscular atrophy, cystic fibrosis, and congenital heart disease are also screened using DBS approaches as part of Newborn Screening (NBS) programs (ACHDNC, 2018). Besides being a patient-centered approach, DBS also offers a simple and practical alternative to traditional phlebotomy with sample collection from a finger/heel prick and delivering cost savings for sample delivery and storage where analyte stability has been demonstrated at ambient temperatures.

The pre-analytical phase during the development of DBS methods for application in quantitative bioanalysis integrates a variety of processes that should be carefully evaluated during the optimization and validation of the DBS method to diminish errors in laboratory testing. First, it is important to consider that DBS is a solid matrix that consists of dried whole blood on a paper card, which is widely different from the commonly used conventional liquid matrices.

The most used paper substrates in DBS are cellulose-based filter paper, such as the Ahlström 226, Whatman® 903 Protein Saver Card, and the Whatman® Fast Transient Analysis (FTA)® Drug Metabolism and Pharmacokinetics (DMPK). The latter has been classified into three classes known as FTA® DMPK-A, FTA® DMPK-B, and FTA® DMPK-C cards. FTA cards are intended for use with blood samples, and their difference relies on the fact that the FTA® DMPK-A and FTA® DMPK-B cards contain chemicals to lyse cells and denature proteins, while the FTA® DMPK-C card contains pure cellulose without any chemicals or denaturing agents, and it could be more suitable for LC-MS analysis for being less prone to cause matrix effects.

Several factors have been identified as potentially affecting the quantitative outcomes obtained from DBS bioanalysis, such as: (a) DBS specimen collection devices (cellulose and noncellulose-based materials); (b) quality of the whole blood collection/spotting on the materials; (c) spot size and volume diffusion; (d) punch location; (e) analyte stability and spot homogeneity during the drying process; (f) stability under transportation and storage of the DBS samples; (g) internal standard addition; (h) matrix effects; (i) target analyte recovery; (j) analysis of a fixed subpunch spot or an entire DBS; (k) drug partitioning between plasma and red blood cells (RBCs); and (l) hematocrit (Hct) levels (Denniff and Spooner, 2010; Henion et al., 2013; Klak et al., 2019; Lenk, et al., 2015a; Malsagova et al., 2020; Moat et al., 2020; Spooner et al., 2019; Volani et al., 2017; Zakaria et al., 2016).

In the literature, review papers overview the impact of DBS-related preanalytical factors on TDM (Capiau et al., 2019; Delahaye et al., 2021; Klak et al., 2019), on inherited metabolic disorders (IMDs) (Moat et al., 2020), and by using mass spectrometry-based methods (Volani et al., 2017; Zakaria et al., 2016). Procedures and practices for the validation of bioanalytical methods using dried whole blood were reported, and the benefits and challenges of DBS-based assays have been extensively discussed (Capiau et al., 2019; Henion et al., 2013; Jager et al., 2014; Londhe and Rajadhyaksha, 2020; Spooner et al., 2019).

The analytical sample preparation workflow for DBS samples involves manual (or semi-automated) punching out of a fixed-size disk (e.g. 2–3 mm diameter) from the center of the DBS, transferring it to a centrifuge tube or well(s) within a microtiter plate and performing the sample extraction with an organic solvent containing an internal standard. Figure 4.2 describes a general workflow for DBS collection, transport, and preparation for analysis. Fully automated approaches for the extraction and analysis of DBS samples have been investigated to overcome the tedious and time-consuming punching step during sample preparation, especially to improve throughput for a large number of samples (Déglon et al., 2009, 2012; Gaugler et al., 2021; Luginbühl and Gaugler, 2020; Oliveira et al., 2014a, 2014b).

Over the past decade, DBS has gained considerable attention from pharmaceutical companies to support Phase I–IV PK studies in children, neonates, critically ill

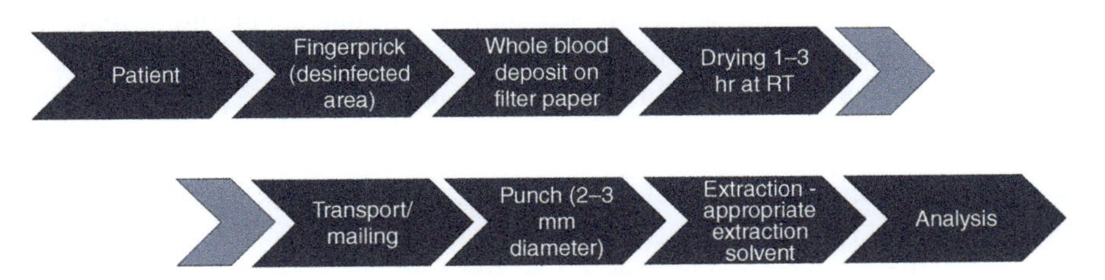

Figure 4.2 DBS workflow for sample collection and analysis. The collection area (finger, heel) is first disinfected. The skin is then punctured with a sterile lancet, and the first blood drop is wiped away and subsequent drops are placed on the collection paper, marked with circles to be filled. Once all the required circles are filled, the collection paper is left to dry for a few hours at room temperature on a nonabsorbent surface. After drying, the DBS paper cards are stored in plastic bags containing desiccant pouches before transport/mailing to the analytical laboratory. For sample preparation, the DBS cards are punched for removal of a disk (2–3 mm), which is then extracted with an appropriate extraction solvent before analysis.

patients, omics approaches (e.g. genomics, proteomics, and metabolomics), toxicological studies, forensic toxicology, and clinical applications as an alternative to conventional sampling for biological fluids (Londhe and Rajadhyaksha, 2020; Nys et al., 2017). Several advantages have been demonstrated for the use of DBS over traditional plasma sampling (Figure 4.1; Table 4.1). However, one of the main analytical challenges associated with the use of DBS on cellulose-based paper cards in clinical settings is the effect of blood Hct (Denniff and Spooner, 2010).

Hct is the ratio of the volume of red cells to that of the whole blood sample, and it varies according to age, gender, and disease state, impacting the viscosity of the blood. A sample with high-Hct blood results in a smaller blood spot and darker color, while low-Hct blood provides a larger blood spot and a lighter color due to a low number of RBCs (Figure 4.3). Therefore, Hct affects the characteristics of a DBS sample resulting in variability in blood spot area, drying time, analyte stability, homogeneity, and extraction efficiency (Denniff and Spooner, 2010). Consequently, Hct impairs the precision and accuracy of the quantitative data derived from the sample. For example, when punching and extracting a fixed diameter subpunch of a low-Hct blood sample, the analyte concentration is underestimated. During sample preparation for quantitative assays, attention should be paid to the preparation of calibration curve and quality controls (QC) samples in which the Hct of the control blood can vary from that of the study samples.

For the generation of high-quality quantitative bioanalysis data from DBS samples, it is important to evaluate the homogeneous dispersion of blood on the DBS cards to assure accurate and precise results. Moreover, the use of DBS microsampling for a PK or TDM study requires comprehensive quantitative bioanalytical validation and bridging studies with traditional blood sampling (FDA, 2013; Kothare et al., 2016). On the other hand, the detection of biomarkers for IMDs that are intended only for qualitative screening might tolerate nonhomogeneous blood sample distribution on DBS cards.

Some analytical alternatives for assessing and reducing/eliminating Hct effects when using cellulose-based paper cards for DBS sample collection include:

- spotting an accurate volume of blood and extraction/analysis of the entire DBS (Oliveira et al., 2014b).
- use of a filtration membrane/device to separate plasma from the RBCs (Hauser et al. 2018; Sturm et al., 2015).
- development of alternative substrates for accurate biofluid collection (Denniff and Spooner, 2014).
- normalization to the measured Hct level of the sample (Capiau et al., 2019) or to another coextracted endogenous analyte, e.g. chloride (Crawford et al., 2020), potassium (Capiau et al., 2013), and hemoglobulin levels (Richardson et al., 2018).
- development of microfluidics devices, as new patient-centered approaches for blood collection (Lei and Prow, 2019; Lenk et al., 2015b).

Table 4.1 General Considerations for Analytical Workflow Using Microsampling Techniques.

Sampling technique	Purpose of the technique	Advantages	Limitations	Potential for self-sampling	Selected references
Common considerations for: DBS; DPS; PDBS; VAPD; VAPD-mini; HemaSpot™; Mitra® VAMS®; Capitainer® qDBS; HemaXis™ DB, hemaPEN®; Tasso-M20	–	• Less invasive than venipuncture since a few microliters of whole blood is needed (<50 µL) • Convenient sample storage and transport, minimal space is required and no temperature regulation • Reduced risk of infectious transmission and cross-contamination • Based on the technique and extraction procedure, often less time-consuming sample preparation than traditional sample extraction procedures. • Drying of blood on a solid phase makes analytes less reactive and more stable in a dried format than in frozen samples • High-throughput analysis by laboratory automation solutions	• Requires a sensitive and specific assay due to small sample volume • Requires elution of the sample from the filter paper (cellulose-based, polymeric membrane, or absorbent tip) • Incorporation of internal standard • Inadequate for air-sensitive or volatile analytes • Stability depends on the class of analytes	–	–

(*Continued*)

Table 4.1 (Continued)

Sampling technique	Purpose of the technique	Advantages	Limitations	Potential for self-sampling	Selected references
Dried blood spot (DBS)	• To provide a microsample of whole blood from a finger/heel prick	• Blood collection on a conventional paper card (~5–10 μL) • Might allow incurred sample reanalysis (4 spots/card) depending on blood homogeneity on paper card • Serial sampling on a single paper card • High-throughput analysis by DBS robotic punching devices and/or on-line extraction/analysis of DBS cards • Different cellulose and noncelullose supports commercially available, e.g. Whatman® 903, FTA® DMPK series, Ahlström 226	• Hematocrit (Hct) level impacts quantitative assays due to the blood spot size and spot homogeneity • Quality of the DBS sample (sample collection and spotting variations) • Accurate volumetric application of blood on cards for quantitative assays requires training • Manual punching of cards limits laboratory throughput • Requires complete drying before storage/transport (~4 hr)	Yes, minimum training is required.	Denniff and Spooner (2010), Henion et al. (2013), and Jager et al. (2014)
Dried plasma spot (DPS)	• To provide a dried plasma spots from a whole blood sample	• Plasma is volumetrically collected in a polymetric membrane or cellulose discs • Fast total sample collection time (~3 min) • Fast drying time (~15 min) • No Hct impact • Plasma is obtained without centrifugation • Commercially available plasma separation membranes for self-sampling, e.g. Vivid™ and Cobetter®.	• DPS are colorless, resulting in the spots not being easily recognized by automated systems • DPS cards require additional layers to perform blood constituent separation • Limited laboratory throughput since the upper membrane has to be peeled away from the lower membrane • Low plasma yield • Plasma composition is different from those produced by centrifugation	• Yes, when using commercially available plasma separation membranes. • No, when the plasma is obtained by centrifugation and then spotted on cards.	Li et al. (2012b), and Nys et al. (2017)

Perforated dried blood spot (PDBS)	• Similar to DBS with precut or perforated spots with larger absorption area for maximum sample usability	• Reduced Hct effect compared to regular DBS • No spot size distribution and volume diffusion variations • Minimize carryover since there is no need to punch the cards • No sample loss	• Requires complete drying before storage/transport (~4 hr) • Relies on spotting an accurate volume of blood (5–10 μL) onto the center of the paper disks • Accurate volumetric application of blood on cards for quantitative assays requires training • Limited laboratory throughput since automated/online extraction/analysis is not available • For research purposes only, not commercially available	No	Li et al. (2011, 2012a)
Volumetric absorptive paper disc (VAPD)	• Combines polymeric and filter paper to minimize the effect of Hct	• Accurate collection of 20 μL of whole blood • Cards with a single hole punch of 5.5 mm • No Hct effect due to sample dispersion • No special labware for accurate blood spotting	• Manual punching of cards limits laboratory throughput • Requires complete drying before storage/transport (~4 hr) • Limited laboratory throughput since automated/online extraction/analysis is not available • For research purposes only, not commercially available	Yes, minimum training is required.	Nakahara et al. (2018)

(*Continued*)

Table 4.1 (Continued)

Sampling technique	Purpose of the technique	Advantages	Limitations	Potential for self-sampling	Selected references
Mini-disc (VAPD-mini)	• Same as VAPD but for collection of a smaller whole blood volume	• Accurate collection of 5 μL of whole blood • Cards with single hole punch of 3 mm • No Hct effect due to sample dispersion • No special labware for accurate blood spotting	• Manual punching of cards limits laboratory throughput • Requires complete drying before storage/transport (~4 hr) • Limited laboratory throughput since automated/online extraction/analysis is not available • For research purposes only, not commercially available	Yes, minimum training is required.	Nakahara et al. (2018)
HemaSpot™	• To improve DBS test reliability by providing uniform blood collection and distribution	• Blood collection on a paper card with three different volume capacities • When dried produces eight sample replicates (HemaSpot HF) • Even blood distribution, reducing Hct effect • Allows incurred sample reanalysis for repetitive and reproducible testing, using replicate spots • Punching not required • No sample loss • Commercially available from Spot On Sciences	• Requires complete drying before storage/transport (~4 hr) • Not designated for quantitative assays unless an accurate volume of blood is spotted onto the center of the paper disks • Limited laboratory throughput since online extraction/analysis of DBS is not available	Yes, minimum training is required.	Brooks et al. (2016) Spot On Sciences, Inc.

VAMS®	• Collection of dried biological samples for quantitative bioanalysis by absorbing a fixed accurate volume of blood onto an absorbent tip	• A fixed accurate volume of blood (10, 20, or 30 μL) • Absence of Hct impact in quantitative assays • Serial sampling • Can be used with different biomatrices (urine, plasma, serum, tears, etc.) • No special labware for accurate blood collection • Minimize carryover since there is no need for punching • No sample loss • Readily amenable to tip-based automation systems • Allows incurred sample reanalysis for repetitive and reproducible testing using replicate tips • Commercially available, Neoteryx, LLC Mitra® VAMS®	• There is no spare sample for reanalysis of the same tip • Manual removal of the absorbent tip limits laboratory throughput • Requires complete drying before storage/transport (~2 hr) • On-line extraction and analysis is not available • Can be under or overfilled with blood	Yes, minimum training is required.	Denniff and Spooner (2014), Protti et al. (2019), Spooner et al. (2015), and Volani et al. (2017) Trajan Scientific and Medical, Inc.

(*Continued*)

Table 4.1 (Continued)

Sampling technique	Purpose of the technique	Advantages	Limitations	Potential for self-sampling	Selected references
Capillary microsampling (CMS)	• Utilization in bioanalysis of certified capillaries for simplified collection and handling of small, exact volumes of liquid matrices	• Filled end-to-end by capillary forces • Enables handling of samples (blood, plasma, and other biofluids) in liquid state, making it compatible with existing technologies for sample treatment before analysis • Collection of an accurate and precise volume of biofluid • Can be diluted and stored for later analysis or spotted on cellulose-based cards for DPS generation • Can perform quantitative assays • High-throughput analysis by laboratory automation solutions • Commercially available, e.g. Vitrex®, Shimadzu Microsampling Wings®	• Longer sample preparation time than dried matrices, since capillary must be centrifuged, washout/diluted, and followed conventional treatment for liquid samples • Precision and accuracy of the capillaries is related to the manufacture's quality and to the training of the person collecting the samples • Capillary filled with biofluid is placed in a sample tube and frozen • Requires appropriate refrigeration for transportation • Requires a sensitive and specific assay due to small sample volume	Patients could be trained but is more challenging for at-home sampling and transport/mailing	Nilsson et al. (2013) and Verhaeghe et al. (2017, 2020) Vitrex Medical A/S, Shimadzu
Capitainer®-qDBS	• Provides a combination of microfluidics and DBS, producing a precise and accurate volume of whole blood in a pre-perforated disc	• Nullify Hct bias due to blood volume dispersion. • Can perform quantitative assays • A precise and accurate volume of blood is added to a pre-perforated paper disc, even if sampling volume is higher	• Sample loss (excess is discarded) • Manual removal of the precut spot limits laboratory throughput • Limited laboratory throughput since automated/online extraction/analysis is not available • Requires complete drying before storage/transport (~4 hr)	Yes, minimum training is required.	Velghe and Stove (2018) and Lenk et al. (2015b) Capitainer AB

Capitainer®-qDBS (continued)		• Ease of use due to pre-punched paper discs • Might allow incurred sample reanalysis (2 spots/card) • Serial sampling on a single paper card • No punching • Gives clear indication when a sample has been collected • Cannot be under or over filled • Commercially available from Capitainer			
HemaXis™ DB	• Combines microfluidic-DBS technology for reproducible whole blood volumetric sampling with standard paper cards	• A fixed accurate volume of blood (5.5 µL) • No Hct impact in quantitative assays • Blood collection on conventional paper card • Might allow incurred sample reanalysis (4 spots/card) • Serial sampling on a single paper card • High-throughput analysis by DBS robotic punching devices and/or on-line extraction/analysis of DBS cards • Commercially available from DBS System	• Manual punching of cards limits laboratory throughput • Requires complete drying before storage/transport (~4 hr)	Yes, minimum training is required.	Verplaetse and Henion (2016) DBS System SA

(*Continued*)

Table 4.1 (Continued)

Sampling technique	Purpose of the technique	Advantages	Limitations	Potential for self-sampling	Selected references
hemaPEN®	• Microcapillary device for volumetric whole blood sampling	• Four accurate and precise volumes of blood (2.74 μL) from a single device • Intuitive pen-like device • Eliminate Hct bias • Might allow incurred sample reanalysis (4 spots/card) • Serial sampling on a single paper card • Faster drying when compared to regular DBS • Cartridge design integrates into microtiter plate formats • High-throughput analysis by laboratory automation solutions • Cannot be under- or overfilled • Capillary K2 EDTA-coated • Commercially available from Trajan	• Sample capacity of just a few μL (<3 μL) of blood per channel requires efficient analyte extraction • On-line extraction and analysis is not available • Difficult to open and obtain sample for analysis	Yes, minimum training is required.	Deprez et al. (2019) Trajan Scientific and Medical, Inc.
Tasso^{+}™ Tasso-SST™	• Microcapillary device for volumetric whole blood sampling	• Attached to the upper arm or thigh for whole blood collection • Tasso^{+}: Different volumes for whole blood collection (200–600 μL) using a single device • Tasso^{+}: different anticoagulant available (K_2EDTA, serum or plasma separator, lithium heparin) • Tasso-SST: blood collection volume of 200–300 μL and no anticoagulant	• Potential adverse environmental (or climatic) conditions • Associated with shipping that could affect the stability of the analyte to be tested • Hemolysis • Sample need to be centrifuged after collection, which represents a challenge for home testing	Yes, minimum training is required.	Hendelman et al. (2021) and Williams et al. (2021) Tasso, Inc.

Tasso^{+TM} Tasso-SST™ (continued)		• A push-button controls lancet activation and vacuum action • Minimal pain • Might allow incurred sample reanalysis • Blood is absorbed and stabilized • Separate drying step not required • User-friendly system for simple and convenient handling • High-throughput analysis by laboratory automation solutions • Ready to use for a variety of downstream applications • Commercially available from Tasso, Inc.			
Tasso-M20	• Microcapillary device for volumetric whole blood sampling	• Four accurate and precise volumes of blood (17.5 μL) from a single device • Attached to the upper arm or thigh for whole blood collection • Eliminate Hct bias • Might allow incurred sample reanalysis (4 spots/device) • Serial sampling • Faster drying when compared to regular DBS • Separate drying step not required. • Commercially available from Tasso, Inc.	• Sample capacity of just a few μL of blood per sample requires efficient analyte extraction • On-line extraction and analysis is not available	Yes, minimum training is required.	Tasso, Inc.

(Continued)

Table 4.1 (Continued)

Sampling technique	Purpose of the technique	Advantages	Limitations	Potential for self-sampling	Selected references
OneDraw™	• Microneedle-based device for volumetric whole blood sampling	• Attached to the upper arm or thigh for whole blood collection • OneDraw collects 150 µL of blood using a single device • A push-button controls lancet activation and vacuum action • Minimal pain • Might allow incurred sample reanalysis • Blood is absorbed and stabilized • Separate drying step not required. • User-friendly system for simple and convenient handling • High-throughput analysis by laboratory automation solutions • Commercially available from Drawbridge Health	• Currently only for blood glucose monitoring • Only available for dried samples • Nonvolumetric	Yes, minimum training is required.	Drawbridge Health, Inc.
TAPII®	• Microneedle-based device for volumetric whole blood sampling	• Attached to the upper arm or thigh for whole blood collection • Collects 100 µL of blood using a single device • A push-button controls lancet activation and vacuum action • Minimal pain • Might allow incurred sample reanalysis	• Small sample volume available • Hemolysis • Dried sample option is not available	Yes, minimum training is required.	Blicharz et al. (2018) and Xing et al. (2020) YourBio Health, Inc.

TAPII® (continued)		• Contains lithium heparin anticoagulant • Separate drying step not required. • User-friendly system for simple and convenient handling • High-throughput analysis by laboratory automation solutions • Commercially available from YourBio Health			
Loop Medical One	• Device for volumetric whole blood sampling	• Attached to the upper arm or thigh for capillary blood collection • High-volume collection (1mL) • Minimal pain • Separate drying step not required. • User-friendly system for simple and convenient handling • High-throughput analysis by laboratory automation solutions	• Not commercially available	Yes, minimum training is required.	Loop Medical SA

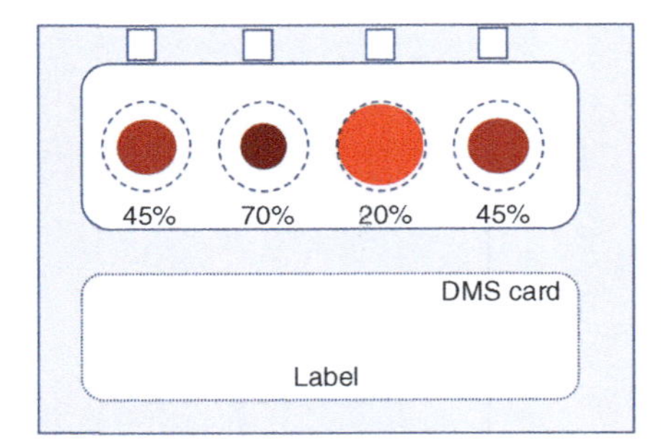

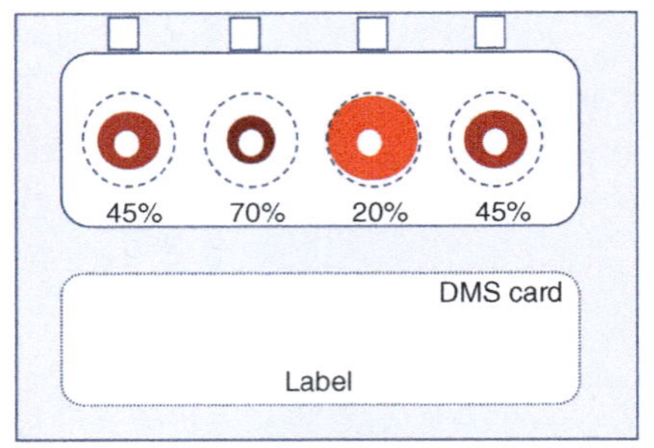

Figure 4.3 Illustration of three hematocrit levels of whole blood (20, 45, and 70%) applied on conventional paper cards used for dried blood spots. Left: Hct of 20% gives the larger blood spot, while 70% results on the smaller one. Right: representation of fixed-sized subpunches (3 mm) taken from DBS samples with blood of different Hct levels.

A recent work (Anderson et al., 2020) used 1522 DBS samples for high-frequency longitudinal tracking of protein biomarkers of inflammation-related acute phase response (APR). In this work, an elegant strategy was used to minimize the effect of variations in plasma volume between DBS from the same individual (typically ±10–15%). The normalization method used a sample-specific scale factor computed from an equally weighted combination of three proteins, namely albumin (Alb), hemopexin (Hx), and immunoglobulin M (IgM), whose abundances are usually constant within individuals over time. The investigators concluded that "the practicality of collecting large scale longitudinal biomarker data using DBS microsamples established a new dimension of dynamic biological 'Big Data', orthogonal and complementary to the established streams of data from genomics, electronic medical records, and wearable sensors." Some alternative procedures for the prediction of the Hct level from a nonvolumetrically applied DBS sample include measurement of potassium by indirect potentiometry (Capiau et al., 2013), chloride (Crawford et al., 2020), and hemoglobin (Richardson et al., 2018) by colorimetric assays. Hemoglobin contents could also by determined by dedicated instrumental techniques such as diffuse reflectance (Capiau et al., 2016), single-wavelength reflectance (Capiau et al., 2018), near-infrared spectroscopy (Osteresch et al., 2016), and image analysis (Del Ben et al., 2019).

HemaSpot™ Building on the well-established DBS technology, HemaSpot™ (Spot On Science, Inc.), was introduced to enable further innovative options for biospecimen collection, storage, and transport of microsamples. According to the manufacturers, the technology also addresses the long-standing Hct problem associated with blood dispersion on DBS (Figure 4.4a1), since the fluid disperses uniformly across the eight blades, reducing the Hct effect. HemaSpot™ HF

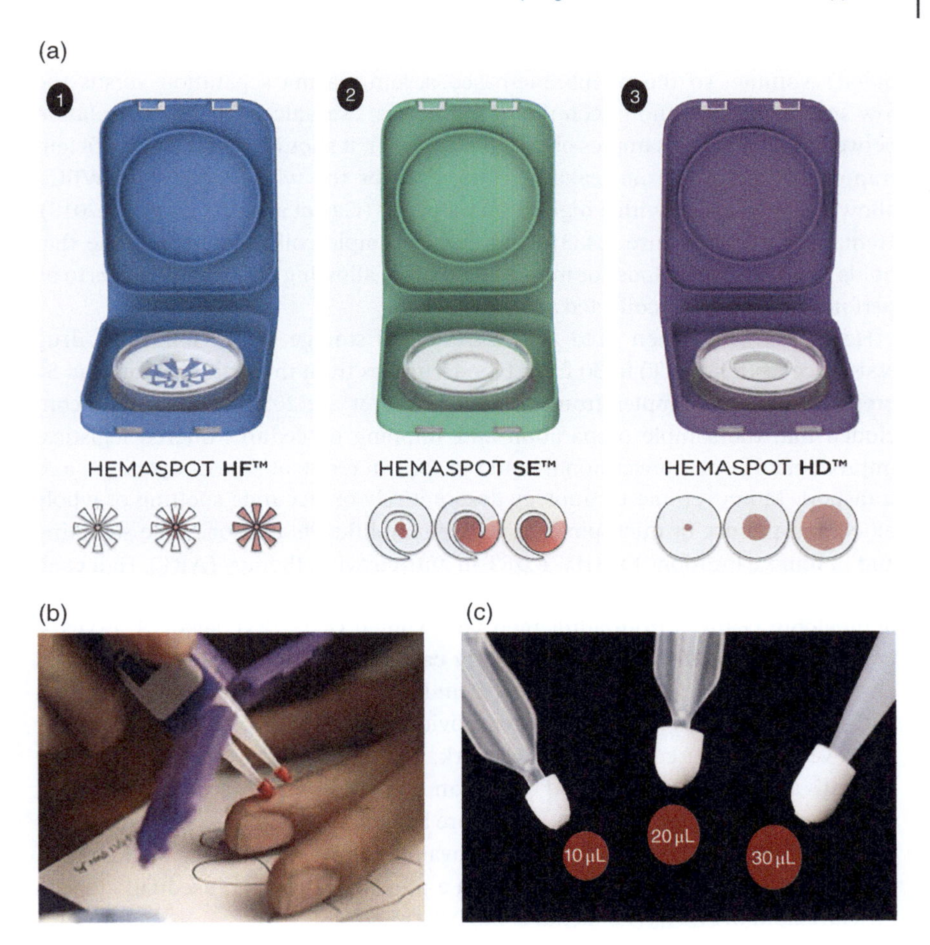

Figure 4.4 A selection of cellulose and noncellulose-based whole blood sampling devices. (a1) HemaSpot HF fan-shaped design enables even distribution across each of the eight blades allowing for reproducible analysis from the sample (no punching); (a2) HemaSpot SE spiral-shaped design allows for separation of whole blood into its constituent parts of whole blood cells and serum, and (a3) HemaSpot HD larger sample collection membrane allows for multiple punches, providing the opportunity to run a variety of essays on a single sample. Courtesy of Spot On Science. (b) Mitra® cartridge devices with two volumetric absorptive microsampling (VAMS)® samplers. (c) VAMS® samplers with blood collection volume of 10, 20, or 30 µL. Courtesy of Neoteryx, LLC.

(Figure 4.4a1) has a fan-shaped design that enables even distribution of whole blood across each of the eight blades, allowing for reproducible analysis from the same sample, without punching (Yamamoto et al., 2020). HemaSpot™ SE allows for the separation of whole blood into whole blood cells and an outflow serum

(Figure 4.4a2). It is important to note that the RBCs accounts for 40–45% of the blood's volume, so the membrane-based serum/plasma separation versus the low-speed centrifugation technique should be evaluated for the correlation between the plasma's samples obtained. Moreover, it should consider the efficient trapping of RBC, and no leakage of the RBC or the white blood cell (WBC), allowing a separation with a high extraction yield (Gao et al. 2020; Su et al., 2018). HemaSpot™ HD (Figure 4.4a3) has a larger sample collection membrane that holds up to 160 μL of blood for multiple punches, allowing a variety of assays to be performed on a single collected sample.

HemaSpot™ has been used as a novel blood storage device for HIV-1 drug resistance testing (DRT) in 30 fresh blood samples from the United States and 54 previously frozen samples from Kenya (Brooks et al., 2016). The authors concluded that the sample preparation and shipping procedures offered logistical improvements over conventional liquid plasma in terms of ease of operation and transport. Moreover, the technology does not rely on accurate spotting of whole blood and the use of micropipettes, which simplifies its use for home sampling and in remote locations for HIV-1 DRT in antiretroviral therapy (ART). Hall et al. (2020) also evaluated the HemaSpot™ device for the determination of glycated hemoglobin (HbA_{1c}) from adult men and women ($n = 128$), aged 18–75 years, with any form of diabetes and regularly carrying out self-monitoring of blood glucose. The HbA_{1c} levels determined from HemaSpot device showed correlation with venous HbA1c results, therefore, providing to patients a potential device for improved glycemic control. In this work, the authors also reported that the patient's acceptance of the blood collection method was 61.7%, and that the participants indicated that they would be more likely to have their testing carried out if this method of blood collection was available. In agreement with this, high patients' satisfaction was also reported in a previous study that used DBS for the measurement of HbA_{1c} (Fokkema et al., 2009).

4.3.1.1.2 Noncellulose-Based Devices for Sample Collection

Volumetric Absorptive Microsampling Recent advances in technologies include the volumetric absorptive microsampling (VAMS®) device, which maintains the benefits of small-volume sampling and the patient-centered aspects associated with DBS sampling. The Neoteryx Mitra® VAMS® device is composed of an absorbent polymeric tip that rapidly wicks up a fixed volume (10, 20, or 30 μL) of whole blood by capillary action and dries in 2 hr or less (Denniff and Spooner, 2014). The VAMS® microsampling collection device collects a fixed volume of blood, regardless of the blood Hct level (Figure 4.4b, c).

VAMS® presents some relevant advantages over DBS sampling, mainly related to accurate and precise blood volumes and Hct-independent recovery (Protti et al., 2019). Moreover, VAMS can be applied to other biological fluids such as plasma, serum, urine, and oral fluids (Marasca et al., 2020). Although some bioanalytical aspects are still under evaluation for clinical applications, VAMS® has been increasingly recognized as a viable alternative to DBS and other dried microsampling techniques.

To provide insights to apply and adapt the VAMS® technique to bioanalysis of whole blood and other matrices (urine and oral fluids), the most important practical aspects related to sampling and pretreatment of VAMS® procedures were discussed in a recent publication (Protti et al., 2019). The evaluated parameters included: touching of the sample surface; drying time; storage at room and controlled temperature; extraction of matrices by device tip detachment or direct device immersion in the extraction solvent; different solvents and mixtures as extraction medium; vortex mixing, shaking, ultrasonication and optional centrifugation; and direct injection versus additional sample pretreatment.

In a recent publication, VAMS® was evaluated as a microsampling approach for the quantitative home monitoring of the tacrolimus area under the concentration versus time curve (AUC) (Gustavsen et al., 2020). The results revealed that the VAMS® approach demonstrated high accuracy and precision when compared with venous AUC, making it an attractive approach for TDM and human clinical trials. As another example, the forensic assessment of cocaine consumption, blood, and plasma were analyzed by a validated LC-MS method, and the pre-analytical procedure adopted VAMS® for blood and plasma sampling (Sciberras et al., 2019). The sample pretreatment investigated the VAMS® volume accuracy (20 μL) at different tip-sample contact times (1–20 s), which demonstrated that the VAMS devices can efficiently sampled whole blood or plasma with high accuracy for all contact times longer than 3 s. Among different extraction solvents and procedures tested for analyte extraction (ultrasound-assisted extraction or vortex-assisted extraction), a low volume of 100 μL for the extraction solvent (methanol/acetonitrile/phosphate) and ultrasound-assisted extraction at 40 kHz for 10 min furnished the best results. The analysis of real samples includes the analysis of macro- and microsamples from cocaine users, and the whole blood and plasma were simultaneously sampled, either by venipuncture or by fingerpricking, and the VAMS® devices prepared for the assay. The LC-MS method was successfully validated and applied to real whole blood and plasma with VAMS® microsampling devices. The results showed high agreement between whole blood and plasma prepared as dried matrices and those of traditional venipuncture samples.

4.3.1.1.3 Capillary Liquid Microsampling

Capillary liquid microsampling (CMS) consists of a simplified approach for the collection and handling of small volumes of biofluids at the point of care with high precision and accuracy (White et al., 2014). Samples are collected with anticoagulant-coated (or uncoated) precision microcapillary tubes in a glass

(Figure 4.5a) or plastic capillary using capillary action (Figure 4.5b, c). The content of the filled capillaries can subsequently be applied to a substrate for the preparation of dried format microsamples, such as DBS, or be transferred to a buffer or dilution solution for storage and further processing. Alternatively, plasma or serum can be generated from the whole blood microsamples by centrifugation of the capillary and used for the preparation of DPS or dried serum/plasma spots (DSS) (Nilsson et al., 2013).

The removal of the whole blood, plasma, or serum (after centrifugation) from the capillary is often facilitated after mixing it with a washout liquid that dilutes the sample, followed by handling and analysis using the same procedures employed for conventional large-volume samples. The result is a homogenous diluted sample that can be handled in the subsequent analytical steps. The washout liquid is usually a buffer that may contain organic modifiers such as acetonitrile or methanol up to at least 20% (Guerra Valero et al., 2019; Nilsson et al., 2013). The use of CMS in quantitative bioanalysis was investigated for the

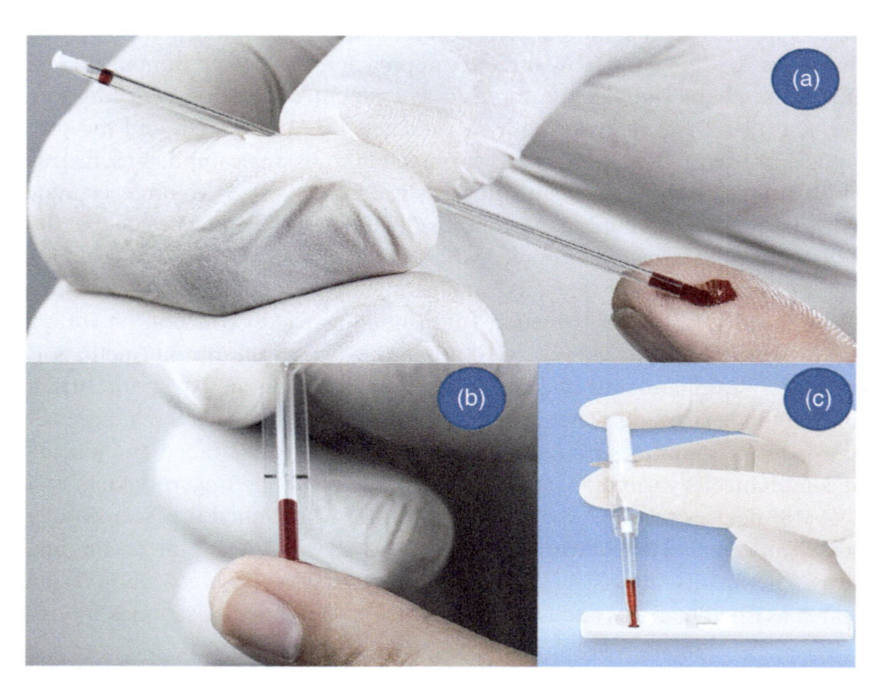

Figure 4.5 Capillary whole blood sampling devices. (a) Vitrex Medical self-sealing hematocrit capillary tubes. (b) Vitrex Medical plastic bulb dispenser for accurate blood volume measurement from 5 to 80 μL. (c) Minivette POCT (Sarstedt) for accurate blood collection for point-of-care tests available from volumes of 10–200 μL. Courtesy of Vitrex Medical.

determination of vancomycin in plasma samples collected from a fingerprick. Conventional arterial plasma samples were also collected from an indwelling cannula for a bridging study (Guerra Valero et al., 2019). The analytical methods were successfully validated, and no significant bias was observed, showing adequate agreement between CMS and the paired plasma samples.

Although CMS is a well-established approach for accurate whole blood collection and has many ethical benefits for animal studies, microcapillaries or calibrated pipettes will only be readily applied for humans in situations where trained and experienced nursing or laboratory personnel are available. In addition, the requirement for training and provision of calibrated pipettes for long-running multisite clinical studies is particularly difficult for this technology.

Capillary microsampling of whole blood (15 µL) from fingersticks or heel sticks was used to collect PK samples from pediatric subjects in two projects (Verhaeghe et al., 2020). The first case was a clinical study for mebendazole in subjects between 1 and 16 years old. The second case was a phase IB study to assess PK, safety, tolerability, antiviral activity, and impact on the clinical course of respiratory syncytial virus (RSV) infection after multiple oral doses of an anti-RSV drug in infants (>1 month to ≤24 months old) hospitalized with RSV infection. For both applications, capillary microsampling was chosen to obtain PK samples because of the young age of the subjects and the remote geographical locations. Furthermore, in the case of the mebendazole study, the absence of centrifuge equipment at locations was also a factor in using this approach. Moreover, bridging studies in healthy volunteers were first conducted to evaluate the feasibility of CMS in adults and to correlate PK data from venous plasma sampling and whole blood capillary microsampling. Both studies were completed, and the potential of CMS microsampling in clinical studies was demonstrated. However, some issues were reported by the authors, including contamination of the samples and improperly filled capillaries (Verhaeghe et al., 2020).

4.3.1.1.4 Microfluidic-DBS Devices

Motivated by the advantages of microsampling, mainly the benefit to patients and the cost-effectiveness of at-home sampling techniques, there has been considerable interest in the development of novel technologies that attain the benefits of DBS microsampling, to obtain high-quality data without the concerns associated with Hct. This has led to the development of technologies based on DBS sampling, such as Capitainer® qDBS (Capitainer AB), HemaXis DB10 (DBS System SA), hemaPEN (Trajan Scientific and Medical), and fluiSPOTTER® (Fluisense ApS).

Capitainer® qDBS uses an approach that automatically measures and stores an exact blood volume, which enables accurate quantitative measurements and does not require assistance by healthcare professionals, providing a great alternative for at-home sampling (Velghe and Stove, 2018). The Capitainer® qDBS device is equipped with an inlet port to which a drop of blood from a fingerprick is added, resulting in the filling of a capillary microchannel with a fixed volume of 10 µL from an undefined volume of a fingerprick blood. Upon complete filling of this capillary channel, a thin film at the inlet dissolves, resulting in the absorption of the excessive amount of blood by a paper matrix, leading to the separation of the excess blood and the filled channel. Finally, upon dissolving of the film at the outlet, the capillary channel is emptied through capillary forces, resulting in the absorption of 10 µL of blood by a pre-perforated paper disk (Ahlstrom 222 filter paper) (Figure 4.6a) (Lenk, et al., 2015b; Velghe and Stove, 2018).

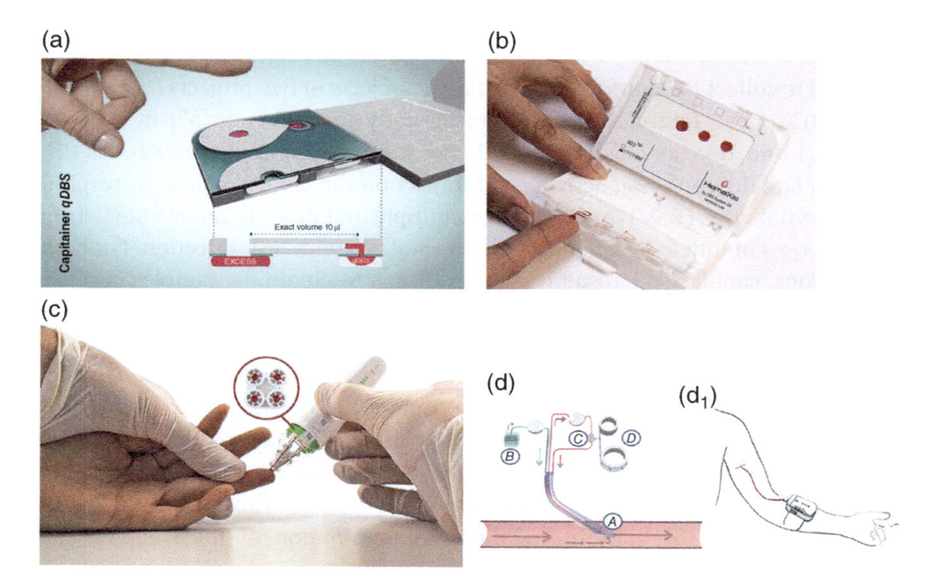

Figure 4.6 Microfluidic-DBS devices. (a) Capitainer qDBS collects an unknown amount of excess blood in DBS paper (Ahlstrom grade 270), while 10 µL of the accurate volume of blood is absorbed in the Capitainer qDBS paper disc (Ahlstrom grade 222). Courtesy of Capitainer. (b) HemaXis DB10 incorporates a microfluidic chip with a standard filter card and a protective case for accurate whole blood collection of four samples from the fingertip and controls an accurate volume of 10 µL. Courtesy of DBS System SA. (c) hemaPEN for the collection of four volumetrically fixed, accurate, and precise whole blood samples from a single source onto a cartridge containing prepunched DBS pads (inset). Courtesy of Trajan Scientific and Medical. (d) fluiSPOTTER, wearable technology for automated blood sampling of 20 serial blood samples over 20 hr or less. (d_1) Sketch of Fluispotter placed on the left lower arm, with the sampling catheter placed in the brachial vein in the upper arm. Reservoir (*B*) contains sodium citrate and an internal standard that is mixed at the time of sampling with venous blood from a catheter placed inside the peripheral vein (*A*). The fortified blood sample is quantitatively spotted via a micro-peristaltic pump (*C*) onto a blood spot paper strip (PerkinElmer 226) (*D*) for sequential blood sampling. Reproduced from Adhikari et al. (2020)/with permission of Future Science Group.

The effect of blood Hct, lipid content, and the volumetric performance of the Capitainer® qDBS device was tested with radiolabelled [^{14}C]-diclofenac (Spooner et al., 2018). The results demonstrated that regardless of the Hct levels (25–65%) or lipid content, the device efficiently deposited a fixed volume of blood (13.5 μL). Moreover, the high volumetric precision and accuracy of the derived dried blood was comparable to that of conventional positive displacement pipettes. The study demonstrated that the Capitainer® qDBS device is suitable for the accurate collection of blood samples. An additional study reported a validated method for the quantification of caffeine and paraxanthine in 133 venous patient samples to evaluate the potential of Capitainer® qDBS as a hematocrit-independent DBS collector and to investigate whether the amount of blood applied had an influence on the device performance (Velghe and Stove, 2018). Venous patient samples with a Hct range of 18.8–55.0 were used, and the results demonstrated that the device nullifies the Hct-based area basis for caffeine and paraxanthine and that the volume added at the inlet of the device does not have an impact on the measured analyte concentrations, being considered a reliable alternative for whole blood collection in existing and emerging applications.

Another approach using microfluidics and a standard DBS card, HemaXis DB10 (DBS System SA), delivers an accurate and precise amount of blood (5.5 μL) with no bias due to Hct (Figure 4.6b). The DBS card (903 Protein Saver cards, Whatman) used within this device is fully compatible with the manual or automated handling process at the laboratories, although other DBS cards are fully compatible with the device too. A comparative study between the use of an automatic pipet and the HemaXis DB10 for the creation of precise volumetric DBS samples has been performed (Verplaetse and Henion, 2016). Samples obtained from both approaches were successfully analyzed using a fully automated sample extraction system and analysis by online LC-MS/MS, allowing detection of five stimulants in fingerprick blood. Analytical data from DBS prepared with the automated pipet and the HemaXis DB10 were comparable, but the latter was easier to operate than the former, making the approach more suitable for sample collection by unskilled persons, especially for patient-centered application for at-home sampling (Verplaetse and Henion, 2016).

Another example of a microfluidic-DBS device is the hemaPEN (Trajan Scientific and Medical). This Hct-independent sampling device collects whole blood via four integrated 2.74 μL microcapillaries, each depositing the blood on a prepunched paper disk (Figure 4.6c). This hemaPEN technique was evaluated through an extensive bioanalytical validation for the quantification of caffeine and its metabolite paraxanthine in human patient samples (Deprez et al. 2019). A comparative analysis of patient samples (Hct range: 0.17–0.53) using the hemaPEN, 3 mm DBS subpunches, and whole blood revealed a limited Hct dependence of the hemaPEN devices. Over a Hct range of 0.20–0.50 Hct, the hemaPEN measurements compared to those obtained from liquid venous blood resulted in a slight Hct-based bias of 6.9 and 5.4% for caffeine and paraxanthine, respectively. However, subpunched DBS resulted in Hct-based bias ≥25% when compared to those from liquid venous blood. When considering an intermediate Hct of 0.40, the bias remained within 6%. Therefore, hemaPEN showed good potential for the precise volumetric collection of DBS microsamples.

A unique wearable automated blood sampling system called fluiSPOTTER® has been developed by Fluisense ApS (Figure 4.6d; Adhikari et al., 2020). The device is used for the collection and storage of 20 serial blood samples over 20 hr or less with minimal blood loss in animals. The device is not yet approved for human use but could be beneficial in studies where the serial concentration of analytes needs to be determined for 20 hr and without disturbing the free roaming of animals or for human studies in the future.

4.3.1.2 Other Biological Matrices

Microsampling strategies are also possible with biological fluids other than whole blood, such as plasma, serum, saliva, urine, and breast milk. When obtained from spotting the biofluid on cellulose-based filter paper or noncellulose materials, these approaches are also called dried matrix spotting (DMS), to reflect its wider applicability. The development of DMS has received great attention of researchers for application in clinical settings for TDM of therapeutic agents and clinical elemental analysis (Resano et al., 2018), since it provides positive attributes previously discussed to DBS and other microsampling devices, such as simplicity, noninvasiveness, cost-effectiveness, and patient-convenience. Moreover, for these matrices, the amount of sample available is typically not a problem. Saliva, urine, tears, sweat, and breast milk contain many analytes of interest for screening, diagnosis, and monitoring, and many of the hazards associated with blood collection do not apply to those matrices. Additionally, collections of these matrices are noninvasive, which have obvious advantages for patients, children, and the terminally ill. Add to that the fact that standardized methods for separating plasma/serum from whole blood require lab settings and the provision of equipment such as centrifugation and laboratory supplies, therefore, circumventing this difficulty.

4.4 New Development Areas

4.4.1 Automated Sample Collectors

Recent developments in microsampling technologies for patient-centered care involves the development of automated sample collection devices that can also facilitate closer patient monitoring and sampling during clinical events, especially in situations like the novel Covid-19 outbreak, offering patients an alternative option for home sampling and helping to reduce in-person visits to clinics and potential exposures (Anderson, 2021).

TASSO⁺, TASSO-M20, TASSO SST (Tasso, Inc., WA, USA), TAPII® (YourBio Health, MA, USA), OneDraw™ (Drawbridge Health, Inc., CA, USA, n.d.), and Loop Medical One (Loop Medical SA, Lausanne, Switzerland) are alternative blood collection methods that are more convenient and less invasive than venipuncture or even fingerstick. These devices are single-use and sterile, allowing painless blood draw and are user-friendly enough to be operated by a layperson without any medical training at home. These devices produce blood liquid or dried samples that are further processed in the same way as venipuncture samples.

Some examples of these currently available technologies (Figure 4.4) for patient-centered blood collection include TASSO-SST™ (collects 200–300 μL) and TASSO⁺ (different volume sizes from 200 to 600 μL of blood) (Tasso, Inc., n.d.). This technology is based on microfluidics that utilizes a slight vacuum to pull blood from capillaries through tiny channels in the skin into a small laboratory tube. These devices may collect up to 150 μL of blood, which is enough to test for cholesterol, infectious diseases, cancer cells, diabetes, and other conditions. Recently, the same company introduced TASSO-M20™, which can deliver four dried blood samples with a precise volume of 17.5 μL each (CV <5%) for PK studies in clinical trials and TDM. Touch-activated phlebotomy (TAPII®), OneDraw™, and Loop Medical One are similar technologies to that of TASSO-SST™. According to YourBio Health, Inc, the TAPII® device collects approximately 270–350 μL of blood. OneDraw™ collects blood samples onto a filter paper, which has been used for the measurement of glycated hemoglobin (HbA1c) in people with diabetes. Loop Medical One is under development and little information about it is currently available, but it collects 1 mL of capillary blood to perform hematology panels and a large number of immunoassays.

Two recent papers describe the use of the TAPII™ device. Doornek et al. reported the monitoring of antibodies levels against SARs-CoV following Covid-19 vaccine boosters. (Doornek et al., 2022). Overall, 844 participants self-conducted at-home blood collection with the TAPII® device. The authors concluded that the innovative device for sample collection sets a precedent for greater research access using a consumer-directed technology for real-world studies. Shook et al. compared the durability of the anti-spike antibodies in infants from vaccine-induced mothers and in infants from mothers tested positive for SARs-CoV-2 during pregnancy. Using the TAPII® device, capillary blood samples were collected from 49 infants from vaccinated mothers, at 6 months (Shook et al., 2022).

The use of Tasso-M20 (former Tasso OnDemand™) was investigated for the development and validation of a bioanalytical method for the analysis of samples collected by the gefapixant drug development program (Roadcap et al., 2020). This novel format of the Tasso device utilizes four integrated 10 μL Neoteryx Mitra® VAMS® sample tips for the collection of dried blood samples. The authors concluded that the device successfully showed the ability of dried blood microsampling to predict the plasma equivalent concentration and PK results for gefapixant, indicating that Tasso-M20 could be a simple at-home blood collector to support telemedicine, wellness, and research.

Neoteryx Mitra® VAMS® and TASSO™ technologies were utilized to collect whole blood and Li-heparin plasma, respectively, while the traditional phlebotomy blood sample collection was used to generate serum and Li-heparing plasma samples from three patients treated with anti-A therapeutic, a humanized IgG1 mAb (Williams et al., 2021). Good recoveries of anti-A therapeutic were obtained using Mitra® VAMS® (90.2–109%) as well as TASSO-M20 (93.4–117%) as the sample collection methods when compared with venipuncture data (92.5–106%). Regarding the PK data generated with all three sample collection methods, comparable PK parameters were obtained with a percent difference ≤11%, when compared to the stablished method of venipuncture. The authors highlight the need for a larger study to evaluate statistical comparability but recognize the utility and advantages of the low-volume sampling technologies for improving patient care.

The Tasso-SST automated microsampling device was also evaluated as a self-collection of capillary blood for the determination of the anti-SARS-CoV-2 IgG antibody (Hendelman et al., 2021). Capillary blood was obtained via unsupervised and supervised utilization of the Tasso-SST. Venous blood was also collected by standard venipuncture. The three investigated methods furnished Sera that were all tested concurrently using the CE-marked EuroImmun anti-SARS-CoV-2 S1 IgG assay. In the first blood draw, the sample self-collection was successfully accomplished by 85 out of 91 patients enrolled in the study (93.4%) and 6 samples required redraw (6.6%). The data obtained demonstrated high concordance for the anti-SARS-CoV-2 IgG results between Tasso-SST capillary and the standard venous blood-derived sera. The same high correlation was observed when samples were stressed by winter and summer simulations to mimic extreme shipping conditions. The authors discussed some noted limitations that may raise concern, such as adequate education of patients is needed to perform unsupervised blood collection; the time needed for shipping prior to specimen processing; potential adverse climatic conditions that can affect the stability of the samples, and the final volume of blood obtained with the Tasso-SST device that was 339 μL, which produced approximately 196 μL of serum. In conclusion, the Tasso-SST device demonstrated accuracy and reliability as a self-collection for the measurement of SARS-CoV-2 IgG antibodies to reduce patient's exposure during the current pandemic.

4.4.2 Microfluidic Point-of-Care Devices

Miniaturized devices for POC medical diagnostics are a rapidly developing area which incorporates patient centric blood sampling technologies (Mejía-Salazar et al. 2020). These new tools are promising for a broad range of applications, from early diagnosis of several diseases, including cancer (Syedmoradi et al., 2021), diabetes (Meetoo and Wong, 2015), pathogen (Nguyen et al., 2020), cardiovascular (Goodacre et al., 2011), HIV/AIDS (Mauk et al., 2017), and COVID-19 (Brendish et al., 2020; Anderson, 2021).

POC devices present some advantages compared to laboratory-based tests, such as rapid and reliable sample-to-answer results for use by nontechnical operators, portability, nonrequirement of specialized equipment, cost efficiency for patients and providers by streamlining workflow and improving resource allocation across transportation, trained operator, and physician follow-up. The core idea of a POC diagnostic is to develop a self-containing miniaturized device that can be used to detect or monitor different end points including blood pressure, biomarkers and pathogens, in complex biological matrices, both in clinical and/or home settings (Primiceri et al., 2018). In addition, a great potential utility of such POC devices includes its use in low infrastructure laboratories with limited resources available. In the era of one-step medical checkout, POC devices can promote rapid actionable information for patient care and physicians, allowing timely therapeutic interventions and improving patients' clinical outcomes.

4.5 Summary of Currently Available Patient Centric Sampling Technologies

Table 4.1 summarizes the general advantages and limitations for the most common microsampling techniques with regard to analytical considerations when utilized for quantitative assays, particularly in a clinical setting.

4.6 Microsampling Analysis by LC-MS—Analytical Considerations

Dealing with small volumes of biosamples (<50 µL) poses an important aspect for the analytical detection technologies to be used, since it may potentially impact the assay sensitivity of the analytical methodology. Miniaturization of sampling devices, driven by the patient centric benefits, gives rise to important questions for the development, selection, and acquisition of miniaturized analytical technologies capable of providing high sensitivity, selectivity, high-throughput, and robustness.

The use of microsamples for LC-MS makes more demands on the sensitivity of quantitative assays than those using standard sample volumes. The adoption of microsampling approaches and technologies should also be evaluated for its ability to collect a homogeneous sample so that a human whole system is represented.

A positive aspect of modern mass spectrometers is that improvements in fast electronics, high vacuum technologies, fast scanning speed, and, mostly, high efficiency in ion transmission assure higher sensitivity, resulting in the detection of smaller concentrations of low sample volumes.

The reliability of quantitative data derived from the use of microsampling techniques is supported by reports showing cross-validation results, which demonstrate that the data correlates well with those data obtained from the analysis of conventional liquid samples (Guerra Valero et al., 2018). Statistical procedures like Passing-Bablok help to provide a valuable estimation of agreement between microsamples and conventional samples and indicate possible systematic bias for comparison of methods based on a robust, nonparametric model, nonsensitive to the distribution of errors, and data outliers (Bilić-Zulle, 2011). While there are no fixed rules for making the comparison between venous blood versus capillary blood sampling, reports discuss different recommendations for using DBS microsampling in clinical development programs (Evans et al., 2015; Kothare et al., 2016; Wickremsinhe et al., 2020). It is important to note that peripheral sampling (fingerpick) is different from venous sampling as no anticoagulant is required for sample collection. Therefore, bridging studies between liquid samples (blood, plasma, or serum) and dried matrices are important to understand the concordance between data. Analysis of bridging studies data using standard statistical analysis such as Pearsons's coefficient, Lin's concordance coefficient, Bland–Altman plots (mean-difference or limits of agreement plots) to compare two measurement methods for the same variable, and Deming or Passing-Bablok for estimation of the analytical method agreement and possible systematic bias between, helps furnish more robust criterion to establish a concordance/correlation of data between dried matrices and plasma concentrations.

4.6.1 Basic Principles of Liquid Chromatography (LC) and Mass Spectrometry (MS) for Bioanalysis of Microsamples

LC-MS has emerged as a powerful technique for its capabilities for separating, identifying, and quantifying substances in complex samples. In clinical laboratories, the application of LC-MS provides unique capabilities for selective and sensitive investigation of drugs, protein, metabolites, lipids, and disease-related biomarkers in clinically relevant biological matrices (Jannetto and Fitzgerald, 2016; Teclemariam et al., 2020).

LC is a separation technique in which the components of a sample mixture are separated based on their different interactions between a mobile phase and a stationary phase (Snyder et al., 2010). In the MS system, the separated analytes are ionized and identified based on their mass-to-charge (m/z) ratio (Watson and Sparkman, 2008). Coupling LC with MS adds another dimension to the analysis of complex biological matrices, leading to a powerful analytical technique that combines the separation power of the LC with the detection specificity of the MS. Therefore, the LC-MS technique is an important tool for the analysis of drugs, metabolites, biomarkers, peptides, and proteins with a wide range of applications, e.g. PK/PD assays, proteomics, metabolomics, disease control, and clinical diagnostics (Lee, 2012). The high degree of specificity offered by the LC-MS systems enables simplified sample preparation and chromatographic separations, leading to simpler processes and higher throughput analysis. LC-MS system also allows multiplexing capabilities in matrix-matched analytical standards, faster than any other analytical method with conventional detectors (Pitt, 2009). Moreover, even with a decrease in the available sample volume for analysis, the technological advances in microspray ionization sources, microflow LC, and microchip-based LC have helped to advance microsampling in the drug discovery industry and clinical settings (Bowen and Barfield, 2019).

4.6.1.1 Microspray Ionization Sources

In clinical laboratories, the vast majority of the LC-MS methods use atmospheric pressure ionization (API), with electrospray ionization (ESI) being the method of choice. ESI is a soft ionization technique capable of analyzing both small (<1000 Da) and large molecules (i.e. proteins up to 200 kDa), and it is extensively used for a broad range of target analytes, particularly those which are polar, non-volatile, and thermally labile. On the other hand, atmospheric pressure chemical ionization (APCI) is generally applied for the analysis of moderately polar and apolar (i.e. lipids), thermally stable compounds with low molecular weight (<1000 Da) (Banerjee and Mazumdar, 2012). Both ESI and APCI can analyze a wide variety of molecules using positive and/or negative polarities for complex biological samples.

In 1989, ESI was introduced by Fenn et al. (1989) and in 2020, John Fenn and Koichi Tanaka were awarded the Nobel Prize in Chemistry for the development of soft ionization techniques for large molecules, ESI, and matrix-assisted laser desorption ionization (MALDI), respectively (Banerjee and Mazumdar, 2012). Fenn developed ESI (Fenn et al., 1989) into an ion source capable of ionizing molecules at atmospheric pressure, directly from the liquid phase, forming an aerosol by the application of a high voltage to the liquid phase. ESI ionizes macromolecules resulting in characteristic multiple charge state distributions for individual molecules, while preserving weaker noncovalent bonds in the gas phase (Przybylski and Glocker, 1996). Due to its ease of use, ESI is one of the most used ionization techniques employed in clinical and research laboratories for clinical diagnostics such as TDM, PK/PD, metabolite identification, and biomarker measurements, as well as in discovery areas like proteomics, lipidomics, and metabolomics for identification of new biomarkers and targets.

The principle of ESI is to generate gas-phase ions from analytes before they are subjected to mass spectrometric analysis. This involves a series of steps including nebulization, evaporation, charged droplet shrinking (charge density increment), repetitive Coulomb fission cycles, and gas-phase ionization (Konermann et al., 2013; Wilm, 2011; Wilm and Mann, 1994).

Ionization efficiency is the effectiveness of producing gas-phase ions from analytes in solution, and it is highly dependent on the physicochemical properties of the analyte, interface design (emitters' geometry, inner diameter (ID), and materials), mobile phase composition, and flow rate (Page et al., 2007; Reschke and Timperman, 2011). There is a clear and direct dependency of the LC flow rate and charged droplet radius (Page et al., 2007). The smaller the charged droplet formed in the spraying process, the more susceptible it is to a greater ionization efficiency, since the solvent evaporation process will be more efficient and fewer coulombic fission events will be required to generate gas-phase ions. Ionization at lower flow rates besides improving sensitivity also leads to a reduction in matrix suppression. The resulting increase in sensitivity for the detection of analytes is a key factor to enable efficient LC-MS analysis of microsamples, where sample volume is limited and alternative approaches for gaining increased sensitivity are not viable, i.e. sample preconcentration.

In the last decades, changes in emitter and capillary geometries have gained attention to promote better ionization efficiency, while using low flow rates. A major breakthrough and performance improvement of miniaturized ESI concept was demonstrated by a microelectrospray emitter nanoelectrospray—nano-ES) that yielded an increase in sensitivity when the internal diameter of a gold-coated capillary was reduced from 100 to 1–2 μm (Wilm and Mann, 1994). In another work, a series of experiments that varied ionization and transmission efficiencies of an ESI interface were evaluated to better understand how those factors affected MS sensitivity. Some of the results demonstrated that analyte peak intensity increased as the flow rate of the ESI decreases. Lower ESI flow rates allow the emitter to be placed closer to the instrument, increasing the transmission efficiency by >90%, while higher flow rates require a greater ESI emitter-to-inlet distance (Page et al., 2007). The advantages offered by miniaturized ESI emitters deliver the benefit of low sample consumption and solvent demands, as well as enhanced sensitivity.

Conventional ESI sources are compatible with flow rates from 10 μL/min to over 1 mL/min. The larger the LC flow rate through the ESI capillary, the higher is the requirement for drying gas (N_2) and thermal heating to form the aerosol and the charged droplets (Ho et al., 2003). This is even more important if the LC mobile phase is highly aqueous. Microelectrospray (μESI) ionization uses flow rates of 100–1,000 nL/min and nanoelectrospray (nano-ESI) ionization uses flow rates of less than 500 nL/min (Gibson et al., 2009). In μESI and nano-ESI, the droplet formation occurs more readily than in conventional ESI (Figure 4.7) due to the improved desolvation at such low flow rates. The robustness of the spray and the position of the emitter are also important factors to ensure efficient evaporation and ionization of liquid samples and to achieve the highest sensitivity. New advances in this field consider tip geometry, changes in the surface to improve spraying, and the way the potential is applied (Gibson et al., 2009).

Ion suppression effects as an indicator of nano-ESI behavior have been studied, and the results indicated that higher flow rates favor enrichment of the surface ion droplets with charges and uneven droplet fission, resulting in a strong suppression of more hydrophilic ions (Schmidt et al., 2003). On the other hand, below 20 nL/min, an increase of all analytes ionization occurs, and ion suppression is almost eliminated, which offers great potential for highly sensitive methods for detecting a small amount of analytes or biomarkers.

4.6.1.2 Microflow Liquid Chromatography

Although there have been recent developments and technological innovations in nano-ESI, conventional high-flow LC systems are not appropriate to direct coupling with nanoscale emitters. Therefore, much effort has been made to develop

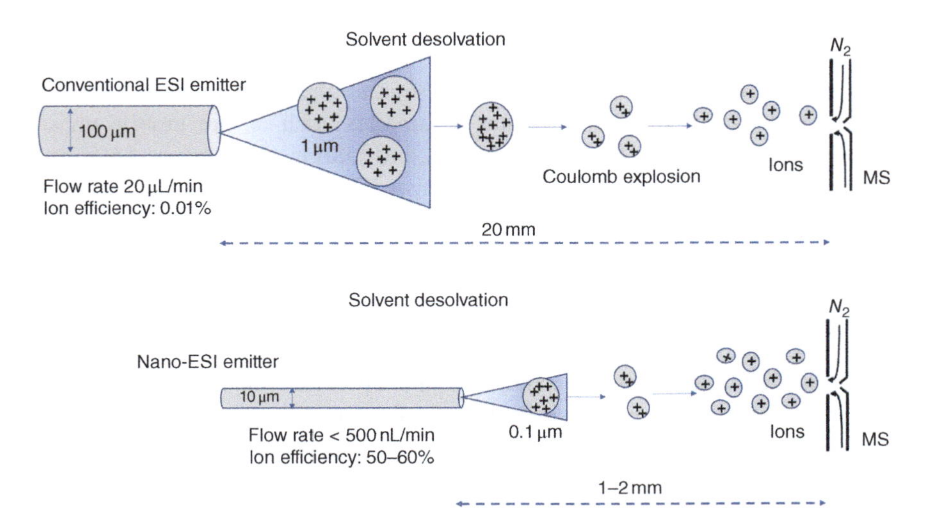

Figure 4.7 Illustration of ESI and nano-ESI: Comparison of ions efficiency.

LC columns with flow rates compatible with nano-/micro-emitters, while achieving high separation capacity for complex samples of limited amounts. The main innovations in LC column development include reduced particle sizes, increased column length, reduced column internal diameter, and the use of integrated microfluidics technologies (Arnold and Needham, 2013; Needham and Valaskovic, 2015).

An essential feature in LC is to achieve highly efficient separations in a fast analysis time. Therefore, much of the innovations that have been made in the LC systems have been driven by advancements in chromatographic column technologies. The goal of a chromatographic separation is to resolve peaks in terms of increasing separation between adjacent peaks and/or decreasing peaks widths. Resolution (R) in chromatography is expressed in terms of chromatographic efficiency (N), selectivity (α), and retention factor (k') through the Resolution Equation (4.1) (Snyder et al., 2010):

To that end, to obtain a higher resolution (R), adjustment of the three parameters of Equation (4.1) will directly influence the separation.

$$R = \frac{\sqrt{N}}{4} * \frac{(\alpha - 1)}{\alpha} * \frac{k'}{1 + k'} \quad (4.1)$$

Selectivity (α): the selectivity describes the separation of two species (X and Y) on the chromatographic column. Selectivity is a chemical factor that can be affected by changing the stationary phase chemistry and the ability of each analyte to interact with the stationary phase (Snyder et al., 2010). For example, by replacing a C18 stationary phase with a C8, or using different separation modes, e.g. reversed phase (RP) and hydrophilic interaction (HILIC). It is also possible to change the separation selectivity with the same column by changing experimental parameters such as the mobile phase composition, elution gradient conditions, and column temperature. Different organic modifiers produce different chromatographic results, and a good chromatographic separation is based on choices of mobile phase and stationary phase chemistries to create different interactions between them and the target analytes.

Retention factor (k'): the retention factor measures the retention of a specific peak relative to an unretained peak. The retention factor is influenced primarily by the mobile phase composition and column temperature (Snyder et al., 2010). It increases when a mobile phase with a weak elution strength is used or vice versa. Optimal k' values are $2 \leq k' \leq 10$, which represents a good balance for acceptable retention and analysis time. If k' value is lower than 1, target analytes may be eluting with many other interferences from the biological matrices and, in many cases, they could cause ion suppression in the ESI ion source of the mass spectrometer. The better the analytes are separated from the void volume, the higher is k', although high k' would also translate into an undesired longer run time.

Column efficiency (N): the efficiency of a column is measured by theoretical plates, and it is primarily influenced by column length, particle size, pore size, mobile phase velocity, and mobile phase viscosity (Snyder et al., 2010). Column efficiency may be increased by using narrower and shorter columns, packed with smaller particle sizes (e.g., 2.1×50 mm; 2.5 μm vs. 4.6×100 mm; 5 μm) that will provide twice the efficiency in the same analysis time. Another great advantage of using smaller particle sizes is the sharper chromatographic peaks obtained with less tailing/peak broadening, with consequently increased sensitivity (S/N: signal intensity/noise variation at baseline).

In general, clinical and analytical laboratories strive for faster results to increase their sample throughput but without sacrificing resolution and, if possible, maintaining the ability to separate as many compounds as possible (peak capacity). This is the chromatographers dilemma (Figure 4.8) that cannot be answered in an absolute way usually researchers tend to choose where they will stand depending on the following analytical question: (a) How many compounds need to be chromatographically resolved? (b) Do all compounds need to be quantified and qualified? (c) Is the biological matrix simple or complex? (d) How is the sample preparation performed? (e) How many samples need to be processed per day/shift/month? (f) Will it be a targeted or a nontargeted LC-MS approach? Usually, the answer is always a compromise between those three parameters: speed, resolution, and peak capacity.

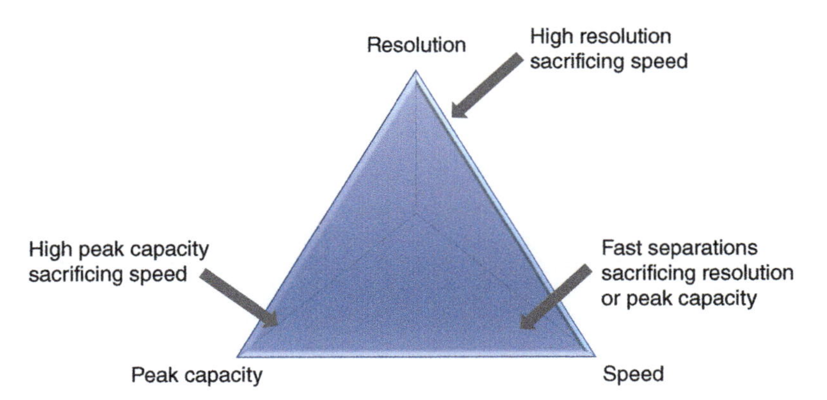

Figure 4.8 Chromatographers dilemma: Correlation between chromatographic speed, peak resolution, and peak capacity.

In recent decades, attempts have been made to improve those parameters and allow users to get faster results, higher peak capacity, and higher peak resolution, in the same analytical run.

The enhanced LC-MS sensitivity (low limit of quantification and detection), required to analyze small sample volumes from microsampling devices, can be achieved by reducing the internal diameter of the LC columns and, consequently, the flow rate of the LC mobile phase delivered to the ionization source. Miniaturization of the LC column diameter has moved from standard 4.6 mm ID to narrow-bore columns of 2.1 mm (ID), microbore (0.5–1.0 mm, ID), capillary (0.3–0.5 mm, ID), and nano (75–100 μm ID.) formats. Considering fixed amounts of injected sample, smaller ID columns produce more intense peaks, which provide better detection limits for MS and other concentration-sensitive detectors, making it feasible for the detection of low concentrations of drugs and/or biomarkers from microsamples.

Table 4.2 summarizes different column designations, taking into consideration the column ID, flow rate, and typical injection volume. The classification of miniaturized LC columns was proposed according to Saito et al. (2004).

Standard ESI sources generally operate at flow rates up to 1 mL/min and at a lower flow rate in the range of 10–200 μL/min, which makes it suitable for use with standard, narrow-bore, and microbore LC chromatographic columns. On the other hand, capillary and nano-LC columns operate at flow rates lower than 10 and 0.5 μL/min, respectively, and require suitable ionization sources, such as capillary and nano-ESI.

The use of miniaturized LC columns offers many advantages: (a) reduction of column ID leads to enhanced sensitivity due to minimized chromatographic dilution and drastic improvement of the signal-to-noise ratio (*S/N*); (b) lower injection volume, make it appropriate for the situation of limited sample amounts and/or for analysis of rare matrices; (c) low solvent consumption, and consequent

Table 4.2 Classification of LC Columns According to the Inner Diameter (ID), Typical Flow Rates, and Injection Volumes.

Column designation	Typical inner diameter (ID) (mm)	Typical flow rate (mL/min)	Typical injection volume (μL)
Standard LC	4.6	1.0–2.0	10–50
Narrow-bore LC	2.1	200–600	1–10
Microbore LC	0.5–1.0	10–300	2–6
Capillary LC	0.3–0.5	0.5–10	1–2
Nano LC	<0.10	<0.005	≤1

reduced environmental pollution; (d) reduced consumption of stationary phases, and better temperature control in response to the lower thermal mass (Saito et al., 2004).

The downscale factor (*f*) is a measurement of the increase in sensitivity due to the decrease in the ID of the columns for the same amount of injected sample, and reduction of the flow rate, when using a concentration-sensitivity detector, such as ESI-MS (Wilson et al., 2015).

This downscale factor (Equation 4.2) represents the sensitivity gain when switching from a conventional LC column with a 4.6 mm ID to a capillary column with 300 μm ID ($ID_1 > ID_2$), which theoretically shows a remarkable 230-fold greater sensitivity than conventional LC-MS. So, the smaller the flow rate the higher is the number of ionized species that enter the mass spectrometer, resulting in increased analyte sensitivity. The sensitivity associated with downscaling of the column ID and low flow rates has also been investigated by the quantification of infliximab, a biotherapeutic molecule (Boychenko et al., 2017). Capillary flow (EASY-Spray™ source, Thermo Scientific) and nanoflow (EASY-Spray® transfer line, Thermo Scientific) provided major sensitivity improvements when compared to analytical flow rates (Ion Max™ source, Thermo Scientific) (Figure 4.9).

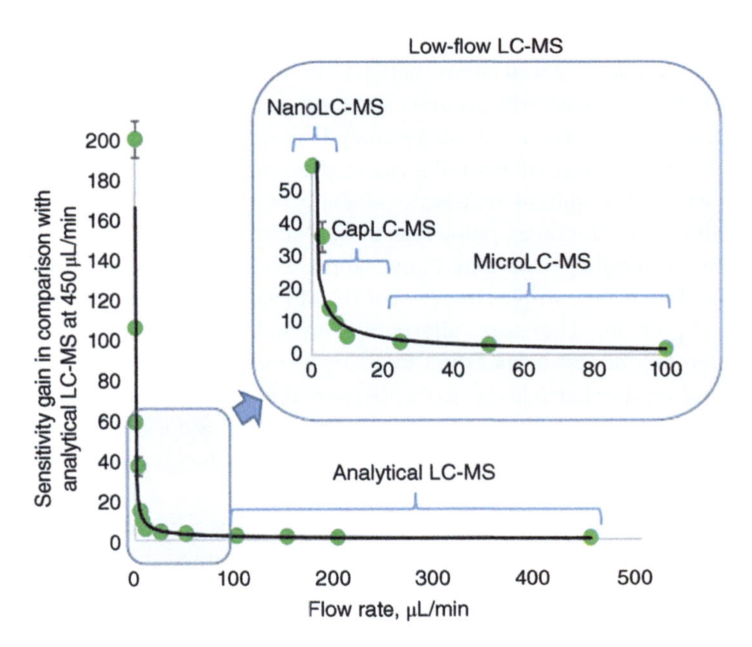

Figure 4.9 Sensitivity gains observed due to change in column diameter and flow rates for two target peptides obtained from a 1 pmol cytochrome C digest. Reproduced from Boychenko et al. (2017)/with permission of ThermoFisher Scientific.

$$f = \left(\text{ID1} / \text{ID2}\right)^2 \quad (4.2)$$

Nanoflow liquid chromatography (nano-LC) and nano-ESI coupled to high-resolution MS (HRMS) were evaluated for profiling of xenobiotics and metabolites and compared to conventional LC-ESI-HRMS (Chetwynd et al., 2014). The results demonstrated that detection limits were improved by between 2- and 2000-fold when using the nanosystem operating at a flow rate of 700 nL/min when compared to those obtained from a ultra-performance liquid chromatography coupled to electrospray (UHPLC-ESI) operated at 200 μL/min. Another analytical system for single-cell metabolomics of mammalian cells showed 3- to 132-fold greater sensitivity when using a nano-LC-ESI-MS system (Nakatani et al., 2020). Several hydrophilic metabolites, including amino acids and nucleic acids related metabolites (posing a challenging detection), were successfully identified in the sub-fmol detection limit range. Although nanoflow platforms offer highly sensitive methods for metabolomics analysis by nano-LC-ESI-MS, the use of these platforms is still very limited but could offer a potential for enhancement of the metabolomic coverage, mainly for low abundant signaling metabolites, such as steroids and eicosanoids, as discussed in a recent review (Chetwynd and David, 2018).

Some drawbacks of working in such small flow rates are the extra-column volumes and gradient delay volume, meaning that the complete system including chromatographic pumps, autosampler, reduced connection devices, ionization interfaces, and columns all need to be optimized to be accurate and reproducible and to avoid peak broadening, reduced peak capacity, and longer analysis times. Today, there are LC systems available with pumping systems and injectors that operate well at these flow rates. This comes with a concomitant increased availability of columns with different dimensions and alternative stationary phase selectivity (Nazario et al., 2015).

Hence, when microsampling is used as a novel approach for sample collection, the limited availability of the samples to perform LC-MS assays requires the best analytical method possible. Therefore, alternatives include considering micro, capillary, and nano-LC modes of operation by using chromatographic columns with smaller ID and smaller particle size as an alternative to higher peak capacity, resolution, and speed. To maximize sensitivity from different LC systems, Table 4.3 presents analytical figures of merit that can be used to compare the LC systems for microsampling analysis.

Comparing the three operational modes in LC-MS, all of them have advantages and drawbacks. Nano-LC is the most sensitive and very well suited for complex samples in untargeted/discovery assays where separation and detection of most peaks are required (e.g. shotgun proteomics experiments). However, this technique could be more difficult to implement for routine clinical use since it requires more experienced operators to optimize the analytical parameters and perform analyses on a routine basis. The potential high peak capacity of the nanoflow

Table 4.3 Analytical Figures of Merit for Standard, Capillary, and Nanoflow LC Systems.

	Standard LC	Capillary LC	Nano LC
Sample consumption	***	*	*
Solvent consumption	***	**	*
Ease of use	***	**	*
ESI ionization efficiency	**	***	***
Sensitivity	**	***	***
LC-MS system robustness	***	**	*
Peak capacity	**	***	***
Targeted assays	***	**	*
Untargeted/discovery assays	**	***	***

*Low; **medium; ***high.

system could lead to reduced matrix suppression. Although this comes with the price of longer run times and a decrease in loading capacity, unless the column length is increased or the analysis is run in conjunction with pre-concentration steps such as multidimensional liquid chromatography (2D-LC), an approach which is commonly used for shotgun proteomics.

Capillary LC (cap-LC) approach is well suited for targeted assays as it uses a few microliters of sample and has lower solvent consumption in comparison to standard LC methods. Although not as sensitive as a nano-LC, cap-LC usually achieves very good detection limits when compared with conventional and microbore LC due to the downscaling factor and better ionization efficiency. Cap-LC with appropriate connection devices to avoid peak broadening and chromatographic dilution fills the bridge between the nLC applications, which shows maximum sensitivity, and the standard/microbore LC with high throughput for targeted assays.

In conclusion, when there is a small amount of sample to be analyzed by LC-MS systems and the analytes of interest are in small concentration, lowering the LC flow rate to a few microliters/min, or even nanoliters/min, could promote the best usage of the sample, allowing replicate injections. Furthermore, it increases the detection sensitivity since a higher proportion of analytes are ionized as droplets in the ion source. Special attention should be taken regarding peak dispersion both in the LC system and in the ESI tip since nonspecific interactions with surfaces may be important when analyzing proteins and peptides. Hence, the material choice of the column, tubing, and emitter should be carefully selected. Nano-LC will be the best choice when there is a need of having higher peak capacity in the LC separation and improved ionization efficiency and sensitivity, particularly for applications in proteomics, metabolomics, and other untargeted assays.

4.6.1.3 Microchip-Based LC

Recently, a range of microchip-based columns has also been developed, which further enhances the sensitivity of the assays. The work of Grinias and Kennedy (2016) has shown also important improvements in using those devices for "omics" application. The microfluidic chip-based systems can be efficiently coupled with LC-MS systems, combining an increased sensitivity, chromatographic performance, and ease-of-use, which together provide significant advantages for the analysis of complex matrices, also bringing the benefit of low flow LC to nonexpert LC-MS users. Some examples of commercially available microchip-based columns include the IonKey™ (iKey™, Waters Corporation), the EASY-Spray™ (ThermoFisher Scientific), the PicoChip® (New Objective), the HPLC-Chips (Agilent Technologies), cHiPLC® (Sciex), and the Chip-Mate™ (Advion, Inc.). The most important technological advantages of the use of such technologies are: (a) to facilitate the analysis of smaller sample volumes by lower flow rate and make it more accessible to nonexpert users, (b) high separation efficiency, (c) reduced consumption of mobile phases and stationary phases, (d) reduced contamination from external sources, such as plasticizers, (e) reduced matrix suppression, and (f) reduced waste production. An example of work toward a miniaturized LC technology is the PicoChip® from New Objective, illustrated in Figure 4.10. The PicoChip® device is assembled with a PicoFrit® nanospray column combining a PicoTip® emitter with an IntegraFrit™ column into a single zero-dead-volume column, a high-voltage union, and a transfer line. There is no need to attach an emitter since it sprays directly from the outlet of the column, while diminishing the delay volume to drastically reduce peak broadening, it also avoids issues with clogging and pressure build-up, which increases system robustness. These technologies are compatible with a small volume of microsamples/patient centric samples and also bring convenience and productivity to the bioanalytical laboratory.

4.7 Conclusions

Patient centric blood sampling has been largely facilitated by new advancements in microsampling techniques. Patients can now take part in clinical trials remotely, in a convenient way, facilitating patients' enrolment, engagement, and retention. There are many scientific and ethical reasons for the interest in microsampling, including its great potential to shift disease diagnosis and monitoring, closer to the point of care. In clinical analysis, microsampling offers a unique opportunity for TDM and PK studies in vulnerable patients' population, including neonates and the elderly, sampling a smaller volume of a blood, and increasing the comfort for patients/volunteers and delivering a patient centric approach. The increasing interest in these technologies indicates that the future of microsampling is promising for home-based self-sampling and clinical analysis.

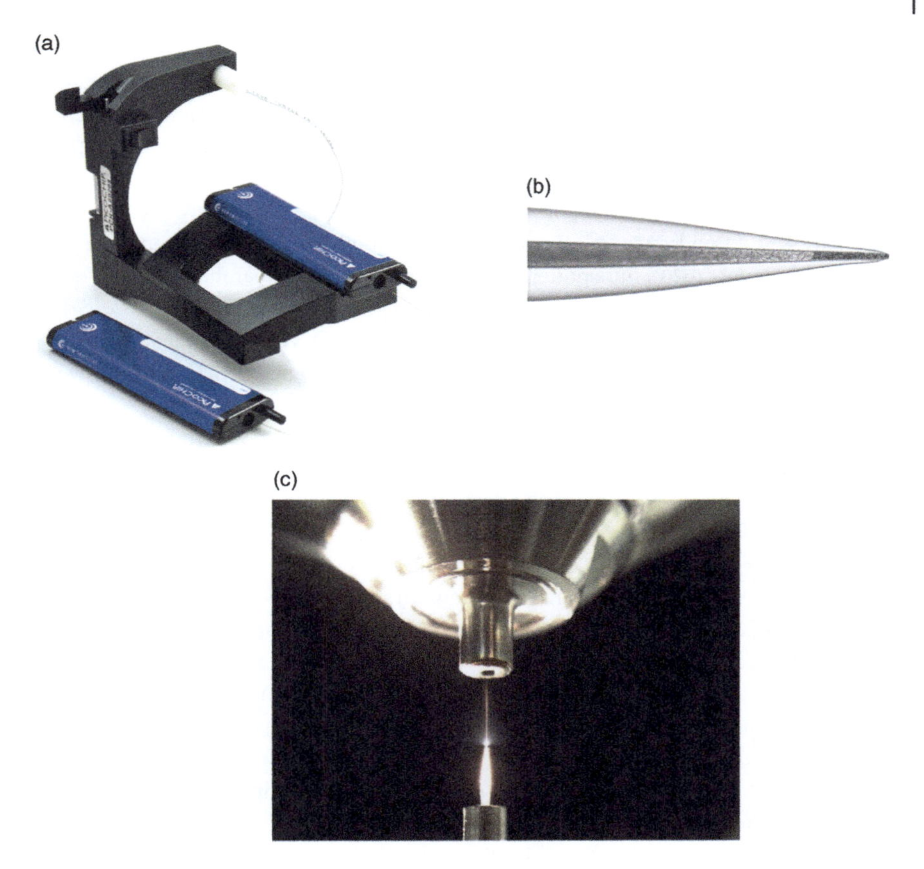

Figure 4.10 PicoChip® technology (a); anatomy of a Pico Frit® column integrated with the ESI emitter (b), and Pico Frit® with formed ESI plume (c). Courtesy of New Objective (New Objective, Inc.).

The performance improvements in the pre-analytical and analytical steps, in particular the consistent collection of blood with different Hct levels by different microsampling devices, and the increased MS sensitivity, have enabled the quantitative determination of different molecules using microsampling devices. In the future, we expect that microsampling techniques will continue to be developed and evaluated for clinical studies in the drug development process, as well as used in routine clinical practice, for the collection of samples in remote locations or at home by self-sampling. Furthermore, with the recent advancement in miniaturization of LC separations and ESI sources, the growth of LC-MS-based bioanalytical methods for quantitative assays is expected to continue. In summary, further efforts in automated sample preparation and high throughput will further enhance microsampling applications in clinical laboratories.

References

ACHDNC. Advisory Committee on Heritable Disorders in Newborns and Children. (2018). *Health Resources and Service Administration*. Retrieved January 20, 2021, from https://www.hrsa.gov/advisory-committees/heritable-disorders/recommendations-reports/reports/index.html.

Adhikari, K. B., Rohde, M., Velschow, S., Feldt-Rasmussen, U., Johannesen, J., & Johnsen, A. H. (2020). Fluispotter, a novel automated and wearable device for accurate volume serial dried blood spot sampling. *Bioanalysis, 12*, 665–681.

Alghamdi, S., Alsulami, N., Khoja, S., Alsufiani, H., Tayeb, H. O., & Tarazi, F. I. (2020). Vitamin D supplementation ameliorates severity of major depressive disorder. *Journal of Molecular Neuroscience, 70*, 230–235.

Anderson, L., Razavi, M., Pope, M. E., Yip, R., Cameron, L. C., Bassini-Cameron, A., & Pearson, T. W. (2020). Precision multiparameter tracking of inflammation on timescales of hours to years using serial dried blood spots. *Bioanalysis, 12*, 937–955.

Anderson, M. (2021). How the COVID-19 pandemic is changing clinical trial conduct and driving innovation in bioanalysis. *Bioanalysis, 13*, 1195–1203.

Arnold, D. W., & Needham, S. R. (2013). Micro-LC-MS/MS: The future of bioanalysis. *Bioanalysis, 5*, 1329–1331

Banerjee, S., & Mazumdar, S. (2012). Electrospray ionization mass spectrometry: A technique to access the information beyond the molecular weight of the analyte. *International Journal of Analytical Chemistry, 2012*, 1–40.

Bateman, K. P. (2020). The development of patient-centric sampling as an enabling technology for clinical trials. *Bioanalysis, 12*, 971–976.

Becker, M., Gscheidmeier, T., Groß, H., Cario, H., Woelfle, J., Rauh, M., Metzler, M., Zierk, J. (2022). Differences between capillary and venous blood counts in children—A data mining approach. *International Journal of Laboratory Hematology, 44*, 729–737.

Bhatnagar, S., Dave, K., Venuganti, V. V. K. (2017). Microneedles in the clinic. *Journal of Controlled Release, 260*, 164–182.

Bilić-Zulle, L. (2011). Comparison of methods: Passing and Bablok regression. *Biochemia Medica, 21*, 49–52.

Blicharz, T. M., Gong, P., Bunner, B. M., Chu, L. L., Leonard, K. M., Wakefield, J. A., Williams, R. E., Dadgar, M., Tagliabue, C. A., El Khaja, R., Marlin, S. L., Haghgooie, R., Davis, S. P., Chickering, D. E., & Bernstein, H. (2018). Microneedle-based device for the one-step painless collection of capillary blood samples. *Nature Biomedical Engineering, 2*, 151–157.

Bowen, C. L., & Barfield, M. (2019). Microsampling applications with LC-MS bioanalysis. Wenkui Li, Wenying Jian and Yunli Fu *Sample preparation in LC-MS bioanalysis* Wiley (pp. 117–127).

Boychenko, A., Meding, S., Decrop, W., Ruehl, M., & Swart, R. (2017). *Capillary-flow LC-MS: Combining high sensitivity, robustness, and throughput*. ThermoFisher Scientific Inc. Technical note 72277.

Brendish, N. J., Poole, S., Naidu, V. V., Mansbridge, C. T., Norton, N. J., Wheeler, H., Presland, L., Kidd, S., Cortes, N. J., Borca, F., Phan, H., Babbage, G., Visseaux, B., Ewings, S., & Clark, T. W. (2020). Clinical impact of molecular point-of-care testing for suspected COVID-19 in hospital (COV-19POC): A prospective, interventional, non-randomised, controlled study. *The Lancet Respiratory Medicine*, *8*, 1192–1200.

Brooks, K., DeLong, A., Balamane, M., Schreier, L., Orido, M., Chepkenja, M., Kemboi, E., D'Antuono, M., Chan, P. A., Emonyi, W., Diero, L., Coetzer, M., & Kantor, R. (2016). HemaSpot, a novel blood storage device for HIV-1 drug resistance testing. *Journal of Clinical Microbiology*, *54*, 223–225.

Capiau, S., Stove, V. V., Lambert, W. E., & Stove, C. P. (2013). Prediction of the hematocrit of dried blood spots via potassium measurement on a routine clinical chemistry analyzer. *Analytical Chemistry*, *85*, 404–410.

Capiau, S., Veenhof, H., Koster, R. A., Bergqvist, Y., Boettcher, M., Halmingh, O., Keevil, B. G., Koch, B. C. P., Linden, R., Pistos, C., Stolk, L. M., Touw, D. J., Stove, C. P., & Alffenaar, J. C. (2019). Official international association for therapeutic drug monitoring and clinical toxicology guideline: Development and validation of dried blood spot-based methods for therapeutic drug monitoring. *Therapeutic Drug Monitoring*, *41*, 409–430.

Capiau, S., Wilk, L. S., Aalders, M. C. G., & Stove, C. P. (2016). A novel, nondestructive, dried blood spot-based hematocrit prediction method using noncontact diffuse reflectance spectroscopy. *Analytical Chemistry*, *88*, 6538–6546

Capiau, S., Wilk, L.S., De Kesel, P. M. M., Aalders, M. C. G., & Stove, C. P. (2018). Correction for the hematocrit bias in dried blood spot analysis using a nondestructive, single-wavelength reflectance-based hematocrit prediction method. *Analytical Chemistry*, *90*, 1795–1804

Capitainer AB. (n.d.). https://capitainer.com

Chapman, K., Chivers, S., Gliddon, D., Mitchell, D., Robinson, S., Sangster, T., Sparrow, S., Spooner, N., & Wilson, A. (2014). Overcoming the barriers to the uptake of nonclinical microsampling in regulatory safety studies. *Drug Discovery Today*, *19*, 528–532.

Chetwynd, A. J., & David, A. (2018). A review of nanoscale LC-ESI for metabolomics and its potential to enhance the metabolome coverage. *Talanta*, *182*, 380–390.

Chetwynd, A. J., David, A., Hill, E. M., & Abdul-Sada, A. (2014). Evaluation of analytical performance and reliability of direct nanoLC-nanoESI-high resolution mass spectrometry for profiling the (xeno)metabolome. *Journal of Mass Spectrometry*, *49*, 1063–1069.

Churiso, G., van Henten, S., Cnops, L., Pollmann, J., Melkamu, R., Lemma, M., Kiflie, A., van Griensven, J., & Adriaensen, W. (2020). Minimally invasive microbiopsies as an improved sampling method for the diagnosis of cutaneous leishmaniasis *Open Forum Infectious Diseases 7*, 1–3.

Crawford, M. L., Collier, B. B., Bradley, M. N., Holland, P. L., Shuford, C. M., & Grant, S. R. (2020). Empiricism in microsampling: Utilizing a novel lateral flow device and intrinsic normalization to provide accurate and precise clinical analysis from a finger stick. *Clinical Chemistry*, *66*, 821–831.

Cross, T. G., & Hornshaw, M. P. (2016). Can LC and LC-MS ever replace immunoassays? *Journal of Applied Bioanalysis*, *2*, 108–116.

Darwish, I. A. (2006). Immunoassay methods and their applications in pharmaceutical analysis: Basic methodology and recent advances. *International Journal of Biomedical Science*, *2*, 217–235.

DBS System SA. (n.d.). https://hemaxis.com

Déglon, J., Thomas, A., Cataldo, A., Mangin, P., & Staub, C. (2009). On-line desorption of dried blood spot: A novel approach for the direct LC/MS analysis of μ-whole blood samples. *Journal of Pharmaceutical and Biomedical Analysis*, *49*, 1034–1039.

Déglon, J., Thomas, A., Mangin, P., & Staub, C. (2012). Direct analysis of dried blood spots coupled with mass spectrometry: Concepts and biomedical applications. *Analytical and Bioanalytical Chemistry*, *402*, 2485–2498.

Del Ben, F., Biasizzo, J., & Curcio, F. (2019). A fast, nondestructive, low-cost method for the determination of hematocrit of dried blood spots using image analysis. *Clinical Chemistry and Laboratory Medicine*, *57*, e81–e82.

Delahaye, L., Veenhof, H., Koch, B. C. P., Alffenaar, J. C., Linden, R., & Stove, C. (2021). Alternative sampling devices to collect dried blood microsamples: State-of-the-art. *Therapeutic Drug Monitoring*, *43*, 310–321.

Denniff, P., & Spooner, N. (2010). The effect of hematocrit on assay bias when using DBS samples for the quantitative bioanalysis of drugs. *Bioanalysis*, *2*, 1385–1395.

Denniff, P., & Spooner, N. (2014). Volumetric absorptive microsampling: A dried sample collection technique for quantitative bioanalysis. *Analytical Chemistry*, *86*, 8489–8495.

Deprez, S., Paniagua-González, L., Velghe, S., & Stove, C. P. (2019). Evaluation of the performance and hematocrit independence of the hemaPEN as a volumetric dried blood spot collection device. *Analytical Chemistry*, *91*, 14467–14475.

Doornek, T., Shao, N., Burton, P., Ceddia, F., & Fraile, B. (2022). *Antibody response rollowing COVID-19 boosters during the omicron wave in the united states: A decentralized, digital health, real-world study*. medRxiv 2022.07.31.22278173.

Dorofaeff, T., Bandini, R. M., Lipman, J., Ballot, D. E., Roberts, J. A., & Parker, S. L. (2016). Uncertainty in antibiotic dosing in critically ill neonate and pediatric patients: Can microsampling provide the answers?. *Clinical Therapeutics*, *38*, 1961–1975.

Drawbridge OneDraw, Inc. (n.d.). https://www.drawbridgehealth.com/onedraw/.

Eklund, J. H., Holmström, I. K., Kumlin, T., Kaminsky, E., Skoglund, K., Höglander, J., Sundler, A. J., Condén, E., & Meranius, M. S. (2019). "Same same or different?" a review of reviews of person-centered and patient-centered care. *Patient Education and Counseling*, *102*, 3–11.

Evans, C., Arnold, M., Bryan, P., Duggan, J., James, C. A., Wenkui, L., Lowes, S., Matassa, L., Olah, T., Timmerman, P., Xiaomin, W., Wickremsinhe, E., Williams, J., Woolf, E., Zane, P. (2015). Implementing Dried Blood Spot Sampling for Clinical Pharmacokinetic Determinations: Considerations from the IQ Consortium Microsampling Working Group. *The AAPS Journal*, *17*, 292–300.

Fenn, J. B., Mann, M., Meng, C. K., Wong, S. F., & Whitehouse, C. M. (1989). Electrospray ionization for mass spectrometry of large biomolecules. *Science*, *246*, 64–71.

Fluisense ApS. (n.d.). https://www.fluisense.com

Fokkema, M. R., Bakker, A. J., de Boer, F., Kooistra, J., de Vries, S., & Wolthuis, A. (2009). HbA1c measurements from dried blood spots: Validation and patient satisfaction. *Clinical Chemistry Laboratory Medicine*, *47*, 1259–1264.

Food and Drug Administration. (2013). *Draft guidance for industry bioanalytical method validation*. Rockville, MD: US Department of Health and Human Services, FDA, Centre for Drug Evaluation and Research.

Freeman, J. D., Rosman, L. M., Ratcliff, J. D., Strickland, P. T., Graham, D. R., Silbergeld, E. K. (2018). State of the science in dried blood spots. *Clinical Chemistry*, *64*, 656–679.

Gao, Q., Chang, Y., Deng, Q., & You, H. (2020). A simples and rapid method for blood plasma separation driven by capillary force with an application in protein detection. *Analytical Methods*, *12*, 2560–2570.

Gaugler, S., Sottas, P., Blum, K., & Luginbühl, M. (2021). Fully automated dried blood spot sample handling and extraction for serological testing of SARS-CoV-2 antibodies. *Drug Testing and Analysis*, *13*, 223–226.

Gibson, G. T. T., Mugo, S. M., & Oleschuk, R. D. (2009). Nanoelectrospray emitters: Trends and perspective. *Mass Spectrometry Reviews*, *28*, 918–936.

Goodacre, S. W., Bradburn, M., Cross, E., Collinson, P., Gray, A., & Hall, A. S. (2011). The randomised assessment of treatment using panel assay of cardiac markers (RATPAC) trial: A randomised controlled trial of point-of-care cardiac markers in the emergency department. *Heart*, *97*, 190–196.

Grinias, J. P., & Kennedy, R. T. (2016). Advances in and prospects of microchip liquid chromatography. *TrAC Trends in Analytical Chemistry*, *81*, 110–117.

Guerra Valero, Y. C., Roberts, J. A., Lipman, J., Fouriea, C., Starr, T., Wallis, S. C., & Parker, S. L. (2019). Analysis of capillary microsamples obtained from a skin-prick to measure vancomycin concentrations as a valid alternative to conventional sampling: A bridging study. *Journal of Pharmaceutical and Biomedical Analysis*, *169*, 288–292.

Guerra Valero, Y. C., Wallis, S. C., Lipman, J., Stove, C., Roberts, J. A., & Parker, S. L. (2018). Clinical application of microsampling versus conventional sampling techniques in the quantitative bioanalysis of antibiotics: A systematic review. *Bioanalysis, 10*, 407–423.

Gustavsen, M. T., Midtvedt, K., Vethe, N. T., Robertsen, I., Bergan, S., & Åsberg, A. (2020). Tacrolimus area under the concentration versus time curve monitoring, using home-based volumetric absorptive capillary microsampling. *Therapeutic Drug Monitoring, 42*, 407–414.

Guthrie, R., & Susi, A. (1963). A simple phenylalanine method for detecting phenylketonuria in large populations of newborn infants. *Pediatrics, 32*, 338–343.

Hall, J. M., Fowler, C. F., Barrett, F., Humphry, R. W., Van Drimmelen, M., & MacRury, S. M. (2020). HbA1c determination from HemaSpot™ blood collection devices: Comparison of home prepared dried blood spots with standard venous blood analysis. *Diabetic Medicine, 37*, 1463–1470.

Harahap, Y., Diptasaadya, R., & Purwanto, D. J. (2020). Volumetric absorptive microsampling as a sampling alternative in clinical trials and therapeutic drug monitoring during the COVID-19 pandemic: A review. *Drug Design, Development and Therapy, 14*, 5757–5771.

Harstad, E., Andaya, R., Couch, J., Ding, X., Liang, X., Liederer, B. M., Messick, K., Nguyen, T., Schweiger, M., Tarrant, J., Zhong, S., & Dean, B. (2016). Balancing blood sample volume with 3Rs: Implementation and best practices for small molecule toxicokinetic assessments in rats. *ILAR Journal, 57*, 157–165.

Hauser, J., Lenk, G., Hansson, J., Beck, O., Stemme, G., & Roxhed, N. (2018). High-yield passive plasma filtration from human finger prick blood. *Analytical Chemistry, 90*, 13393–13399.

Hendelman, T., Chaudhary, A., LeClair, A. C., van Leuven, K., Chee, J., Fink, S. L., Welch, E. J., Berthier, E., Quist, B. A., Wald, A., Wener, M. H., Hoofnagle, A. N., & Morishima, C. (2021). Self-collection of capillary blood using Tasso-SST devices for Anti-SARS-CoV-2 IgG antibody testing. *PLOS ONE, 16*, e0255841

Henion, J., Oliveira, R. V., & Chace, D. H. (2013). Microsample analyses via DBS: Challenges and opportunities. *Bioanalysis, 5*, 2547–2565.

Ho, C. S., Lam, C. W. K., Chan, M. H. M., Cheung, R. C. K., Law, L. K., Lit, L. C. W., Ng, K. F., Suen, M. W. M., & Tai, H. L. (2003). Electrospray ionisation mass spectrometry: Principles and clinical applications. *Clinical Biochemistry Reviews, 24*, 1–10.

Huttner, A., Harbarth, S., Hope, W. W., Lipman, J., & Roberts, J. A. (2015). Therapeutic drug monitoring of the β-lactam antibiotics: What is the evidence and which patients should we be using it for? *The Journal of Antimicrobial Chemotherapy, 70*, 3178–3183.

International Conference on Harmonization. (1995a). *ICH Topic S3A—guidance on toxicokinetics: The assessment of systemic exposure in toxicity studies (I).*

International Conference on Harmonization. (1995b). *ICH Topic S3B—pharmacokinetics: Guidance for repeated dose tissue distribution studies.*

International Conference on Harmonization. (2020). *ICH Topic S11—Nonclinical safety testing in support of development of paediatric medicines.*

International Conference on Harmonization and European Medicines Agency. (2019). *ICH Guideline M10 on bioanalytical method validation.*

Jager, N. G. L., Rosing, H., Schellens, J. H. M., & Beijnen, J. H. (2014). Procedures and practices for the validation of bioanalytical methods using dried blood spots: A review. *Bioanalysis*, *6*, 2481–2514.

Jannetto, P. J., & Fitzgerald, R. L. (2016). Effective use of mass spectrometry in the clinical laboratory. *Clinical Chemistry*, *62*, 92–98.

Keevil, B. G., Fildes, J., Baynes, A., & Yonan, N. (2009). Liquid chromatography-mass spectrometry measurement of tacrolimus in finger-prick samples compared with venous whole blood samples. *Annals of Clinical Biochemistry*, *46*, 144–145.

Klak, A., Pauwels, S., & Vermeersch, P. (2019). Preanalytical considerations in therapeutic drug monitoring of immunosuppressants with dried blood spots. *Diagnosis*, *6*, 57–68.

Knudsen, R. C., Slazyk, W. E., Richmond, J. Y., & Hannon, W. H. (1993). Guidelines for the shipment of dried blood spot specimens. *Infant Screening*, *16*, 1–3.

Konermann, L., Ahadi, E., Rodriguez, A. D., & Vahidi, S. (2013). Unraveling the mechanism of electrospray ionization. *Analytical Chemistry*, *85*, 2–9.

Kothare, P. A., Bateman, K. P., Dockendorf, M., Stone, J., Xu, Y., Woolf, E., & Shipley, L. A. (2016). An integrated strategy for implementation of dried blood spot in clinical development programs. *The AAPS Journal*, *18*, 519–527.

Krleza, J. L., Dorotic, A., Grzunov, A., Maradin, M., & Croatian Society of Medical Biochemistry and Laboratory Medicine. (2015). Capillary blood sampling: National recommendations on behalf of the Croatian Society of Medical Biochemistry and Laboratory Medicine. *Biochemia Medica*, *25*, 335–358

Lacombe, J., Rifai, O., Loter, L., Moran, T., Turcotte, A. F., Grenier-Larouche, T., Tchernof, A., Biertho, L., Carpentier, A. C., Prud'homme, D., Rabasa-Lhoret, R., Karsenty, G., Gagnon, C., Jiang, W., & Ferron, M. (2020). Measurement of bioactive osteocalcin in humans using a novel immunoassay reveals association with glucose metabolism and β-cell function. *American Journal of Physiology—Endocrinology and Metabolism*, *318*, E381–E391.

Lavallee, D. C., Chenok, K. E., Love, R. M., Petersen, C., Holve, E., Segal, C. D., & Franklin, P. D. (2016). Incorporating patient-reported outcomes into health care to engage patients and enhance care. *Health Affairs*, *35*, 575–582.

Lee, J., & Lee, S. (2013). Lab on a chip in situ diagnosis: From blood to point o care. *Biomedical Engineering Letters*, *3*, 59–66.

Lee, M. S. (Ed.). (2012). *Mass spectrometry handbook* (pp. 1–1368). Wiley.

Lei, B. U. W, & Prow, T. W. (2019). A review of microsampling techniques and their social impact. *Biomedical Microdevices*, *21*, 81–110.

Lenk, G., Hansson, J., Beck, O., & Roxhed, N. (2015a). The effect of drying on the homogeneity of DBS. *Bioanalysis*, *7*, 1977–1985.

Lenk, G., Sandkvist, S., Pohanka, A., Stemme, G., Beck, O., & Roxhed, N. (2015b). A disposable sampling device to collect volume-measured DBS directly from a fingerprick onto DBS paper. *Bioanalysis*, *7*, 2085–2094.

Li, F., Ploch, S., Fast, D. M., & Michael, S. (2012a). Perforated dried blood spot accurate microsampling: The concept and its applications in toxicokinetic sample collection. *Journal of Mass Spectrometry*, *47*, 655–657.

Li, F., Zulkoski, J., Fast, D., & Michael, S. (2011). Perforated dried blood spots: A novel format for accurate microsampling. *Bioanalysis*, *3*, 2321–2333.

Li, Y., Henion, J., Abbott, R., & Wang, P. (2012b). The use of a membrane filtration device to form dried plasma spots for the quantitative determination of guanfacine in whole blood. *Rapid Communications in Mass Spectrometry*, *26*, 1208–1212.

Li, Z., Yi, Y., Luo, X., Xiong, N., Liu, Y., Li, S., Sun, R., Wang, Y., Hu, B., Chen, W., Zhang, Y., Wang, J., Huang, B., Lin, Y., Yang, J., Cai, W., Wang, X., Cheng, J., Chen, Z., Sun, K., Pan, W., Zhan, Z., Chen, L., & Ye, F. (2020). Development and clinical application of a rapid IgM-IgG combined antibody test for SARS-CoV-2 infection diagnosis. *Journal of Medical Virology*, *92*, 1518–1524.

Lin, L. L., Prow, T. W., Raphael, A. P., Harrold Iii, R. L., Primiero, C. A., Ansaldo, A. B., & Peter Soyer, H. (2013). Microbiopsy engineered for minimally invasive and suture-free sub-millimetre skin sampling. *F1000Research*, *2*, 120.

Londhe, V., & Rajadhyaksha, M. (2020). Opportunities and obstacles for microsampling techniques in bioanalysis: Special focus on DBS and VAMS. *Journal of Pharmaceutical and Biomedical Analysis*, *182*, 113102–113118.

Losiak, W., & Losiak-Pilch, J. (2020). Cortisol awakening response, self-reported affect and exam performance in female students. *Applied Psychophysiology and Biofeedback*, *45*, 11–16.

Loop Medical SA (n.d.). https://www.loop-medical.com

Luginbühl, M., & Gaugler, S. (2020). The application of fully automated dried blood spot analysis for liquid chromatography-tandem mass spectrometry using the CAMAG DBS-MS 500 autosampler. *Clinical Biochemistry*, *82*, 33–39.

Malsagova, K., Kopylov, A., Stepanov, A., Butkova, T., Izotov, A., & Kaysheva, A. (2020). Dried blood spot in laboratory: Directions and prospects. *Diagnostics*, *10*, 248–263.

Marasca, C., Protti, M., Mandrioli, R., Atti, A. R., Armirotti, A., Cavalli, A., De Ronchi, D., & Mercolini, L. (2020). Whole blood and oral fluid microsampling for the monitoring of patients under treatment with antidepressant drugs. *Journal of Pharmaceutical and Biomedical Analysis*, *188*, 113384.

Mauk, M., Song, J., Bau, H. H., Gross, R., & Bushman, F. D. (2017). Miniaturized devices for point of care molecular detection of HIV. *Lab on a Chip*, *17*, 382–394.

Mbughuni, M. M., Stevens, M. A., Langman, L. J., Kudva, Y. C., Sanchez, W., Dean, P. G., & Jannetto, P. J. (2020). Volumetric microsampling of capillary blood spot vs whole blood sampling for therapeutic drug monitoring of tacrolimus and cyclosporin A: Accuracy and patient satisfaction, *The Journal of Applied Laboratory Medicine*, *5*, 516–530.

Meetoo, D.D., & Wong, L. (2015). The role of point of care testing in diabetes management. *British Journal of Health Care Management*, *21*, 63–67.

Mejía-Salazar, J. R., Rodrigues Cruz, K., Materón Vásques, E. M., & Oliveira-Junior, O. N. (2020). Microfluidic point-of-care devices: New trends and future prospects for eHealth diagnostics. *Sensors*, *20*, 1951–1969.

Moat, S. J., George, R. S., & Carling, R. S. (2020). Use of dried blood spot specimens to monitor patients with inherited metabolic disorders. *International Journal of Neonatal Screening*, *6*, 26–42.

Nakahara, T., Otani, N., Ueno, T., & Hashimoto, K. (2018). Development of a hematocrit-insensitive device to collect accurate volumes of dried blood spots without specialized skills for measuring clozapine and its metabolites as model analytes. *Journal of Chromatography B*, *1087–1088*, 70–79.

Nakatani, K., Izumi, Y., Hata, K., & Bamba, T. (2020). An analytical system for single-cell metabolomics of typical mammalian cells based on highly sensitive nano-liquid chromatography tandem mass spectrometry. *Mass Spectrometry*, *9*, A0080–A0088.

Naylor, S. (2003). Biomarkers: Current perspectives and future prospects. *Expert Review of Molecular Diagnostics*, *3*, 525–529.

Nazario, C. E. D., Silva, M. R., Franco, M. S., & Lanças, F. M. (2015). Evolution in miniaturized column liquid chromatography instrumentation and applications: An overview. *Journal of Chromatography A*, *1421*, 18–37.

Needham, S. R., & Valaskovic, G. A. (2015). Microspray and microflow LC-MS/MS: The perfect fit for bioanalysis. *Bioanalysis*, *7*, 1061–1064.

New Objective, Inc. (n.d.). https://www.newobjective.com

Nguyen, T., Chidambara, V. A., Andreasen, S. Z., Golabi, M., Huynh, V. N., Linh, Q. T., Bang, Q. Q., & Wolff, A. (2020). Point-of-care devices for pathogen detections: The three most important factors to realise towards commercialization. *TrAC—Trends in Analytical Chemistry*, *131*, 116004–116016.

Nilsson, L. B., Ahnoff, M., & Jonsson, O. (2013). Capillary microsampling in the regulatory environment: Validation and use of bioanalytical capillary microsampling methods. *Bioanalysis*, *5*, 731–738.

Nys, G., Kok, M. G. M., Servais, A.-C., & Fillet, M. (2017). Beyond dried blood spot: Current microsampling techniques in the context of biomedical applications. *TrAC Trends in Analytical Chemistry*, *97*, 326–332.

Oliveira, R. V., Henion, J., & Wickremsinhe, E. (2014a). Fully-automated approach for online dried blood spot extraction and bioanalysis by two-dimensional-liquid chromatography coupled with high-resolution quadrupole time-of-flight mass spectrometry. *Analytical Chemistry, 86*, 1246–1253.

Oliveira, R. V., Henion, J., & Wickremsinhe, E. R. (2014b). Automated direct extraction and analysis of dried blood spots employing on-line SPE high-resolution accurate mass bioanalysis. *Bioanalysis, 6*, 2027–2041.

Osteresch, B., Cramer, B., & Humpf, H.U. (2016). Analysis of ochratoxin A in dried blood spots—Correlation between venous and finger-prick blood, the influence of hematocrit and spotted volume. *Journal of Chromatography B, 1020*, 158–164.

Page, J. S., Kelly, R. T., Tang, K., & Smith, R. D. (2007). Ionization and transmission efficiency in an electrospray ionization-mass spectrometry interface. *Journal of the American Society for Mass Spectrometry, 18*, 1582–1590.

Pitt, J. J. (2009). Principles and applications of liquid chromatography-mass spectrometry in clinical biochemistry. *The Clinical Biochemist Reviews, 30*, 19–34.

Primiceri, E., Chiriacò, M., Notarangelo, F., Crocamo, A., Ardissino, D., Cereda, M., Bramanti, A. P., Bianchessi, M. A., Giannelli, G., & Maruccio, G. (2018). Key enabling technologies for point-of-care diagnostics. *Sensors, 18*, 3607–3640.

Prior, H., Sewell, F., & Stewart, J. (2017). Overview of 3Rs opportunities in drug discovery and development using non-human primates. *Drug Discovery Today: Disease Models, 23*, 11–16.

Protti, M, Mandrioli, R, & Mercolini, L. (2019). Tutorial: Volumetric absorptive microsampling (VAMS). *Analytica Chimica Acta 1046*, 32–47.

Przybylski, M., & Glocker, M. O. (1996). Electrospray mass spectrometry of biomacromolecular complexes with noncovalent interactions—New analytical perspectives for supramolecular chemistry and molecular recognition processes. *Angewandte Chemie International, 35*, 806–826.

Resano, M., Belarra, M. A., García-Ruiz, E., Aramendía, M., & Rello, L. (2018). Dried matrix spots and clinical elemental analysis. Current status, difficulties, and opportunities. *TrAC—Trends in Analytical Chemistry, 99*, 75–87.

Reschke, B. R., & Timperman, A. T. (2011). A study of electrospray ionization emitters with differing geometries with respect to flow rate and electrospray voltage. *Journal of The American Society for Mass Spectrometry, 22*, 2115–2124.

Richardson, G., Marshall, D., Keevil, B. G. (2018). Prediction of haematocrit in dried blood spots from the measurement of haemoglobin using commercially available sodium lauryl sulphate. *Annals of Clinical Biochemistry, 55*, 363–367.

Roadcap, B., Hussain, A., Dreyer, D., Carter, K., Dube, N., Xu, Y., Anderson, M., Berthier, E., Vazvaei, F., Bateman, K., & Woolf, E. (2020). Clinical application of volumetric absorptive microsampling to the gefapixant development program. *Bioanalysis, 12*, 893–904.

Roberts, J. A., de Waele, J. J., Dimopoulos, G., Koulenti, D., Martin, C., Montravers, P., Rello, J., Rhodes, A., Starr, T., Wallis, S. C., & Lipman, J. (2012). DALI: Defining antibiotic levels in intensive care unit patients: A multi-centre point of prevalence study to determine whether contemporary antibiotic dosing for critically ill patients is therapeutic. *BMC Infectious Diseases, 12*, 152–158.

Rossiter, C., Levett-Jones, T., & Pich, J. (2020). The impact of person-centred care on patient safety: An umbrella review of systematic reviews. *International Journal of Nursing Studies, 109*, 103658–103677.

Saito, Y., Jinno, K., & Greibrokk, T. (2004). Capillary columns in liquid chromatography: Between conventional columns and microchips. *Journal of Separation Science, 27*, 1379–1390.

Scaglione, F., Esposito, S., Leone, S., Lucini, V., Pannacci, M., Ma, L., & Drusano, G. L. (2009). Feedback dose alteration significantly affects probability of pathogen eradication in nosocomial pneumonia. *European Respiratory Journal, 34*, 394–400.

Schmidt, A., Karas, M., & Dülcks, T. (2003). Effect of different solution flow rates on analyte ion signals in nano-ESI MS, or: When does ESI turn into nano-ESI? *Journal of the American Society for Mass Spectrometry, 14*, 492–500.

Sciberras, D., Otoul, C., Lurquin, F., Smeraglia, J., Lappert, A., De Bruyn, S., & Jaap van Lier, J. (2019). A pharmacokinetic study of radiprodil oral suspension in healthy adults comparing conventional venous blood sampling with two microsampling techniques. *Pharmacology Research & Perspectives, 7*, e00459.

Sharma, A., Jaiswal, S., Shukla, M., & Lal, J. (2014). Dried blood spots: Concepts, present status, and future perspectives in bioanalysis. *Drug Testing and Analysis, 6*, 399–414.

Shook, L. L., Atyeo, C. G., Yonker, L. M., Fasano, A., Gray, K. J., Alter, G., Edlow, A. G. (2022). Durability of Anti-Spike Antibodies in Infants After Maternal COVID'19 Vaccination or Natural Infection. *JAMA, 327*, 1087–1089.

Shimadzu (n.d.). https://www.shimadzu.com

Simmonds, M. J., Baskurt, O. K., Meiselman, H. J., & Marshall-Gradisnik, S. M. (2011). A comparison of capillary and venous blood sampling methods for the use in haemorheology studies. *Clinical Hemorheology and Microcirculation, 47*, 111–119.

Snyder, L. R., Kirkland, J. J., & Glajch, J. L. (2010). *Introduction to modern liquid chromatography* (3rd ed, pp. 1–960). Wiley.

Spooner, N., Anderson, K. D., Siple, J., Wickremsinhe, E. R., Xu, Y., & Lee, M. (2019). Microsampling: Considerations for its use in pharmaceutical drug discovery and development. *Bioanalysis, 11*, 1015–1038.

Spooner, N., Denniff, P., Michielsen, L., De Vries, R., Ji, Q. C., Arnold, M. E., Woods, K., Woolf, E. J., Xu, Y., Boutet, V., Zane, P., Kushon, S., & Rudge, J. B. (2015). A device for dried blood microsampling in quantitative bioanalysis: Overcoming the issues associated with blood hematocrit. *Bioanalysis, 7*, 653–659.

Spooner, N., Olatunji, A., & Webbley, K. (2018). Investigation of the effect of blood hematocrit and lipid content on the blood volume deposited by a disposable dried blood spot collection device. *Journal of Pharmaceutical and Biomedical Analysis, 149*, 419–424.

Spooner, N., Ramakrishnan, Y., Barfield, M., Dewit, O., & Miller, S. (2010). Use of DBS sample collection to determine circulating drug concentrations in clinical trials: Practicalities and considerations. *Bioanalysis, 2*, 1515–1522.

Spot On Sciences, Inc. (n.d.). https://www.spotonsciences.com

Sturm, R., Henion, J., Abbott, R., & Wang, P. (2015). Novel membrane devices and their potential utility in blood sample collection prior to analysis of dried plasma spots. *Bioanalysis, 7*, 1987–2002.

Su, X., Zhang, S, Ge, S., Chen, M., Zhang, J., Zhang, J., & Xia, N. (2018). A low cost, membranes based serum separator modular. *Biomicrofluidics, 12*, 024108.

Syedmoradi, L., Norton, M. L., & Omidfar, K. (2021). Point-of-care cancer diagnostic devices: From academic research to clinical translation. *Talanta, 225*, 122002.

Taddeo, A., Prim, D., Bojescu, E. D., Segura, J. M., & Pfeifer, M. E. (2020). Point-of-care therapeutic drug monitoring for precision dosing of immunosuppressive drugs. *The Journal of Applied Laboratory Medicine, 5*, 738–761.

Tanna, S., Ogwu, J., & Lawson, G. (2020). Hyphenated mass spectrometry techniques for assessing medication adherence: Advantages, challenges, clinical applications and future perspectives. *Clinical Chemistry and Laboratory Medicine, 58*, 643–663.

Tasso, Inc. (n.d.). https://www.tassoinc.com/.

Teclemariam, E. T., Pergande, M. R., & Cologna, S. M. (2020). Considerations for mass spectrometry-based multi-omic analysis of clinical samples. *Expert Review of Proteomics, 17*, 99–107.

Teubel, J., & Parr, M. K. (2020). Determination of neurosteroids in human cerebrospinal fluid in the 21st century: A review. *Journal of Steroid Biochemistry and Molecular Biology, 204*, 105753.

Tey, H. Y., See, H. H. (2021). A review of recent advances in microsampling techniques of biological fluids for therapeutic drug monitoring. *Journal of Chromatography A, 1635*, 461731.

Touw, D. J., Neef, C., Thomson, A. H., & Vinks, A. A. (2005). Cost-effectiveness of therapeutic drug monitoring: A systematic review. *Therapeutic Drug Monitoring, 27*, 10–17.

Trajan Scientific and Medical, Inc. (n.d.). https://www.neoteryx.com

van Amsterdam, P., & Waldrop, C. (2010). The application of dried blood spot sampling in global clinical trials. *Bioanalysis, 2*, 1783–1786.

Velghe, S., & Stove, C. P. (2018). Evaluation of the Capitainer-B microfluidic device as a new hematocrit-independent alternative for dried blood spot collection. *Analytical Chemistry, 90*, 12893–12899.

Verhaeghe, T., Dillen, L., Stieltjes, H., Zwart, L., Feyen, B., Diels, L., Vroman, A., & Timmerman, P. (2017). The application of capillary microsampling in GLP toxicology studies. *Bioanalysis*, *9*, 531–540.

Verhaeghe, T., Meulder, M. D., Hillewaert, V., Dillen, L., & Stieltjes, H. (2020). Capillary microsampling in clinical studies: Opportunities and challenges in two case studies. *Bioanalysis*, *12*, 905–918.

Verplaetse, R., & Henion, J. (2016). Hematocrit-independent quantitation of stimulants in dried blood spots: Pipet versus microfluidic-based volumetric sampling coupled with automated flow-through desorption and online solid phase extraction-LC-MS/MS bioanalysis. *Analytical Chemistry*, *88*, 6789–6796.

Vitrex Medical A/S. (n.d.). https://www.vitrexmedical.com

Volani, C., Caprioli, G., Calderisi, G., Sigurdsson, B. B., Rainer, J., Gentilini, I., Hicks, A. A., Pramstaller, P. P., Weiss, G., Smarason, S. V., & Paglia, G. (2017). Pre-analytic evaluation of volumetric absorptive microsampling and integration in a mass spectrometry-based metabolomics workflow. *Analytical and Bioanalytical Chemistry*, *409*, 6263–6276.

Watson, I. D. (2005). Clinical analysis overview. In P. Worsfold, A. Townshend, & A. Polle (Eds.), *Encyclopedia of analytical science* (2nd ed, pp. 126–132). Elsevier.

Watson, J. T., & Sparkman, O. D. (2008). *Introduction to mass spectrometry: instrumentation, applications and strategies for data interpretation* (4th ed, pp. 1–819). Wiley.

White, S., Hawthorne, G., Dillen, L., Spooner, N., Woods, K., Sangster, T., Cobb, Z., & Timmerman, P. (2014). EBF: Reflection on bioanalytical assay requirements used to support liquid microsampling. *Bioanalysis*, *6*, 2581–2586.

Wickremsinhe, E. R., Ji, Q. C., Gleason, C. R., Anderson, M., & Booth, B. P. (2020). Land O'Lakes workshop on microsampling: Enabling broader adoption. *The AAPS Journal*, *22*, 135–141.

Williams, K. J., Lutman, J., McCaughey, C., & Fischer, S. K. (2021). Assessment of low volume sampling technologies: Utility in nonclinical and clinical studies. *Bioanalysis*, *13*, 679–691.

Wilm, M. (2011). Principles of electrospray ionization. *Molecular and Cellular Proteomics*, *10*, 1–8.

Wilm, M. S., & Mann, M. (1994). Electrospray and Taylor-cone theory, Dole's beam of macromolecules at last? *International Journal of Mass Spectrometry and Ion Processes*, *136*, 167–180.

Wilson, S. R., Vehus, T., Berg, H. S., & Lundanes, E. (2015). Nano-LC in proteomics: Recent advances and approaches. *Bioanalysis*, *7*, 1799–1815.

World Health Organization, OECD, International Bank for Reconstruction and Development. (2018). *Delivering quality health services: A global imperative for universal health coverage*. World Health Organization, World Bank Group and OECD.

Xing, J., Loureiro, J., Patel, M. T., Mikhailov, D., Gusev, A. I. (2020) Evaluation of a novel blood microsampling device for clinical trial sample collection and protein biomarker analysis. *Bioanalysis, 12*, 919–935.

Yamamoto, C., Nagashima, S., Isomura, M., Ko, K., Chuon, C., Akita, T., Katayama, K., Woodring, J., Hossain, S., Takahashi, K., & Tanaka, J. (2020). Evaluation of the efficiency of dried blood spot-based measurement of hepatitis B and hepatitis C virus seromarkers. *Scientific Reports, 10*, 3857–3866.

Yeoman, G., Furlong, P., Seres, M., Binder, H., Chung, H., Garzya, V., & Jones, R. R. M. (2017). Defining patient-centricity with patients for patients and caregivers: A collaborative endeavour. *BMJ Innovations, 3*, 76–83.

Youhnovski, N., Bergeron, A., Furtado, M., & Garofolo, F. (2011). Pre-cut dried blood spot (PCDBS): An alternative to dried blood spot (DBS) technique to overcome hematocrit impact. *Rapid Communications in Mass Spectrometry, 25*, 2951–2958.

YourBio Health, Inc. (n.d.). https://yourbiohealth.com

Zakaria, R., Allen, K. J., Koplin, J. J., Roche, P., & Greaves, R. F. (2016). Advantages and challenges of dried blood spot analysis by mass spectrometry across the total testing process. *The Journal of the International Federation of Clinical Chemistry and Laboratory Medicince, 27*, 288–317.

Zhang, X. X., Xu, D., Guo, D., Han, H. X., Li, D. W., & Ma, W. (2020). Enzyme-free amplified SERS immunoassay for the ultrasensitive detection of disease biomarkers. *Chemical Communications, 56*, 2933–2936.

Zochowska, D., Bartłomiejczyk, I., Kamińska, A., Senatorski, G., & Pączek, A. (2006). High-performance liquid chromatography versus immunoassay for the measurement of sirolimus: Comparison of two methods. *Transplantation Proceedings, 38*, 78–80.

5

Automation in Microsampling: At Your Fingertips?

Sigrid Deprez[1,#], *Liesl Heughebaert*[1,#], *Nick Verougstraete*[1,2], *Veronique Stove*[2,3], *Alain G. Verstraete*[2,3], *and Christophe P. Stove*[1]

[1] *Laboratory of Toxicology, Faculty of Pharmaceutical Sciences, Ghent University, Ghent, Belgium*
[2] *Department of Laboratory Medicine, Ghent University Hospital, Ghent, Belgium*
[3] *Department of Diagnostic Sciences, Faculty of Medicine and Health Sciences, Ghent University, Ghent, Belgium*

 # Equally contributed as co-first authors.

5.1 Introduction

5.1.1 Identifying the Current Bottlenecks for Routine Implementation of Microsampling in Clinical Practice

In recent years (and especially since the start of the COVID-19 pandemic), collecting samples in a more patient centric way, compared to the standard practice, is receiving increased attention. One of the patient centric sampling approaches that are available is microsampling, which includes the collection of blood (or other biological matrices) in a manner that is different from conventional venous phlebotomy (e.g. after a fingerstick). Microsampling has now become a well-known technique in bioanalysis since its introduction as dried blood spot (DBS) some 60 years ago for newborn screening. Important benefits, including the minimal invasiveness, the small sample volume, and the ease of transport and storage, were part of the driving force for the implementation of this technique in other fields such as for phenotyping, therapeutic drug monitoring (TDM), doping analysis, or toxicology (Cox & Eichner, 2017; De Kesel et al., 2014b; Edelbroek et al., 2009; Sadones et al., 2014; Stove et al., 2012; Wilhelm et al., 2014). Another advantage of this patient centric sampling technique is the possibility to perform home-sampling studies. In particular, this has been an important incentive for its

Patient Centric Blood Sampling and Quantitative Bioanalysis, First Edition.
Edited by Neil Spooner, Emily Ehrenfeld, Joe Siple, and Mike S. Lee.

implementation in clinical trials and can contribute in the future to the evolution of wellness (health) monitoring from patients' homes, instead of "sick monitoring." However, a major gap remains between the large number of research articles on this subject and the rather limited adoption in clinical laboratories and by the bioanalysis field in general. How can we bridge this gap? In the past decade, an important solution has become available, namely the possibility of performing (semi-)automated DBS sample analysis. Since a large part of the analytical workflows are already automated in clinical laboratories, this was one of the main hurdles to be overcome to facilitate routine adoption. In the first part of this chapter, we identify the remaining bottlenecks for the routine implementation of (automated) DBS analysis in clinical practice, with a particular focus on clinical applications in toxicology and TDM. The second part of the chapter is focused on the available options to (semi-)automate the analysis of microsamples (including both DBS and its alternatives). Finally, the third part discusses important aspects that have to be considered when aiming to implement automated DBS analysis into a clinical laboratory.

The routine implementation of microsampling-based strategies into clinical practice has been hampered by a number of aspects. The use of these techniques is associated with both practical and technical challenges compared to established liquid plasma analysis, making the adoption of DBS analysis in a clinical laboratory more complicated. One of the most frequently described issues associated with DBS sample analysis is the hematocrit (Hct) effect (De Kesel et al., 2013; Velghe et al., 2019a). This effect has both an analytical and physiological nature and is often referred to as a major hurdle for the implementation of DBS analysis in clinical practice. The main analytical challenge is the varying spot size of a DBS, generated when applying blood with varying Hcts on filter paper, due to a difference in viscosity. It is typical in DBS analysis for a fixed-size subpunch to be taken. Hence, for a subpunch of a high-Hct DBS sample, the analyte concentration will typically be overestimated because of the larger blood volume contained in that subpunch, derived from a smaller blood spot. The opposite occurs for low Hct samples. Physiologically, the Hct mainly influences the blood–plasma distribution of the analyte of interest, and also capillary-venous differences can be observed (Capiau et al. 2016b). Consequently, the Hct affects data accuracy and interpretation compared to historical plasma, or whole blood data (Abu-Rabie et al., 2015; De Kesel et al., 2013; Velghe et al., 2019a). Importantly, the spreading effect of the Hct on DBS filter paper can be overcome by volumetrically depositing a fixed volume onto the paper in combination with whole spot analysis. However, from a practical point of view, this approach is more complicated in a home-sampling context, as it requires the deposition of a fixed volume of blood, e.g. by using (volumetric) microcapillaries. Alternatively, devices have been developed that allow volumetric microsampling (Delahaye et al., 2021b). By analyzing the complete dried (volumetrically collected) microsample, the impact of spreading on DBS-based quantitation can be overcome. Several of these devices have become commercially available (using DBS or other formats), such as the Mitra® volumetric absorptive microsampling (VAMS) device (Neoteryx, USA) (De Kesel et al., 2015; Denniff & Spooner, 2014), the HemaXis device (DBS system SA, Switzerland) (Leuthold et al., 2015), the Capitainer® qDBS device (KTH, Sweden) (Velghe & Stove, 2018a), and the HemaPEN® (Trajan, Australia) (Deprez et al., 2019). Furthermore, non-fingerstick dried blood microsampling devices have been developed, based on (micro)needle technologies, e.g. the Tasso-M20 (Tasso, USA) (Tasso INC., 2021). Additionally, dried plasma spot devices have entered the market, such as the Noviplex™ device (Shimadzu, Japan), with the aim to address the blood–plasma conversion issues (Shimadzu & Novilytic, 2021). A summary of the advantages and disadvantages of each of these (as well as other) devices has recently been published in a review by Delahaye et al. (2021b). Another possibility to overcome the issues associated with the spreading effect imposed by the Hct is the use of conventional DBS in combination with approaches to predict the Hct of the DBS sample. Several techniques have become available to realize this in a nondestructive way, using ultraviolet–visible spectroscopy (UV/Vis) or near-infrared (NIR) spectroscopy (Capiau et al., 2016a; Delahaye et al., 2021a; van de Velde et al., 2021). Importantly, these techniques have also been incorporated in automated DBS extraction units, allowing fully automated Hct prediction and analysis (as will be discussed in Section 5.2). Hence, it can be concluded that the Hct issue can be overcome by applying recent developments in this field, notably volumetric collection devices and Hct prediction strategies. However, other challenges associated with DBS analysis are still hampering implementation and need to be tackled before routine integration into the routine practice of clinical laboratories is feasible (Abu-Rabie, 2013; Capiau et al., 2016b; Velghe et al., 2016).

Compared to liquid plasma and whole blood methods (standard practice), additional validation steps are required for microsampling-based methods, e.g. the evaluation of the influence of the Hct effect on the recovery, making the validation more time-consuming (Capiau et al., 2019). A recent publication on the validation of microsampling methods for TDM purposes gives useful guidance on this and other issues (Capiau et al., 2019). Another bottleneck for the routine adoption of patient centric blood sampling approaches is that existing therapeutic intervals for analytes are typically expressed as serum or plasma concentrations, meaning that either new intervals or a conversion factor needs to be set up for the interpretation of DBS results. To accomplish this, additional bridging studies are needed, which are also time-consuming and expensive to perform.

A further challenge lies in the requirement for very sensitive analytical equipment, as only a few microliters of sample are available for analysis. Although liquid chromatography tandem mass spectrometry (LC-MS/MS) equipment and methods have increased in sensitivity, these systems come with a significant cost, which might not be affordable for smaller-scale laboratories. Moreover, a high technical expertise is required to set up and perform LC-MS/MS bioanalytical methods. Additionally, sample quality may be more of a concern, as quality control (QC) is less easy to perform on sample collection and storage when using microsampling for home-sampling studies, e.g. in the context of TDM. Until recently, this was indeed true, since a quality check of the sample could only be performed by trained personnel upon arrival of the samples at the laboratory. However, currently, applications are available that enable a quality check of the DBS, either at the laboratory or by the patient themselves, at home (Dantonio et al., 2014). One of those applications is the web-based app developed by Veenhof et al. at the University of Groningen within the framework of a doctoral dissertation. However, this app is no longer available to be used in clinical laboratories (Veenhof et al., 2019). Moreover, the feasibility of getting good quality dried blood microsamples following adequate patient instruction has been demonstrated by Van Uytfanghe and co-workers in the context of a study quantifying phosphatidylethanol (PEth; a direct alcohol marker) using Mitra devices (Van Uytfanghe et al., 2021). Here, by using video demonstrations and written instructions, and emphasizing the importance of correct sampling, 92% of the samples were of sufficient quality and suitable for analysis (no signs of over- or underfilling after a quality check by two independent assessors). Regrettably, clinical validation studies and studies on the added value of the use of alternative matrices on patient outcome are largely lacking.

Another key factor to be understood is the economics of implementation and routine use of microsampling technologies. One study has demonstrated that implementing DBS in a clinical setting is (or can be) cost-efficient, when considering all the associated costs (Martial et al., 2016). This aspect will be further discussed in Section 5.3. Another requirement for routine clinical implementation is the participation in proficiency testing (PT) programs for quality assessment. For microsampling methods, the setup of QC programs is challenging, as standard kits (for LC-MS/MS QC) are typically plasma/serum-based. Furthermore, these QC samples may not be suitable for methods using dried matrices because of matrix-related issues. Often, lyophilized blood is used, which differs substantially in viscosity compared to fresh blood, which will influence the spreading on DBS filter paper and may also result in a different extractability (Francke et al., 2022). Therefore, dedicated PT programs are needed for DBS methods. Currently, only one such program is available, for the quantification of immunosuppressants (tacrolimus, sirolimus, everolimus, cyclosporin A and mycophenolic acid), as an initiative of the Dutch Association for Quality Assessment in Therapeutic Drug Monitoring and Clinical Toxicology (KKGT) (Robijns et al., 2014). Participants of this program send their microsampling devices to the KKGT, where DBS are generated, and the sampled device is subsequently sent back to the participating laboratory for analysis. The fact that different substrate types are used to collect DBS samples (in terms of type of filter paper, alternative collection device, etc.) causes an additional level of complexity for the setup of such programs and for the interpretation of the results. From this point of view, it will be important to harmonize DBS analysis in the future. One potential approach when only plasma- or serum-based PT or external QC samples are available is that these standards could be used to replace a given plasma or serum volume from a blank blood sample after centrifugation, hence allowing the generation of blood samples. These could be used to generate DBS, which could be analyzed, with subsequent back-calculation of the original concentration in the PT sample, which could be reported. This approach has already been successfully applied in our lab (Velghe & Stove, 2018b). Another point of attention is that, in routine follow-up of patients, there is often a requirement for multiple parameters to be determined from a single blood sample, e.g. liver and kidney function or hematology parameters. This can be a limiting factor for microsampling, since, typically, only one or a limited set of parameters can be determined from one microsample. Hence, the benefit of smaller volumes and convenience could be overruled if an additional blood draw would be required for the evaluation of other parameters.

A further challenge associated with the routine implementation of microsampling approaches is the increased complexity required to analyze these samples, which may involve more manual handling and a significant increase in hands-on time, compared to the handling of conventional blood/serum/plasma tubes. With the introduction of automated solutions for the different steps in the sample preparation process, the throughput can be increased, while the hands-on time and risk of human errors are reduced. As the possibility for automation is one of the major requirements for the implementation of the microsampling technique in clinical practice, this will be extensively discussed in Section 5.2.

5.1.2 The Importance of Analytical Automation for Different Application Fields

As mentioned in Section 5.1.1, microsampling is increasingly being used in a number of different fields. For TDM and several other clinical applications, the ultimate goal is to monitor patients from their homes when needed, complying with a patient centric view on healthcare. Since DBS samples are considered to be less biohazardous than routine liquid samples and often do not require cooled transport, they can be sent via regular postal services to the laboratory.

In this way, the patients' quality of life can be significantly improved by avoiding regular hospital visits, solely for the purpose of a venous blood draw. To make this approach feasible for clinical laboratories, sample handling and preparation of these samples should ideally not be more laborious compared to the current standard approach. From that point of view, automated extraction of DBS or microsampling devices would fit perfectly within the increasingly automated workflow present in many clinical laboratories for conventional blood tubes. Hence, the availability of dedicated, less labor-intensive sample preparation systems could be beneficial, which can be achieved by (partially) automating sample extraction.

In routine sports doping analysis, urine has been the preferred matrix for the detection of analytes of interest, as typically (a) a higher volume is available than blood, (b) concentrations in urine are higher than in blood, and (c) urine offers a longer detection window. In a similar way as outlined for TDM, DBS can provide a complementary approach for existing methods in doping analysis. In practice, a conceivable scenario would consist of paired urine and blood sampling for in-competition doping controls. The blood sample could then serve to quantify an active analyte, which could be beneficial as this allows more individualized case management, e.g. determined urine levels could be verified or refuted by the presence of pharmacologically relevant blood levels (Thevis et al., 2020). However, the facts that traditional blood collection is invasive and collection/transport is costly are major limitations. Consequently, the use of matched pairs of urine and DBS samples rather than traditional blood samples in routine sports doping control would offer a substantial added value. This was recently recognized by the World Anti-Doping Agency (WADA), as they accepted DBS as an alternative matrix because of the numerous associated advantages (e.g. the low invasiveness, the fact that less adulteration is possible compared to urine, etc.) (World Anti-Doping Agency, 2021). However, for DBS analysis to be implemented, effective high-throughput testing is one of the major requirements. In other words, for the doping analysis field, automation may be a crucial aspect that will facilitate the implementation into routine practice. To date, some methods have already become available, as will be outlined further in Section 5.2.5 (Dib et al., 2017; Lange et al., 2020; Luginbuhl et al., 2020a; Ryona and Henion, 2016; Tretzel et al., 2016).

Finally, in the field of newborn screening, in which microsampling has been established for almost six decades, automation of DBS methods may facilitate sample processing as these samples are currently mostly processed using the labor-intensive manual "punch-and-elute" methodology. However, recently, some efforts have been made to develop and transfer automated methodologies to the routine laboratory environment (see Section 5.2.5) (Gaugler et al., 2018; Luginbuhl and Gaugler, 2020).

5.2 Automation of Dried Blood Microsampling Analysis Coupled to (LC-)MS/MS: What's Available?

5.2.1 Amenability of DBS Samples for Automation

In this section, we will focus on the automation potential for DBS samples at the different stages of the sample preparation process, including (a) the generation of calibrators and QCs on filter paper, (b) the punching of the DBS card, and (c) the extraction process itself.

The spotting of DBS calibrators, QCs, and blanks can be automated, via the use of liquid handling systems (Capiau et al., 2016b; Velghe et al., 2016). These instruments are already available or are easily accessible in a lot of bioanalytical laboratories. Hence, this should not cause any (large) additional costs. An example of one such handlers is the Freedom EVO robotic liquid handler from Tecan® (Männedorf, Switzerland), as described by Yuan et al. (2012). These authors concluded that automated spotting of DBS is as accurate as manual spotting, with the benefits that it is more efficient and consistent for the laboratory technicians involved in this process. However, some challenges were identified during the setup of the automated process. Since these systems are typically used to handle plasma or serum, one should be aware of the higher viscosity of whole blood and make some adaptations to the pipetting process to avoid the formation of air bubbles and the settling of the red blood cells due to gravitational forces (Yuan et al., 2012). Other liquid handler systems have also been described to automate spotting of DBS, such as the DBS2 system from Instech (Pennsylvania, USA) (now a more recent version, the ABS2™ has become available (Instech, 2021)) or the Accusampler® from Dilab® (VeruTech AB, Sweden) (Fan et al., 2014; VeruTech AB, 2019). Importantly, the Instech systems described here are only approved for laboratory animal research and are not suited for use in a clinical setting.

A second step in the analytical process of DBS sample analysis that can be (semi-)automated to increase efficiency is the punching of the DBS to obtain a 3–8 mm diameter subpunch for analysis (Capiau et al., 2016b; Fan et al., 2014; Velghe et al., 2016). A distinction can be made between semi-automated robotic punchers, where the operator still needs to pay attention to the process (these are typically based on hand- or foot-operated switches) and automated punchers making use of robotic arms (Henion et al., 2013). A semi-automated setup is provided by Analytical Sales and Services (New Jersey, USA), called the DBS Pneumatic Card Punch (Analytical Sales and Services, 2021). Based on a pneumatic foot pedal-controlled "power punch," DBS punches are generated and directly loaded into collection plates or tubes via a glass capillary. This setup is available for 3–6 mm diameter punch sizes and does not need an electricity connection. The downside of this approach is that the operator still needs to manually select the correct position of the area to be punched. Another similar device is the Tomtec manual DBS punching station (Connecticut, USA), which is also a semi-manually operated instrument (Tomtec, 2021). The benefit of these punchers is that they are less costly compared to larger, more automated systems, and at the same time their use is still beneficial compared to handheld manual card punchers, such as the Harris Micro punch (Fisher Scientific, Massachusetts, USA) with respect to efficiency and hand-fatigue of the operator. However, while these approaches may be useful for smaller-scale applications, or in a research setting, more automated systems are preferred (not requiring any intervention by an operator) when aiming at high-throughput DBS sample analysis. Examples of smaller, more automated systems are the Dried Blood Spot Processor (Hudson Robotics, New Jersey, USA) (Hudson Robotics, 2021), the AutoDBS-1, 3 and 4 Dried Blood Spot Processing Workstation (Tomtec) (Tomtec, 2021), and the BSD-300, BSD-600 Ascent and BSD Nova M4 (BSD Robotics, Brisbane, Australia) (BSD Robotics, 2021b). In addition, Perkin Elmer (Massachusetts, USA) has the Wallac DBS Punch and the Panthera-Puncher 9 (Perkin Elmer, 2004; Perkin Elmer, 2022). Some of these devices contain cameras to select the correct punching location without any guidance of an operator. BSD Robotics has also manufactured systems such as the BSD-1000 GenePunch that have a larger capacity (no longer commercially available), which could handle the unattended card punching of up to 300 card samples after manual loading of the cards, or the BSD Galaxy A9, which is also equipped with a camera system (BSD Robotics) (BSD Robotics, 2021a). Some examples of (semi-)automated punching and pipetting devices are displayed in Figure 5.1.

The third and final step in the sample preparation process of DBS is extraction of the compounds of interest, which, when performed manually, can be more time-consuming compared to plasma or whole blood extraction. Over the years, efforts have been made to automate one or multiple steps in the extraction of DBS samples. Overall, these efforts can be subdivided in two categories, namely surface sampling techniques (including Direct Analysis In Real Time (DART), Desorption Electrospray Ionization [DESI], etc.), and online extraction approaches. Since multiple surface sampling techniques exist, each with their own strengths and limitations, these will be discussed separately in Section 5.2.7.1. In what follows, the different online extraction approaches will be discussed.

The first use of flow-through desorption for online extraction of DBS was reported by Déglon et al. (2009). These authors set up a configuration based on an inox cell in which a punched disk from a DBS sample can be placed. By coupling this configuration to a 12-port LC valve, up to six DBS samples could be analyzed in a semi-automated manner by manually switching the position of the valve from one inox cell to another after the end of each run. A successful application utilizing

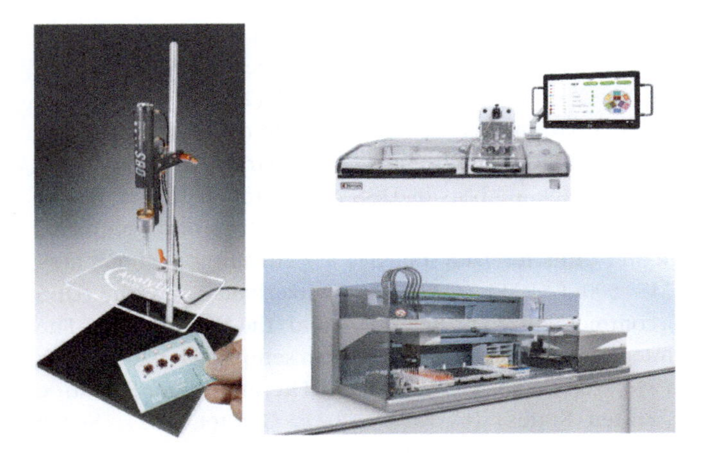

Figure 5.1 Commercially available (semi-)automated punching and pipetting instruments. From top-left to bottom-right: the DBS Pneumatic Card Punch (Analytical Sales and Services), the BSD GalaxyA9 (BSD Robotics) and the Freedom EVO Robotic Handler (Tecan). Reproduced with permission from Analytical Sales and Services, BSD Robotics and Tecan.

this setup for the detection of both polar and nonpolar compounds was achieved by Thomas et al. (2010). The same group also developed a fully automated configuration that was able to analyze up to 30 DBS samples using a rotating inox plate (Déglon et al., 2011). However, a disadvantage of this approach was that cutting or punching out DBS punches was still necessary. Simultaneously in 2009, other groups reported the evaluation of techniques that would allow online DBS extraction as well as elimination of the DBS punching step. Both Van Berkel et al. and Abu-Rabie et al. evaluated the use of the CAMAG thin layer chromatography–mass spectrometry (TLC–MS) interface for the online extraction of DBS coupled to (LC-)MS/MS (Abu-Rabie & Spooner, 2009; Van Berkel & Kertesz, 2009). The TLC–MS interface, controlled either manually or using a switching valve, consists of a 4-mm diameter extraction plunger (i.e. the prototype of the extraction head currently incorporated in the DBS-MS 500, see Section 5.2.2.1) which is used to seal a fixed area of the DBS for extraction. In further work, Abu-Rabie and colleagues reported on an updated version of this interface, the CAMAG DBS-MS 16 prototype (Abu-Rabie, 2011; Abu-Rabie & Spooner, 2011). A third approach, developed by Miller et al. and comparable to the configuration setup by Déglon et al. (2009), employs two end caps from a precolumn filter holder and two 0.5 μm frits in between which a DBS card can be sealed for extraction (Miller et al., 2011). Finally, a fourth reported approach is the direct elution of DBS samples followed by automated online solid phase extraction (SPE) prior to LC–MS/MS analysis. Based on this technique, Spark Holland developed a prototype DBSA-1 device that was previously evaluated by Ooms et al. (2011) and Li et al. (2012).

All the efforts mentioned within this section led to the development of the currently commercially available fully automated DBS extraction units. These systems provide complete online analysis of DBS cards, from inserting a card into a rack to the quantitative result, with no hands-on intervention required. Two automated extraction units for DBS analysis are currently commercially available: the DBS-MS 500 (-HCT) (CAMAG, Switzerland) and the DBS Autosampler (DBSA), commercialized by Spark Holland (the Netherlands). The DBSA system is also integrated in a sample preparation setup by Gerstel (Multi-Purpose Sampler (MPS)-based DBSA, Germany) and ThermoFischer Scientific (Transcend DSX-1, United States) (CAMAG, 2021; Gerstel, 2021; Spark Holland, 2021; ThermoFischer Scientific, 2021b). The Sample Card and Prep System (SCAP, Prolab, Switzerland), which has been used in some publications, is no longer commercially available but will also be discussed within this section (Prolab Instruments GmbH, 2021). The following sections will focus on how these systems work, the main differences between them, and some points to pay attention to during method validation and implementation of DBS-based analytical methods using these instruments.

5.2.2 Commercially Available Automated DBS Extraction Instruments

5.2.2.1 Automated Extraction of DBS: Workflow

The general workflow of an automated extraction unit is illustrated in Figure 5.2. Using a robotic arm, DBS cards are moved toward different working stations. A first station is the optical recognition system which locates the position of the spots on the card (both before and after extraction) and enables card identification through barcode reading. Second, the card is moved toward the extraction module after the internal standard (IS) is sprayed or loaded onto the DBS card in the IS module. After DBS extraction, the extract is guided into a sample loop (DBS-MS 500) or onto the SPE cartridge (DBSA and SCAP). The system is equipped with a wash station to avoid cross-contamination between different samples (Luginbuhl & Gaugler, 2020; Velghe et al., 2019b). Finally, to be able to perform fully automated DBS analysis, these automated extraction units have to be coupled to an LC-MS/MS instrument. Nevertheless, if full automation is not an absolute requirement, both systems can also be coupled to a sample collector allowing the generation of isolated extracts (Gaugler et al., 2021; Spark Holland, 2021). For example, for the analysis of Sars-Cov-2 antibodies, Gaugler et al. reported on the deposition of DBS extracts into 96-well plates using the DBS-MS 500 for automated extraction and a CTC Pal as sample collector (Gaugler et al., 2021). As for throughput,

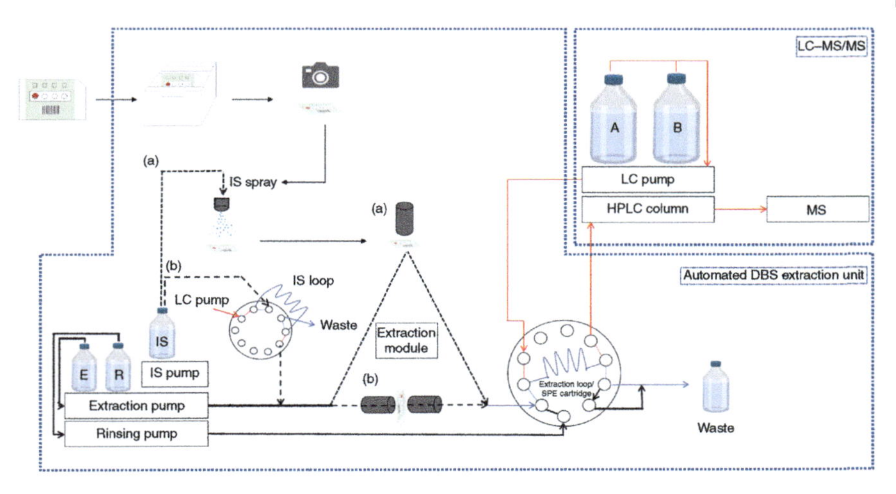

Figure 5.2 Schematic workflow of an automated extraction instrument. Following sampling and drying, DBS cards are placed in racks to allow fully automated extraction. A digital picture is taken of each spot before and after extraction. Before extraction, the IS is either sprayed (a, DBS-MS 500) or added to the extraction solvent using an IS loop (b, DBSA and SCAP). In the extraction module, the extraction takes places either horizontally (a, DBS-MS 500) or vertically (b, Flow-Through Desorption (FTD™), DBSA and SCAP). After DBS extraction, the extract is guided onto the extraction loop or the SPE cartridge/trapping column before injection onto the LC-MS/MS system.

the DBSA, SCAP, and DBS-MS 500 have a capacity of up to 96, 160, and 500 cards per run, respectively (CAMAG, 2021; Gerstel, 2021; Prolab Instruments GmbH, 2021; Spark Holland, 2021). When comparing the general workflow of the systems mentioned here, three other main differences can be found with respect to the extraction process, the extract processing strategy, and the IS application, which are discussed in the following sections (Luginbuhl & Gaugler, 2020).

5.2.2.2 Extraction Process

During extraction, and irrespective of the extraction process used, the DBS card is fixed between clamps containing the extraction head, which consists of a circular plunger that seals a fixed diameter on the blood spot for extraction. In contrast to manual or semi-automated DBS analysis, this setup eliminates the need for physically punching out disks of a DBS card. Instead, the area of the DBS to be extracted is determined by the inner diameter of the circular plunger. For the DBSA and SCAP systems, multiple extraction heads are available, with diameters ranging from 2 to 8 mm, allowing both partial and whole spot analysis (Prolab Instruments GmbH, 2021; Spark Holland, 2021). For the DBS-MS 500, a 4 and 6 mm extraction head is available. After the correct positioning of the extraction head, DBS samples can be extracted either vertically using a flow-through desorption technology (FTD™, patented by Spark Holland) or horizontally (as illustrated in Figure 5.2). The former is used by the DBSA and SCAP system, while the latter is used by the DBS-MS 500 (Velghe et al., 2019b). Both extraction process technologies are performed using high pressure ("active extraction"), aimed at yielding higher recoveries when compared to manual DBS handling ("passive extraction") (Li et al., 2012; Oliveira et al., 2014b; Ooms et al., 2011; Velghe et al., 2019b). Comparing the SCAP and the DBS-MS 500 platform, the DBSA system has the additional feature that a heater is placed upstream of the extraction module, allowing heating of the extraction solvent up to 80–100 °C (Oliveira et al., 2014b; Tretzel et al., 2016). The usefulness of this feature was demonstrated by Hempen et al., as they showed that the use of heated solvent resulted in an increased, Hct-independent recovery of immunosuppressants from DBS (Hempen et al., 2015). Tretzel et al. also used heated solvent extraction to enhance the elution strength of the used solvents (Tretzel et al., 2016). Although the use of heated solvent may improve the recovery, it can also cause degradation of thermolabile analytes and deterioration of chromatographic peaks, two scenarios in which extraction at lower (below 80 °C) or ambient temperature may be preferred (Oliveira et al., 2014b). These factors should be explored as part of the assay method development process.

5.2.2.3 Extract Processing Strategy

Both the DBSA and SCAP systems perform online sample clean-up via the use of online SPE. The DBS-MS 500, on the other hand, makes use of either a direct elution technique (variable extraction loop volumes are available) or a trapping column (Velghe et al., 2019b). The direct elution approach results in a simple and quick sample extraction method, while the use of online SPE or a trapping column allows more extensive sample clean-up when deemed necessary (e.g. due to matrix effects of co-extracted compounds). Using a trapping column, Luginbühl et al. were able to remove interfering phospholipids in the automated DBS analysis of PEth (Luginbuhl et al., 2019, 2021a). From a method development point of view, the use of an online DBS-SPE system provides considerable flexibility, as the SPE sample clean-up and enrichment can be accomplished with different and independent mixtures of solvents, eliminating the need to use a single-solvent system (as is the case when using a direct elution technique) (Oliveira et al., 2014b). In addition, the use of a suitable SPE cartridge can eliminate the need for an additional analytical column and allow direct coupling to the MS/MS, as reported by Oliveira et al. (2014b), Ooms et al. (2011), and Hempen et al. (2015). Nevertheless, the direct elution technique used by the DBS-MS 500 has shown adequate flexibility for optimization of the extraction process, as evidenced by a multitude of reports (Lin et al., 2020; Luginbuhl et al., 2020a; Velghe et al., 2019b). Moreover, the combination of a certain extraction loop and extraction volumes enables the circumvention of certain analytical issues. For example, Duthaler et al. were able to increase method precision by combining a 20-μL extraction loop with 25-μL extraction solvent, with only a minimal loss in sensitivity (as the first 5 μL is directed to waste) (Duthaler et al., 2017). When encountering saturation issues of the MS detector during automated extraction of DBS for quantifying antiepileptics, Velghe et al. coped with this issue by combining a 20-μL loop with an extraction volume of 60 μL. This approach decreased the %CV of the IS area of carbamazepine and its metabolite from >33 to <12.5%. However, because of the higher extraction volume and the differential extraction behavior of valproic acid and its IS (i.e. the IS being extracted first and in a more efficient way compared to valproic acid itself), the latter was lost via the 40 μL that was directed to waste and thereby not detectable anymore. Consequently, the built-in IS spray had to be excluded and all IS were added directly to the extraction solvent to allow detection of all IS used in this method (Velghe et al., 2019b).

5.2.2.4 Internal Standard Application

In a conventional liquid blood or plasma assay, the IS is added directly to the sample, allowing it to compensate for any variability or losses during extraction and to correct for variable LC-MS/MS performance. In contrast, due to the nature of DBS samples, the standard practice in manual DBS processing is addition of the IS to the extraction solvent. Consequently, the extraction behavior of the IS may differ from that of the analyte, as the IS will not be integrated into the matrix and, therefore, cannot correct for any variation in recovery. This could potentially lead to a Hct-based or sample age-based recovery bias. Abu-Rabie et al. extensively evaluated the use of an IS spraying unit which applies IS onto a DBS sample prior to extraction, demonstrating its ability to nullify the Hct-based recovery bias as a practical alternative to conventional IS addition (Abu-Rabie et al., 2011, 2015). In summary, these authors highlighted three points of attention: (a) it is of utmost importance to understand both the Hct-based area bias and recovery bias in DBS-based methods as there can be competing effects, (b) when comparing the overall assay bias using manual versus automated extraction of DBS, the recovery bias seems to be a more dominant factor when automated extraction is used, and (c) addition of the IS onto a DBS via spraying is able to nullify the Hct-based recovery bias.

To date, the DBS-MS 500 is the only system equipped with a built-in IS spray. For the two other systems (DBSA and SCAP), the IS is added during sample extraction (using a separate IS loop, Figure 5.2) (Ganz et al., 2012; Ooms et al., 2011). The latter can be a disadvantage, especially if the absolute recovery is low, as was also demonstrated by Abu-Rabie et al. (2015). Although the built-in IS spray has the ability to eliminate any Hct-based recovery bias and it has been demonstrated that the process is reproducible (Lin et al., 2020; Luginbuhl et al., 2020a), this appears not to be universal (Deprez and Stove, 2021). Hence, it is important to evaluate the performance of the IS spray during method optimization. For the automated DBS analysis of four immunosuppressants, the use of the built-in IS spray did not provide a benefit versus adding the IS to the extraction solvent: high IS area CVs (19.5–21.9%) were seen for the extraction of replicate (blank blood) DBS samples when using the spray, while only minimal CVs (2.7–4.8%) were seen when the IS was added to the extraction solvent (Deprez and Stove, 2021). At this point, it isn't clear whether this was related to this specific application or was caused by suboptimal performance of the spraying module. As a result, addition of the IS to the extraction solvent was preferred, despite the presence of a Hct-dependent recovery bias (see Section 5.2.4). Another example which led to the exclusion of the IS spray was reported by Velghe et al. (2019b). However, here, exclusion of the IS spray was not related to its performance but was a consequence of coping with saturation issues, as outlined in Section 5.2.2.4 (Velghe et al., 2019b). Overall, it can be concluded that the availability of a built-in IS spray offers the additional option to evaluate whether this approach may aid in alleviating or even nullifying the Hct-based recovery bias. However, this also requires evaluation and optimization (i.e. different ways of adding the IS, different IS volumes, and different IS drying times) during method development and validation.

5.2.3 Points of Attention During Method Validation

As DBS analysis typically involves several manual steps, it is not surprising that the introduction of automated extraction units enables a reduction in workload and an increased throughput of DBS-based methods, two major advantages offered by these instruments. However, one must bear in mind that the method validation of a manual DBS sample processing method is not completely interchangeable with the method validation of an automated extraction unit. Therefore, some points of attention specific for the validation of a method employing automated DBS extraction are discussed below.

5.2.3.1 Matrix Effects and Recovery

Generally, matrix effects and recovery are evaluated according to the procedure proposed by Matuszewski et al. (2003). This involves comparison of the absolute peak areas of the analytes (and IS) in a pre- and post-extraction spiked sample to assess recovery and comparison of the absolute peak areas of the analytes (and IS) spiked after extraction with the absolute peak areas of the analytes (and IS) spiked in neat solvent to assess matrix effects (Matuszewski et al., 2003). However, due to the online extraction setup of the automated extraction units, post-extraction spiking is not possible. Therefore, alternative approaches have been proposed, mimicking the abovementioned procedures as much as possible. For the determination of the matrix effect, both offline and online approaches have been used. The offline approach is in essence the determination of the matrix effect using a manual DBS processing method instead of the developed automated method. This approach was used by, e.g. Luginbühl et al. in a procedure to determine the matrix effect of tramadol in DBS (Luginbuhl et al., 2020a). The online approach uses the online extraction setup, whereby an extraction solution containing all analytes and IS (= "spiked extraction solvent") is used to extract blank DBS samples (= DBS cards on which blank blood was spotted) and blank DBS cards (on which no blood was spotted) and comparing the signals obtained for the former with those obtained for the latter (Abu-Rabie et al., 2011; Duthaler et al., 2017; Tretzel et al., 2016; Velghe et al., 2019b).

For the recovery, consecutively extracting and analyzing the same spot multiple times has been proposed as a possible strategy (Duthaler et al., 2017; Ooms et al., 2011; Verplaetse and Henion, 2016a, 2016b). The recovery is then calculated by dividing the signal of the first analysis by the sum of all, assuming that the latter represents a recovery close to 100% (Ooms et al., 2011). However, the number of consecutive extractions differs between published methods. For example, Verplaetse and Henion repeated the extraction 5 or 10 times, while Duthaler et al. performed 3 or 6 repeated extractions for the lower and upper limit of quantification (LLOQ and ULOQ) QCs, respectively (Duthaler et al., 2017; Verplaetse & Henion, 2016a, 2016b). In contrast, Velghe et al. observed perforation of the DBS card after four consecutive extractions, despite programming a drying time of 15 min (Velghe et al., 2019b). Therefore, for both the LLOQ and ULOQ QC samples, a maximum of three repeated extractions was performed to determine recovery. It is also important to note that the use of a particular extraction unit may influence the final outcome of the recovery for a particular analyte, due to differences in the extraction process. First, vertically or horizontally extracting DBS may have an impact. However, to the best of our knowledge, this has not been evaluated yet in a side-by-side comparison. Second, the procedure of performing multiple extractions may be different. For example, when using the SCAP or DBSA system, the DBS card is not removed from the clamp module between two (or more) consecutive extractions of the same spot (Ganz et al., 2012). This is in contrast to the DBS-MS 500 system, where the DBS card is moved from and to the extraction module after each consecutive extraction, leading to a possible underestimation of the recovery caused by minimal differences in extracted spot location (Velghe et al., 2019b). An alternative, less frequently used approach to determine recovery is the comparison of the results obtained by analyzing spiked DBS samples ("pre-spiked") and the results obtained by analyzing blank DBS samples extracted with the corresponding amount of spiked neat solution applied via the IS loop or added directly to the extraction solvent ("post-spiked") (Hempen et al., 2015; Tretzel et al., 2016). Another potential approach, which actually determines the process efficiency (combination of recovery and matrix effects), is the comparison of the results obtained by extracting both a regular DBS sample and a matrix-free (neat solvent) DBS sample pre-spiked with the analytes.

5.2.3.2 The Hct Effect

As outlined in Section 5.1.1, an important aspect during the development and optimization of DBS-based procedures is the evaluation of the Hct effect, as this parameter may impact the validity of the concentrations obtained by DBS methods (Oliveira et al., 2014a). Consequently, the effect of the Hct should be thoroughly evaluated during method validation, and this is also important to investigate when using automated DBS procedures. In short, for the evaluation of the Hct effect on the recovery, DBS should be generated from blood with varying Hcts (at least three levels, reflecting the Hct range expected in the target patient population), at two concentration levels (i.e. low and high QC) (Capiau et al., 2019). In order to eliminate the effect of the Hct-based area bias, this experiment, when conducted using automated DBS analyzers, should be performed by pipetting a fixed amount of blood onto filter paper, followed by whole spot analysis.

When a Hct-related bias is observed during method validation of an automated DBS procedure, multiple approaches can be used to alleviate this bias. These are similar to those for manual DBS extraction (De Kesel et al., 2013, 2014a; Velghe et al., 2019a). A first approach is whole spot analysis, provided that the diameter of the extraction head used is large enough to extract the entire spot. However, this approach relies on the accurate and precise volumetric spotting of blood onto filter paper using micro-dispenser capillary devices (Oliveira et al., 2014b; Tretzel et al., 2016). Oliveira and co-workers evaluated this approach by comparing the Hct-based area bias of partial (2 mm) and whole spot (6 mm) analysis using the DBSA system. Using partial spot analysis, a bias was seen at the Hct extremes (30 and 60%, as compared to 45%), exceeding the predefined 15% acceptance criterion for all three compounds studied (chloroquine, desipramine, and midazolam). In contrast, this bias substantially decreased when using whole spot analysis, proving its usefulness to cope with a Hct-based area bias (Oliveira et al., 2014b). Although a popular and successful approach in a laboratory context, the requirement for volumetric sample application is more complex for the collection of DBS samples in home-sampling settings and remote regions. In addition, this approach only alleviates the Hct-based area bias and does not tackle any other Hct-related effects, such as the recovery bias or the blood–plasma ratio. For example, Deprez and Stove observed a Hct-related recovery bias using the DBS-MS 500 for the fully automated analysis of immunosuppressants in DBS (Deprez & Stove, 2021). In this case, whole spot analysis would not alleviate this phenomenon since the Hct effect observed was mainly related to extractability and not solely to the area bias. Therefore, another approach, the use of a Hct correction model was successfully applied to cope with the observed bias (Deprez & Stove, 2022). However, a disadvantage of the latter approach is that the correction model was based on the Hct determined from corresponding liquid whole blood samples. Of course, ideally, the Hct should be measured or predicted using the DBS samples themselves. A number of noncontact methodologies have been developed that can estimate the Hct of DBS samples, using NIR or UV/Vis spectroscopy, as mentioned in Section 5.1.1 (Capiau et al., 2018; Delahaye et al., 2021a; Oostendorp et al., 2016; van de Velde et al., 2021). In line with this, CAMAG recently embedded a Hct module (based on the methodology developed by Capiau et al. using UV/Vis spectroscopy) in its DBS-MS 500 platform (Capiau et al., 2016a, 2018; Luginbuhl and Gaugler, 2021; Luginbuhl et al., 2020b). Using this approach, Luginbühl et al. performed a proof-of-concept study for the correction of the Hct effect observed in PEth determinations (Luginbuhl et al., 2021b). Using data of four participants, a Hct correction factor was set up and all but two of the measured concentrations were within the acceptance limits of ±15% after applying the correction. In another study, Lange et al. performed a preliminary assessment of the coupling of NIR-based Hct prediction to a Gerstel-DBSA (Lange et al., 2020). In this report, a NIR-based setup developed by Büchi (Essen, Germany) and Oostendorp et al. and recently evaluated by Delahaye and Heughebaert et al. was used (BÜCHI Labortechnik, 2021; Delahaye et al., 2021a; Lange et al., 2020; Oostendorp et al., 2016). However, the application of NIR for Hct prediction and correction for a clinical study has at the time of writing not been performed. Furthermore, to the best of our knowledge, this combined setup is not yet commercially available. Overall, in contrast to whole spot analysis, knowledge of the Hct enables correction for the Hct-based area and recovery bias, as well as facilitating the conversion of DBS-based results to plasma- or serum-based results, which are often the matrices for which reference intervals are available. Nevertheless, more research is needed, effectively demonstrating that the technology is capable of yielding an adequate correction and adequate conversion to plasma results (the latter also requiring a consistent blood–plasma ratio).

5.2.3.3 Calibration Curve and Dilution Integrity

The span of the calibration range is usually determined by the expected concentration breadth in patient samples and is ideally chosen such that most of the samples will fall within this predetermined range. For example, in the field of TDM, a calibration range minimally ranging from half of the lower end of the therapeutic interval to twice the upper end of the therapeutic interval should suffice (Capiau et al., 2019). However, the concentration of some samples may exceed the ULOQ. If the concentration of these samples is to be determined, a dilution integrity experiment must be incorporated in the assay method validation. While this is straightforward for liquid samples and also feasible for manual DBS approaches, this is not readily possible for automated DBS extraction (Henion et al., 2013). Consequently, independent of the application field, this means that the ULOQ should be chosen high enough to minimize the number of samples that cannot be quantified (Duthaler et al., 2017). However, it may not always be relevant to have an exact quantitative result for concentrations above the ULOQ. For example in the field of TDM, in the absence of indications of intoxication, trough concentrations above the ULOQ are not often expected and may rather indicate that the blood was collected at the wrong time point (in which case the measured concentration would not be suited for follow-up anyway), or that a contamination issue occurred (Deprez and Stove, 2021).

5.2.4 Cross-Validation of Automated DBS Procedures

Besides the advantages discussed in Section 5.2.3, the use of an automated online extraction approach for DBS analysis may also offer an increased sensitivity compared to offline extraction of subpunched DBS samples. Oliveira et al. reported that the direct extraction and analysis of DBS samples resulted in a fivefold improvement in assay sensitivity compared to conventional offline extraction of punched DBS samples (Oliveira et al., 2014a). This is realized by direct transfer of the DBS extract to the LC-MS/MS setup, thereby minimizing analyte losses due to sample transfers. Additionally, the entire extract can be sent to the LC column, in contrast to a fixed injection volume (which is typically only a fraction of the total extract) when performing manual sample preparation. An important prerequisite to the replacement of manual extraction of DBS samples by automated (online) extraction is that comparable results can be achieved in terms of accuracy and precision. Few studies are available that directly compared manual with automated DBS extraction. A limited comparison was performed by Tretzel et al., reporting a sevenfold improvement in spot-to-spot cycle time when compared to manual DBS extraction. Furthermore, both the precision and the extraction efficiency of the automated DBS method were superior to its manual counterpart. However, a complete cross-validation was lacking in this report (Tretzel et al., 2016). A study by Martial and colleagues compared manual extraction of voriconazole with automated DBS extraction using the DBSA instrument and to a reference method in plasma (Martial et al., 2018). Based on the cross-validation performed in this study, the authors concluded that both DBS methods complied with all validation parameters, based on FDA and EMA guidelines (European Medicines Agency, 2011; US Department of Health and Human Services Food and Drug Administration, 2018), and that a proportional bias was present between both methods. Hence, a correction factor was needed to obtain interchangeable voriconazole concentrations between both DBS methods. The manual DBS method required correction of a proportional bias with the plasma method, while the DBSA and plasma assay showed good agreement. These authors also remarked that the variability in paired DBS/DBSA and plasma concentration measurements was considered high for TDM purposes, and that this limitation should be balanced against the advantages of DBS sampling of voriconazole and the time gain using the automated DBS approach. The time benefit of using the automated method was a gain in sample preprocessing time, which was reduced down to 3 min, compared to 1.5 hr for the manual approach.

Using the DBS-MS 500, Duthaler et al. reported on the validation of an automated DBS method for antiretroviral drugs. As part of the validation, a method comparison ($n = 30$ per analyte) was performed with the corresponding manual DBS extraction method. The authors reported an acceptable agreement, with a mean bias of automated to manual extraction of −10.5, −8.9, and −3.0% for nevirapine, efavirenz, and lopinavir, respectively. Although it was reported that 70% (or more) of the samples were within 20% difference of the manual DBS method, for efavirenz, the 95% limits of agreement were quite broad (ranging from −38 to +20%). In addition, a fivefold improvement in sensitivity was reported for the automated approach, which was explained by the smaller extraction volume used in the automated setup (25 vs. 150 μL) (Duthaler et al., 2017). In a follow-up study, the same authors performed a cross-validation to compare the results of the automated (capillary) DBS extraction to a fully validated plasma method. To demonstrate the applicability of the method in remote settings, both a bridging study and a field study were set up. Initial results yielded an unacceptable mean bias for all three analytes, i.e. 18.4, −47.4, and −48.1% for nevirapine, efavirenz, and lopinavir, respectively. However, for efavirenz and lopinavir similar differences were observed when comparing whole blood and plasma results (for nevirapine only a small mean bias of 5.7% was reported). Using the Hct and the blood–plasma ratio as a correction factor, the authors reported that the bias between DBS and plasma was nullified for the samples collected in the bridging study. However, for nevirapine the mean bias was still 12.4% (Duthaler et al., 2018). Using the same correction factor, a decrease in mean bias was also observed for the samples collected in the field study for efavirenz (from −52.0 to −16.1%). For nevirapine, good agreements were found 3.6 and −1.7% with and without correction factor respectively. In contrast, for lopinavir, corrected DBS concentrations were still 31% lower than plasma concentrations in the field study, which could be attributed to degradation of lopinavir under the climate conditions in Tanzania. The same group also performed a preliminary method comparison of automated versus manual extraction of ivermectin, yielding a mean bias of 5.8% between both methods. However, since only three participants ($n = 26$ samples, as DBS were generated at multiple time points for each participant) were included, the reported agreement is only preliminary and should be confirmed in a larger method comparison study

(Duthaler et al., 2019a). In previous work, a limited comparison of plasma versus automated capillary DBS concentrations was also performed, which yielded a mean bias of 34% ($n = 119$). The authors hypothesized that the origin of this bias was probably related to the high-protein binding of ivermectin to albumin (Duthaler et al., 2019b). Therefore, as a correction factor, the ratio DBS:plasma was used; however, no data were provided demonstrating actual improvement of this bias. Luginbühl et al. also performed multiple cross-validation studies for their automated procedures. A first study encompassed the comparison of both whole blood and (whole spot) manual DBS extraction with automated DBS extraction for the determination of the direct alcohol marker PEth 16:0/18:1 (Luginbuhl et al., 2019). To compensate for a potential extraction bias in the manual DBS method, the IS was sprayed onto the DBS samples with the CAMAG DBS-MS 500 autosampler prior to the manual sample extraction. The comparison was conducted for 28 DBS samples, and a mean deviation of approximately 2 and 8% and a correlation coefficient of 0.9768 and 0.9666 was obtained for the comparison with whole blood and the manual DBS extraction method, respectively (Luginbuhl et al., 2019). Fatch et al. performed a follow-up study and, using categorization of DBS samples depending on PEth levels, reported a "good agreement" between the different categories following manual extraction versus automated extraction. Furthermore, using log transformation of all (non-categorized) PEth results, a good agreement was reported by these authors. However, the median manual PEth testing result was 98.8 ng/mL, while the median automated result was 155 ng/mL. These results indicate that the automated procedure yielded higher PEth levels as compared to the manual procedure. Additional limitations of this study were the time difference of 3–6 years between manual (performed first) and automated extraction, and the difference in (purity of) the calibrators that were used in both procedures (Fatch et al., 2021). These (and other) factors complicate the interpretation of the results, and therefore the conclusions made should be interpreted with caution. Similarly, a cross-validation study was performed for the automated analysis of tramadol and its metabolite *O*-desmethyltramadol (Luginbuhl et al., 2020a). In this report, 26 authentic samples were used to compare the results obtained using automated (venous) DBS extraction and a manual whole blood extraction method. Mean agreements of 90% ± 19% and 94 ± 14% were reported for tramadol (23/26 samples were used) and *O*-desmethyltramadol (9/26 samples were used), respectively. It should be remarked, though, that all samples were used to set up the regression model, while only a subset of the samples was used to verify the model and calculate mean biases. In addition, the setup and validation of the regression model were done using the same samples. Ideally, this should have been performed with samples distinct from those that were used to set up the regression model. Moreover, only a limited set of samples was used to perform the evaluation, and the automated DBS extractions were performed about 6–16 months after the original whole blood analysis (Luginbuhl et al., 2020a).

A side-by-side comparison of the results obtained from whole blood and DBS generated thereof (using automated extraction) was performed by Lin et al. for the determination of riboflavin (vit B_2). In this article, the authors report for the first time on the use of automated extraction for endogenous compounds. Using 133 whole blood samples and DBS derived thereof, a method comparison was performed in which DBS concentrations were plotted against whole blood concentrations, resulting in a Pearson correlation coefficient of 0.9774. However, when looking at the correlation plot, there seems to be a bias present between both methods (detailed data are lacking in this report), which is an important limitation (Lin et al., 2020). Recently, we published a limited method comparison as part of the method validation of an automated extraction method for four immunosuppressants. Here, 10 venous whole blood samples per analyte were collected and analyzed as venous DBS and in whole blood. Comparison of the results showed a reasonable agreement as 8, 6, 9, and 9/10 samples were within the acceptance limit of ±20% for tacrolimus, sirolimus, everolimus, and cyclosporin A, respectively (Deprez and Stove, 2021). However, since part of the differences between both methods could be attributed to an impact of the Hct on the recovery (as explained in Section 5.2.3.2), a follow-up study was performed to set up and implement a Hct correction model. This allowed to further improve the agreement (Deprez and Stove, 2022). In short, the analysis of a larger set of samples showed that for all analytes, prior to Hct correction, a major overestimation of the DBS results of low Hct samples was present compared to whole blood (because of a higher relative recovery from low-Hct DBS), and vice versa for high Hct samples. Based on a large dataset of paired DBS and whole blood samples for one of the analytes (tacrolimus), a Hct correction factor could be determined using linear regression analysis. The validity of this correction was verified by applying it to an independent set of samples. Additionally, the Hct correction was also able to alleviate the Hct effect for the three other immunosuppressants, indicating that the recovery issue was similar for all analytes. Following Hct correction for all analytes, at least 80% of the automated Hct-corrected-DBS results were within 20% difference of the whole blood concentration. In conclusion, automated DBS analysis seems to be capable of yielding results that are comparable to those obtained by manual processing of DBS or liquid blood samples. However, more studies are needed to demonstrate agreement on a compound-to-compound basis.

5.2.5 Current Applications of Automated Online DBS Extraction

In the past few decades, a wealth of applications have been published using DBS or other alternative microsampling approaches. Simultaneously, the use of automated extraction approaches has emerged with the publication of a substantial number of applications, covering various fields, including newborn screening, TDM, (forensic) toxicology, and (pre)clinical studies. Table 5.1 provides a summary

Table 5.1 Overview of published applications employing automated DBS extraction units.

Instrument and analyte	References	Publication year	Extract processing strategy	Clamp size (inner diameter)	Filter paper type	Application field	Comparison with whole blood/ manual DBS method	Implementation status	Comments
DBSA—Spark Holland									
Propranolol, haloperidol, amitriptyline, imipramine, and verapamil	Ooms et al. (2011)	2011	On-line SPE: HySphere C18HD, 7 μm	2 mm	Whatman FTA® DMPK-C cards	NA[a]	ND	Proof-of-concept	Prototype DBSA No analytical column
Guanfacine	Li et al. (2012)	2012	On-line SPE: HySphere C18HD, 7 μm, 2 × 10 mm	4 mm	Whatman FTA® DMPK-C cards	NA	ND	Proof-of-concept	Prototype DBSA
Chloroquine, desipramine, and midazolam	Oliveira et al. (2014b)	2014	On-line SPE: Hysphere™ C_{18}SE, 7 μm	2 and 6 mm	Whatman FTA® DMPK-C cards	NA	ND	Proof-of-concept	Prototype DBSA No analytical column
Immunosuppressants	Hempen et al. (2015)	2015	On-line SPE: HySphere C8HD, 7 μm, 2 × 10 mm	6 mm	Perkin Elmer 226 cards	TDM	ND	Unknown	No analytical column Temperature-enhanced desorption at 100 °C

Nicotine and metabolites	Tretzel et al. (2016)	2016	On-line SPE: HySphere Resin GP, 5–20 μm	6 mm	Whatman FTA® DMPK-C cards	Antidoping	Limited comparison with manual DBS method	Unknown	Gerstel Temperature-enhanced desorption at 100 °C
Opioids	Verplaetse and Henion (2016b)	2016	On-line SPE: HySphere C18HD, 7 μm, 2×10 mm	2 mm (partial spot)	Perkin Elmer Ahlstrom 226	DOA testing	ND	Unknown	
Stimulants	Verplaetse and Henion (2016a)	2016	On-line SPE: HySphere C2 HD, 7 μm, 2×10 mm	6 mm (whole spot)	Perkin Elmer Ahlstrom 226	DOA testing	ND	Unknown	Temperature-enhanced desorption at 100 °C
Opioids and stimulants	Ryona and Henion (2016)	2016	On-line SPE: HySphere C8HD, 7 μm, 2×10 mm	4 mm	Book type DBS card (Perkin Elmer 226)	Antidoping	ND	Unknown	
AdipoRon and 112254 (adiponectin receptor agonists)	Dib et al. (2017)	2017	On-line SPE: HySphere C8-EC-SE	6 mm	Whatman FTA® DMPK-C cards	Antidoping	ND	Inclusion in routine screening is planned	Gerstel
Voriconazole	Martial et al. (2018)	2018	On-line SPE: HySphere C18HD, 7 μm, 2×10 mm	2 mm	Whatman 903 filter paper	TDM	Automated versus manual DBS method	Unknown	

(*Continued*)

Table 5.1 (Continued)

Instrument and analyte	References	Publication year	Extract processing strategy	Clamp size (inner diameter)	Filter paper type	Application field	Comparison with whole blood/manual DBS method	Implementation status	Comments
(Non-)peptide doping agents	Lange et al. (2020)	2020	On-line SPE (SCX)	6 mm	Whatman FTA® DMPK-C cards	Antidoping	ND	Unknown	NIR-based Hct prediction module coupled to a Gerstel-DBSA
Anabolic steroid esters	Jing et al. (2022)		On-line SPE: Cyclone-P TurboFlow™ column, 0.5 × 50 mm	6 mm	QIAcard FTA® DMPK-C cards	Antidoping	ND	Unknown	
DBS-MS 500—CAMAG									
Acetaminophen, sitamaquine	Van Berkel and Kertesz (2009)	2009	ND	4 mm	Ahlstrom grade 237 paper	Proof-of-concept	ND	Proof-of-concept	Prototype DBS-MS 500
Ibuprofen, 4-nitrophthalic acid, paracetamol, simvastatin, sitamaquine, benzethonium chloride, and proguanil	Abu-Rabie and Spooner (2009)	2009	ND	4 mm	Ahlstrom grade 237 paper	Proof-of-concept	Preliminary comparison of manual versus semi-automated extraction	Proof-of-concept	Prototype DBS-MS 500

Acetaminophen, sitamaquine	Abu-Rabie and Spooner (2011)	2011	20 μL extraction solvent	4 mm	Ahlstrom 226 cards	Proof-of-concept	Preliminary comparison of manual versus automated extraction	Proof-of-concept	Prototype DBS-MS-500
Drug candidate	Heinig et al. (2010)	2010	Trapping column: Chromolith RP-18e 2×5 mm	4 mm	Ahlstrom 226 paper	(Pre-) clinical studies	ND	Proof-of-concept	Prototype DBS-MS 500
Oseltamivir and oseltamivir carboxylate	Heinig et al. (2011)	2011	Trapping column: Nucleosil 100-5 C18HD 4×8 mm	4 mm	Ahlstrom 226 DBS cards and Whatman FTA DMPK-B cards	Population studies, TDM	Limited comparison with manual DBS method	Proof-of-concept	Prototype DBS-MS 500
Amino acids and acylcarnitines	Fingerhut et al. (2014)	2014	Loop (100 μL)	4 mm	Ahlstrom 226 filter paper	Newborn screening	Preliminary comparison with routine NBS method	Unknown	
Midazolam and alfa-hydroxymidazolam	Oliveira et al. (2014c)	2014	Loop (20 μL)	4 mm	Whatman FTA® DMPK-C cards	(Pre-) clinical studies	ND	Unknown	
Antiretroviral drugs	Duthaler et al. (2017, 2018)	2017	Loop (20 μL) + 25 μL extraction volume	4 mm	Grade 226 filter paper cards	TDM	Automated DBS versus manual DBS/plasma method	Unknown	

(Continued)

Table 5.1 (Continued)

Instrument and analyte	References	Publication year	Extract processing strategy	Clamp size (inner diameter)	Filter paper type	Application field	Comparison with whole blood/ manual DBS method	Implementation status	Comments
Drugs screening	Gaugler and Cebolla (2018)	2018	Loop (20 µL) (200 µL/min)	4 mm	Ahlstrom TFN filter paper cards	DOA	ND	Method transferred to routine laboratory and implemented	
	Gaugler et al. (2019a)	2019	Loop (20 µL) + 22 µL extraction volume (50 µL/min)						
Amino acids and acylcarnitines	Gaugler et al. (2018)	2018	Loop (20 µL)	4 mm	903 TFN cards	Newborn screening	ND	Yes	
	Gaugler and Cebolla (2019b)	2019	Loop (20 µL) + 60 µL extraction volume						
Antiepileptics	Velghe et al. (2019b, 2020)	2019	Loop (20 µL) + 60 µL extraction volume	4 mm	Perkin Elmer 226 cards	TDM	Automated DBS versus manual VAMS method	Applied in a remote context (Velghe et al., 2020)	
Phosphatidylethanol	Fatch et al. (2021) and Luginbuhl et al. (2019, 2021a, 2021b, 2021c)	2019	Trapping column: polar-RP 20 × 2 mm, 4 µm	4 mm	Ahlstrom BioSample TFN cards	DOA testing	Automated DBS versus manual DBS/whole blood method	Unknown	

Ivermectin	Duthaler et al. (2019a, 2019b)	2019	Loop (20 μL)	4 mm	Whatman 903 cards	TDM	Automated DBS versus manual DBS/plasma	Applied in clinical study	
Riboflavine (vit B_2)	Lin et al. (2020)	2020	Loop (10 μL) + 15 μL extraction volume	ND	ND	Vitamin analysis	Automated DBS versus whole blood	Unknown	
Tramadol and *O*-desmethyltramadol	Luginbuhl et al. (2020a)	2020	Loop (20 μL)	4 mm	BioSample TFN cards	Antidoping	Automated DBS versus whole blood	Unknown	
SAR-CoV-2 antibodies	Gaugler et al. (2021)	2021	Elution into a 96-well plate	4 mm	TFN filter paper cards	Serological testing	ND	Unknown	
Immunosuppressants	Deprez and Stove (2021)	2021	Loop (100 μL)	4 mm	Perkin Elmer 226 cards	TDM	Automated DBS versus whole blood	Application study planned	
SCAP—Prolab Instruments GmbH									
Oseltamivir and oseltamivir carboxylate	Heinig et al. (2011)	2011	Trapping column: Nucleosil 100-5 C18HD 4 × 8 mm	3.2 mm	Ahlstrom 226 DBS cards and Whatman FTA DMPK-B cards	(Pre-) clinical studies	Limited comparison with manual DBS method	Unknown	SCAP system no longer commercially available

(Continued)

Table 5.1 (Continued)

Instrument and analyte	References	Publication year	Extract processing strategy	Clamp size (inner diameter)	Filter paper type	Application field	Comparison with whole blood/ manual DBS method	Implementation status	Comments
Bosentan and metabolites	Ganz et al. (2012)	2012	Trapping column (using two precolumns (PC)): PC1 LiChrospher 100 RP18 ADS, 25 mm × 2 mm, 25 μm and PC2 CC 8/4 Nucleosil 100-5 C18 HD, 5 μm	2 mm	Whatman DMPK-A cards	TDM	ND	Unknown	
Paracetamol	Taylor et al. (2013)	2013	Loop: volume ND	ND	Whatman 903 Protein Saver Cards	TDM	Limited comparison with plasma and CSF and manual DBS	Unknown	
Midazolam and desipramine	Oliveira et al. (2014a)	2014	Trapping column: Ascentis RP-Amide Supelgard Guard Cartridge 20 × 4 mm, 5 μm	2 mm	Whatman FTA DMPK-C cards	(Pre-) clinical studies	Limited comparison with manual DBS method	Unknown	

[a] CSF, cerebrospinal fluid; DBS, dried blood spot(s); DBSA, dried blood spot autosampler; DOA, drugs of abuse; NA, not applicable; ND, not determined; NIR, near infrared; SCAP, sample card and prep; SPE, solid phase extraction; TDM, therapeutic drug monitoring; VAMS, volumetric absorptive microsampling.

of those analytes for which automated procedures have been published, along with the application field and implementation status

5.2.6 Approaches for Automating Analysis with Other Microsampling Devices

So far, this chapter has focused on the amenability of automating conventional DBS sample analysis. However, equally important are the automation possibilities for the processing of samples obtained with other microsampling devices. As these devices enable the collection of a fixed volume of blood, whole spot analysis can be performed, and the punching step is avoided. In contrast to conventional DBS, the blood spots generated using volumetric devices are often integrated inside a device to avoid contamination. Because of this, an additional step is needed to open the device prior to extraction. This is the case for the HemaPEN, the Capitainer qDBS device, and the micro-needle-based devices (e.g. the Tasso-M20 device). To the authors' knowledge, no automated platforms exist to facilitate this step for any of the above-mentioned devices. This is a point of attention when envisaging the use of these devices at a high-throughput scale. However, for the analysis of breath samples, robots already exist facilitating the opening of the breath sampling cartridges (Munkplast AB, 2021). This could be a starting point for the development of a similar robot which can automatically open multiple types of alternative microsampling devices, thereby eliminating the additional step of manually opening these devices before extraction. For other devices such as the HemaXis device, automation is far more straightforward as this device is readily made in a DBS-card format, which makes it amenable for automation and ensures compatibility with most currently available DBS extraction platforms. However, because of the small volume of 10 μL that is collected, a 6-mm extraction clamp (as is the maximal extraction diameter for the DBS-MS 500) is too small to ensure extraction of the whole 10 μL spot. This means that partial spot extraction should be performed, which would negate part of the benefit offered by these volumetric devices. In contrast, the DBSA enables extraction up to an 8 mm diameter area, which would be sufficient to enable extraction of the entire spot. For the extraction of other volumetric microsampling devices, an additional manual step is required to bring the DBS samples into a 96-well plate format before extraction can be semi-automated using conventional liquid handlers, as outlined in Section 5.2.1.

Another volumetric blood collection technology that—at least in principle—is readily compatible with automated processing is the Mitra device. This differs from DBS samples; in that, it has a polymeric tip for blood collection attached to a plastic handle. Consequently, an alternative approach for sample analysis is required compared to conventional DBS sample analysis. At the Society for

Laboratory Automation and Screening (SLAS) conference in 2019, the implementation of a Tecan automation setup for Mitra devices was presented (Yeung et al., 2019). The system outlined in that presentation was not an off-the-shelf solution, but a custom-built (modified) Tecan system. It was used to automatically sample a large number of VAMS tips for positive and negative controls as well as blank samples needed for the assay. To allow easy automation of VAMS tips extraction, Neoteryx manufactured two versions (manual use and automation use) of VAMS tips in 96-well racks (Yeung et al., 2019). One version contains a finned version of VAMS tips, which are intended to help avoid touching the side of the blood tube during manual preparation. A second version fits any standard "Society for Biomolecular Sciences" (SBS) format. Moreover, the plastic handlers are smooth, enabling them to easily slide in and out of tubes and racks (the latter is no longer commercially available). Unfortunately, the authors reported that neither version was directly compatible with a commercially available Tecan system. Therefore, they added Tecan disposable tips automatically into the smooth VAMS tips for direct utilization by Tecan Liquid-handling. In this way, they were able to demonstrate that it is feasible to set up a system to automatically process a large number of VAMS tips. While this was a custom-made solution, not being commercially available, in the future, commercial alternatives may become available for this purpose. In a tutorial in *Analytica Chimica Acta*, Protti et al. also readily elaborated on the possibilities of automating VAMS extraction (Protti et al., 2019). Because of the small size of the sampling tip, it is possible to perform extraction in microtiter plates. This setup is easily automatable using standard bioanalytical instrumentation with minimal additional effort. To the authors' knowledge, only one method has been published which explicitly reports on the use of automated VAMS extraction (van den Broek et al., 2017). In this article, van den Broek et al. describe an automated, high-throughput sample preparation method coupled to MS detection for the determination of protein biomarkers from Mitra microsampling devices. The authors report the use of a Biomek NX^{P} workstation to automate VAMS extraction. This workstation is a conventional liquid handler, available from Beckman Coulter (2021). However, the removal of the Mitra tip from the microsampler body (the plastic handle) was still performed manually by gently pushing it with the long side of a needle, while holding the tip to the edge of the destination well. Neoteryx, the manufacturer of the Mitra tips, also explored the use of a dedicated, automated liquid handling instrumentation, the Mitra PAL RTC instrument (in collaboration with Brechbühler AG) (Neoteryx, 2015). This instrument was compatible with the 96-rack format to store VAMS and would enable automated addition of extraction solvent and shaking of VAMS in a 96-plate format. However, to the authors' knowledge, this configuration is at present not commercially available. Another possibility to decrease the manual handling using Mitra devices is the use of the autoMitra™ workflow (INTEGRA

VIAFLO® 96). This is a semi-automatable solution with programmable liquid transfer (Neoteryx, 2021). Although the semi-automated system requires some manual handling to control the workflow, the setup is quite compact, is in a bench-top format, and is less costly compared to fully automated systems. Depending on the envisaged application, such a semi-automated solution might be sufficient.

Despite the efforts made to provide automated solutions for the analysis of Mitra devices, very few publications are available describing an automated workflow. When using these devices in large-scale clinical studies (or other high-throughput applications), a fully automated workflow would obviously be preferred. We consider it likely that the possibility of automation will be further explored, preferentially coupling automated extraction with subsequent LC-MS/MS analysis (similar to the online systems that are available for DBS), as this will create many more opportunities for the application of large-scale Mitra device usage.

5.2.7 Alternative Approaches for the Analysis of DBS Samples

5.2.7.1 Direct Analysis of DBS

An alternative approach to reduce or eliminate the additional workload associated with DBS sample analysis is their direct analysis using ambient ionization techniques, including for example DESI and DART (as already briefly mentioned in Section 5.2.1). The former is based on the electrospray ionization (ESI) technique, while the latter relates to atmospheric pressure chemical ionization (APCI). One of the main advantages of these approaches is that both the sample preparation (punching and extraction) and the chromatography are eliminated. However, several issues have also been observed inherent to these techniques. These include (a) a lack of sensitivity due to matrix effects and (b) interferences from isobaric species and in-source fragmentation, which are both related to the absence of chromatographic separation. Moreover, analyte extraction is performed in a different way compared to conventional DBS extraction and is specific to the technique used. Consequently, knowledge of the impact of important DBS characteristics such as the Hct effect is limited or even nonexistent. Since the Hct can impact the recovery and it has been reported that these techniques do not provide equivalent analyte desorption, it is of utmost importance that this is further evaluated. Another important aspect is the irregular nature of the analyte signals in the chronogram, which may complicate the accurate and reliable quantification of the compounds of interest (Manicke et al., 2011; Wagner et al., 2016).

Regardless of these challenges, a multitude of reports using these techniques has been published in the last decade (Crawford et al., 2011; Wiseman et al., 2010). In this section, we will only briefly focus on the use of DESI and paper spray ionization for direct DBS analysis, since it has been reported that

DART-MS is inefficient for the analysis of classical paper-based DBS with respect to analyte recovery (Wagner et al., 2016). Paper spray is an ESI-based MS approach for analyzing dried blood samples. In short, a drop of blood is spotted onto a small triangular piece of paper and allowed to dry. To analyze the sample, a solution (typically an acidified mixture of an organic solvent [>90%] and water) is added near to the base of the paper which is positioned with its sharp vertex facing the MS orifice (Manicke et al., 2011; Yannell et al., 2017). Ionization is achieved by applying a high voltage onto the paper through a copper clip at the base of the paper triangle. This technique was recently commercialized by ThermoFisher Scientific called the Verispray™ Paperspray ion source (ThermoFischer Scientific, 2019). In addition, cartridges containing pre-cut triangles are also available eliminating the time-consuming step of precisely cutting out triangular pieces of paper (ThermoFischer Scientific, 2019; Yannell et al., 2017). Also for this technique, the same challenges and limitations as already mentioned for DESI and DART remain. Finally, since full automation and practical use for quantitative DBS analysis have not yet been proven, we will not further discuss these techniques in detail. For a more detailed overview of the (dis)advantages and applications of each technique, we refer to the reviews of Wagner et al. (2016) and Frey et al. (2020).

5.2.7.2 Coupling DBS Analysis to Automated Immuno-Analyzers

DBS samples are also used in combination with (automated) immunoassays, mainly for diagnostic virology, for example for hepatitis B and C diagnosis. For the diagnosis of these infectious diseases, detection of antigens (e.g. HBsAg), antibodies (e.g. anti-HBs), or the viral genome is necessary. Here, after drying the DBS sample, the dried matrix is redissolved with buffers to result in aqueous extracts compatible with conventional, automated immunoassays used for serological testing (Gruner et al., 2015). In this field, the use of DBS as an alternative to serum samples offers the advantage of easy collection and safe transportation to reference laboratories, without requirement for refrigeration (Villar et al., 2019). Recently, this combined setup was also used in seroprevalence studies in the context of the COVID-19 pandemic. The use of microsampling was advantageous as this allowed to conduct these studies remotely, which was necessary due to the effect of lockdowns. In addition, the (automated) immunoassays allowed high-throughput and efficient analysis, a prerequisite when analyzing large sets of samples. Initially, volumetric sampling was used for this purpose (e.g. using the Mitra device), as discussed in the review by Rudge et al. (Rudge & Kushon, 2021). This was mainly due to the fact that volumetric sampling eliminates the Hct-based area bias in contrast to regular DBS. Currently, reports on the use of DBS for this purpose have also been published combined with, e.g. a modified Roche Elecsys Anti-Sars-CoV-2 assay (Fontaine & Saez, 2021).

5.3 Integration Into a Clinical Laboratory

5.3.1 Requirements, Challenges, and Advantages of Implementation

As outlined in Section 5.1.1, some challenges that were hitherto limiting implementation of DBS sampling have been tackled recently. One of the major bottlenecks using conventional DBS samples was the Hct issue, causing inaccurate results. Since the establishment of Hct prediction technologies from DBS, in combination with volumetric DBS collection, this issue can be resolved. Additionally, guidelines have become available for the validation of DBS-based methods (Capiau et al., 2019). In a clinical laboratory setting, sufficient QC is another important aspect, certainly in the framework of ISO accreditation. In this context, the setup of a PT program for immunosuppressant quantification from dried blood microsamples is a first good step (Stichting Kwaliteitsbewaking Medische Laboratoriumdiagnostiek, 2021). Because of the above-mentioned evolutions in the field of microsampling, some Dutch hospitals now provide DBS-based TDM services to their patients. While regular TDM is usually performed for four distinct drug classes: tricyclic antidepressants (TCA), antibiotics, anticonvulsants, and immunosuppressants (Velghe et al., 2016). DBS-based TDM has hitherto only been adopted for immunosuppressants. Yet, this is relevant, as this drug class represents the largest patient population for TDM, and these patients can really benefit from home-sampling (for other analytes, TDM is partially indicated for hospitalized patients). Several hospitals in the Netherlands have put major efforts to implement DBS for TDM of immunosuppressants in clinical routine such as the UMC Utrecht (2021), the UMC Groningen (2021), the UMC Maastricht (tacrolimus; UMC Maastricht (2018), the Erasmus MC (2021), and the UMC Leiden (2021)). In these hospitals, DBS is offered as a complementary approach to conventional TDM for patients that prefer to sample at home. Also, in cases requiring repeated measurement of drug levels, e.g. for the abbreviated area under the curve estimation for the follow-up of tacrolimus treatment, microsampling can be of added value. Despite the increasing resources and experience, the number of hospitals where DBS is offered as part of the clinical routine has remained rather limited. As far as we are aware, the services currently (end 2021) offered in Dutch Hospital Pharmacies all involve manual DBS extraction and analysis with LC-MS/MS. Only one dedicated, thoroughly validated (and applied) method has been published for the fully automated TDM of immunosuppressants from DBS, by our group, although this automated method is currently not routinely applied (Deprez & Stove, 2021).

Depending on the throughput of the samples, and the priorities of the local hospital laboratory, online automated DBS analysis, as described in Section 5.2, can be beneficial. This especially holds true when envisaging the analysis of large numbers of samples (in the framework of studies or high-volume TDM parameters such as immunosuppressants) in a routine setting. From this point of view, automated units can provide a solution to decrease the hands-on time and increase the throughput, two factors currently limiting clinical implementation on a larger scale. Additionally, only limited training would be needed to deposit the cards into the sample racks, in contrast to when manual extraction is to be performed. However, online DBS extraction is mostly coupled to LC-MS/MS analysis, as outlined in Section 5.2, which requires some expertise to be used (the same argument holds true for conventional LC-MS/MS-based TDM on whole blood/plasma for immunosuppressants, antibiotics, anticonvulsants, and TCA). Current MS software packages are not particularly user-friendly, and operating these instruments requires specialized training. Ideally, an automated online DBS extraction unit could be coupled to a MS analyzer adapted for clinical use. A lot of progress has been made in this area. ThermoFisher Scientific has developed the Cascadion™ SM Clinical Analyzer, enabling automated processing of whole blood or plasma in combination with automated LC-MS/MS analysis (ThermoFischer Scientific, 2021a). In addition, automated processing of the results is possible, with very limited intervention of a laboratory technician. In this way, it is compatible with the automated workflow present in many clinical labs, and it can be integrated in the core lab which is usually operated 24/7. Several clinical labs have already implemented this system for the monitoring of vitamin D and/or immunosuppressants, albeit not yet starting from dried blood microsamples (Barquin-Delpino et al., 2021; Benton et al., 2020; Horber et al., 2021). Shimadzu has also been working on a fully automated sample preparation module to be coupled to LC-MS/MS, the CLAM-2030 (Shimadzu, 2021). At the 14th EBF meeting in November 2021, Franck Saint-Marcoux presented a method using the CLAM-2030 for the automated measurement of tacrolimus starting from HemaPEN samples (Saint-Marcoux, 2021).

In the future, fully automated DBS analyzers compatible with clinical MS analyzers may provide the change needed in the field to convince clinical chemists to integrate this approach in their core laboratories. Even if these technological advances become available, more studies will be needed to demonstrate the equivalence of the results obtained by these automated setups for DBS with those obtained by the current standard of care (plasma or whole blood). Secondly, a thorough cost–benefit analysis will be needed as these fully automated online DBS extraction units (but also clinical MS analyzers) come at a substantial cost and are a significant investment for (smaller-scale) clinical laboratories. This might also be one of the reasons why in the Netherlands, most laboratories still opt for manual (or semi-automated) analysis and no implementation examples of fully automated DBS analysis can be found (to the authors' knowledge). For laboratories with a higher throughput, a fully automated DBS approach might be preferred to create a future-proof laboratory enabling DBS analysis in an automated way. However, for smaller laboratories or laboratories in countries where the access to healthcare is easily accessible and widespread (such as in Belgium), the decision to invest in automation for DBS analysis will depend on whether DBS will remain a minor fraction of the TDM samples or the interest will grow in the upcoming years. Currently, it might be more beneficial (cost-wise) to implement manual DBS analysis (in combination with semi-automated sample extraction), despite the numerous advantages of automation, as is the case in many laboratories in the Netherlands. Nevertheless, if in the future the available systems can be made even more compatible with the current workflow of clinical laboratories, more opportunities can be present in the upcoming years, as outlined below. An extra consideration here is that the proposed solutions should ideally offer random access capabilities, i.e. the possibility to perform different analyses in random access mode, as opposed to batch analysis, which is less compatible with the workflow in clinical laboratories. Whether or not the new *in vitro* diagnostic medical device (IVD) Regulation ((EU) 2017/746), making it more complicated to maintain lab-developed tests will be an accelerator or brake on the penetrance of automated DBS analysis: only future can tell (EUR-Lex, 2017).

5.3.2 Cost-Effectiveness of Implementation of (Automated) DBS Analysis

When considering DBS analysis, we often tend to focus on the patient benefits associated with this approach. Indeed, patient centricity is of utmost importance, particularly during the current pandemic. However, one important point that is often neglected (and on this topic very few studies exist) is the laboratory cost associated with this approach, compared to conventional whole blood or plasma analysis. For a smaller-scale laboratory, the implementation of automated DBS services for TDM purposes would be translated into a substantial additional cost, as already mentioned in Section 5.1.1. Importantly, it is not because the overall picture might give cost-savings that this will automatically mean easy adoption of this approach, as demonstrated by Martial et al. for the TDM of pediatric patients under tacrolimus treatment in the Netherlands, (Martial et al., 2016). More particularly, the calculations by Martial et al. did not consider several laboratory-associated costs that are different for DBS in comparison with conventional whole blood analysis. The authors considered an equal cost for laboratory analysis. When comparing both LC-MS/MS analysis' costs for DBS and whole blood, this price is indeed identical. However, other factors should also be considered. It is evident that patients will save certain costs related to transportation and parking at the hospital when replacing their venous blood draws by home-sampled DBS sent to the lab via postal services. Nevertheless, for example for patients in Belgium, the use of microsampling devices is not currently reimbursed by the public healthcare system, which may lead to an additional cost. Only the analysis itself is reimbursed, regardless the state (dried of liquid) of the blood. Possibly, national healthcare will recognize the benefits of microsampling technologies in certain disease areas in the future and approve a (partial) reimbursement. This is currently the case for lancets, glycaemic control strips, and a glucometer for diabetic patients in Belgium. In addition, depending on the setting, the savings, versus additional costs, may not be evenly spread over all the stakeholders involved. When the added value is not present for the affected party that is the major payer, this results in a barrier. For example, in the current healthcare setting in Belgium, the cost of the blood tube is covered by the laboratory. When in a similar model, the cost of the DBS card, as well as the shipping of the DBS card to the laboratory, would need to be paid by the laboratory, and this would increase the price per sampling device (not even considering a higher cost for subsequent pre-analytical steps). However, when also taking into account the sampling itself (by medically trained personnel) in addition to the sampling material (such as a needle) for a regular blood draw, the total cost for the laboratory is comparable. In addition, the extensive method validation that should be performed for microsample-based methods should also be taken into account in cost estimations.

At the clinical laboratory of Ghent University Hospital, samples arrive via three main transportation ways: (a) a Tempus system (this is a dedicated point-to-point transport line for blood/urine 25 mm tubes, internal sample transportation) (Sarstedt, 2021), (b) a pneumatic transportation system via tubes (internal sample transportation), and (c) courier and postal services (samples arriving from outside the clinic). For the samples arriving in the lab with the latter system, manual sorting and registration of the samples is required, which increases the costs of such samples because of personnel working hours. If DBS would be sent to the laboratory, they would also be subject to manual registration (regardless the automation status of the analysis itself). A solution to avoid this additional cost is to work with pre-barcoded DBS, but this should be compatible with the laboratory information system (LIS). In the framework of SARS-CoV-2 testing, pre-barcoded samples are getting more established in most labs as these are frequently used for this purpose. Alternatively, (also inspired by the current COVID-19 pandemic) an independent, electronic prescribing system can be set up as this is also employed to automatically generate barcodes for SARS-CoV-2 tests which are requested online (Rijksinstituut voor ziekte-en invaliditeitsverzekering (RIZIV), 2020). It should be noted that whole blood samples also arrive via courier or postal services and are subject to the same manual registration process. Additionally, the fact that these

(DBS or whole blood samples) arrive at a fixed time point (i.e. batch-wise) can also be considered an advantage for their subsequent analysis in a batch format. Moreover, to date, DBS or other microsamples are not compatible with pre-analytical automated systems installed in labs for transporting the samples to the connected analyzers or for sorting the samples to perform offline analyses. Analytically, the cost of a laboratory technician can be saved by automating DBS extractions, as this is one of the costliest elements of the laboratory expenses. However, this saving can only be performed when first a significant investment (of several 100,000 euros) is made to purchase an automated extraction unit and a dedicated (clinical) MS analyzer. Obviously, when DBS extraction robots and clinical MS analyzers are more widely used in practice, the acquisition cost will continue to descend over time. When using the DBS-MS 500 for online extraction, the coupled LC-MS/MS system is partially limited to DBS analysis, as switching from the DBS-MS 500 to regular LC must be performed manually. Therefore, quick switching from one analysis to another is not possible. Additionally, at this point, the currently available online DBS extraction units (commercialized by CAMAG, Spark Holland, Gerstel and ThermoFischer Scientific) are not compatible with the above-mentioned automated laboratory tracks as present in many core labs. So, more efforts should be made to make the already available automated extraction units for DBS compatible with a fully automated clinical laboratory setting. A minimum requirement is that DBS samples can be transported via the available laboratory tracks for blood tubes. In that way, they could be sorted and transported through the laboratory in the same automated way as conventional samples. The last step, the placement into the racks of a DBS extraction unit, can then be performed manually. A last point of attention for DBS implementation in general is the fact that often standard laboratory parameters are also requested together with TDM parameters, as already mentioned in Section 5.1. To maintain the benefit associated with the use of DBS for TDM purposes, these standard parameters should also be determined in DBS. Typically, this comes with an increased variability compared to the standard of care (e.g. variability introduced by using a different matrix (dried whole blood vs. plasma), by performing self-sampling, by the Hct effect, etc.). Clinicians should consider whether it is worth to accept this additional analytical uncertainty on the result in exchange for more patient comfort, particularly in countries (such as Belgium) where the healthcare services are quite extensive and easily accessible. In more sparsely populated areas, with a lower availability of hospitals or specialized centers, this benefit might outweigh the potential loss in accuracy and/or precision of the result.

To conclude, the feasibility to implement (automated) DBS analysis in clinical labs depends on multiple factors: (a) the country's healthcare situation, (b) the throughput of the lab, and (c) the envisaged strategy. An overview of the abovementioned advantages and challenges is given in Table 5.2. If only small

Table 5.2 Overview of the potential advantages and hurdles for implementation of (automated) DBS analysis in clinical practice.

Potential advantages and hurdles for implementation of (automated) DBS analysis	
Advantages	**Comments**
Availability of clinical MS[a] analyzers	Specialized training not needed and only very limited intervention of a laboratory technician. Currently, these systems are only compatible with conventional liquid blood/plasma methods.
Cost-savings	Depending on the stakeholder, (automation of) DBS analysis will result in a cost-saving: • The patient: – Costs related to transportation and parking at the hospital versus home-sampled DBS sent to the lab via postal services. – Costs/benefits related to the loss of productivity by time in the hospital versus being able to go to work/school combined with DBS home-sampling • Clinical laboratory: – The total cost of a sampling (using either microsampling devices or a regular blood draw) is comparable (situation in Belgium). – The costs related to the laboratory technicians will be saved using automated DBS extraction.
Decreased hands-on time and increased throughput	Implementation will depend on the throughput of the samples and the priorities of the local hospital laboratory.
Patient centricity	The use of DBS has proven to be beneficial in terms of patient experience.
Hurdles	**Comments**
Administrative practicalities	• Arrival of DBS samples via postal or courier services would result in additional cost for personnel. However, this can easily be overcome by the use of pre-barcoded microsampling devices. • Microsampling devices are currently not compatible with pre-analytical automated systems present in clinical laboratories.
Batch analysis versus random access	An LC-MS/MS system coupled to an automated DBS extraction unit is partially limited to DBS analysis, as switching from automated DBS extraction to conventional LC has to be performed manually. However, since DBS samples will mostly arrive via postal services, batch analysis may be preferred hereby overcoming the former issue.

Table 5.2 (Continued)

Potential advantages and hurdles for implementation of (automated) DBS analysis	
Hurdles	**Comments**
Determination of regular TDM parameters	In routine follow-up of patients, often multiple TDM parameters are determined from a single blood sample. To maintain the benefit associated with the use of DBS for TDM purposes, these standard parameters should also be determined in DBS.
Extensive method validation	Guidelines have become available for the validation of DBS-based methods to facilitate the validation process.
Investments	Depending on the stakeholder, (automation of) DBS analysis will result in additional costs: • The patient: The use of microsampling devices is currently not reimbursed by the public healthcare system (situation in Belgium). • Clinical laboratory: A significant investment is needed to purchase an automated DBS extraction unit and a (clinical) MS analyzer.
IVD Regulation (EU) 2017/746	More complicated to maintain lab-developed tests.
PT programs specific for microsampling applications	E.g. the PT program set up by the KKGT for immunosuppressant quantification from dried microsamples.
Specialized training needed when using LC-MS/MS	This also holds true for conventional LC-MS/MS-based TDM on whole blood/plasma.
The Hct effect	Hct prediction technologies and/or volumetric collection of DBS can be used to overcome this effect.

[a] DBS, dried blood spot(s); EU, European Union; Hct, hematocrit; IVDR, in vitro diagnostic regulation; KKGT, Association for Quality Assessment in Therapeutic Drug Monitoring and Clinical Toxicology; LC-MS/MS, liquid chromatography–tandem mass spectrometry; MS, mass spectrometer; PT, proficiency testing; TDM, therapeutic drug monitoring.

numbers of DBS analyses are envisaged, automation will result in a cost increase and will therefore not be favorable. In contrast, when large numbers of DBS are to be analyzed daily, as in the context of (clinical) studies or in the framework of a "central microsampling laboratory," automation can provide a significant saving in personnel cost and time.

5.4 Conclusions and Future Perspectives

Since the introduction of dried blood microsampling by Guthrie and Susie in the 1960s, the technique has been successfully applied in multiple application fields including screening for metabolic diseases, TDM, toxicology, or doping analysis. Although the advantages are undeniable, a few remaining challenges have hindered the implementation of microsampling in clinical practice, including the Hct effect which complicates the interpretation of the results. The recent introduction of Hct prediction modules (either separate or built-in) for DBS samples enables the correction of any observed Hct-based bias and allows simpler interpretation of DBS-based results. Therefore, future research should focus on the implementation of these modules to confirm their usefulness. Another bottleneck for the routine implementation of DBS sample analysis is the often labor-intensive sample handling and preparation processes. Nevertheless, with the introduction of automated pipetting, punching, and extraction systems, this important bottleneck has also been addressed by reducing or eliminating the workload associated with DBS analysis. To further decrease the manual handling associated with DBS extraction, fully automated online extraction units are also available (DBS-MS 500 HCT and DBSA). Despite their introduction a decade ago, the implementation of such units in clinical practice remains limited. The main hurdles are the significant investment, together with the limited user-friendliness of the coupled LC-MS/MS equipment for implementation in a routine setting. With the increasing availability of clinical MS analyzers, the latter issue could be resolved in the future. For clinical laboratories interested in implementing DBS sample analysis, depending on the throughput and other parameters, it will remain principally a financial consideration whether they settle for manual/semi-automated solutions or whether they embrace fully automated (more costly) solutions. Regardless of the automation status, clinical labs should be encouraged and supported to implement microsampling technologies in general, as this can be of major benefit for their patients. As other alternative microsampling devices also emerge (e.g. VAMS), continuous improvements and efforts will be needed on this end as well, to allow automated analysis of these devices and to be able to make full use of the advantages associated with these alternatives. To summarize, over recent years, advances have been made that brought us a big step closer to achieving a more widespread implementation of patient centric sampling and its associated analytical workflows. However, continuous progress in this field will be needed to reach our ultimate goal of widespread adoption.

Acknowledgments

S.D. would like to thank the Research Foundation-Flanders (FWO) for granting her a PhD fellowship (application number: 11F3121N). L.H. would like to thank the Special Research Fund (BOF) from Ghent University (application number: 01D05220) for granting her a PhD fellowship.

References

Abu-Rabie, P. (2011). Direct analysis of DBS: Emerging and desirable technologies. *Bioanalysis*, *3*, 1675–1678.

Abu-Rabie, P. (2013). New bioanalytical technologies and concepts: Worth the effort? *Bioanalysis*, *5*, 1975–1978.

Abu-Rabie, P., Denniff, P., Spooner, N., Brynjolffssen, J., Galluzzo, P., & Sanders, G. (2011). Method of applying internal standard to dried matrix spot samples for use in quantitative bioanalysis. *Analytical Chemistry*, *83*, 8779–8786.

Abu-Rabie, P., Denniff, P., Spooner, N., Chowdhry, B. Z., & Pullen, F. S. (2015). Investigation of different approaches to incorporating internal standard in DBS quantitative bioanalytical workflows and their effect on nullifying hematocrit-based assay bias. *Analytical Chemistry*, *87*, 4996–5003.

Abu-Rabie, P., & Spooner, N. (2009). Direct quantitative bioanalysis of drugs in dried blood spot samples using a thin-layer chromatography mass spectrometer interface. *Analytical Chemistry*, *81*, 10275–10284.

Abu-Rabie, P., & Spooner, N. (2011). Dried matrix spot direct analysis: Evaluating the robustness of a direct elution technique for use in quantitative bioanalysis. *Bioanalysis*, *3*, 2769–2781.

Analytical Sales and Services. (2021). *DBS-dried blood spot card punch*. Retrieved December 28, 2021, from https://www.analytical-sales.com/product-category/dried-blood-spot/dbs-dried-blood-spot-card-punch/.

Barquin-Delpino, R., Villena-Ortiz, Y., Vima-Bofarull, J., & López-Hellín, J. (2021). *Evaluation of the measurement of immunosuppressants in small volume samples using the Thermo Scientific™ Cascadian™ SM clinical analyzer*. Munich, Germany: EuroMedLab.

Beckman Coulter. (2021). *Biomek NX^P Automated Workstation*. Retrieved December 28, 2021, from https://www.beckman.com/en/liquid-handlers/biomek-nxp?_ga=2.231611039.1458643903.1637591135-1788387561.1637591135.

Benton, S. C., Tetteh, G. K., Needham, S. J., Mucke, J., Sheppard, L., Alderson, S., Ruppen, C., Curti, M., Redondo, M., & Milan, A. M. (2020). Evaluation of the 25-hydroxy vitamin D assay on a fully automated liquid chromatography mass spectrometry system, the Thermo Scientific Cascadion SM Clinical Analyzer with the Cascadion 25-hydroxy vitamin D assay in a routine clinical laboratory. *Clinical Chemistry and Laboratory Medicine*, *58*, 1010–1017.

BSD Robotics. (2021a). *BSD Galaxy A9*. Retrieved December 28, 2021, from https://www.bsdrobotics.com/galaxy9.html.

BSD Robotics. (2021b). *Semi-automated punch instruments*. Retrieved December 28, 2021, from https://www.bsdrobotics.com/.

Büchi Labortechnik. (2021). *NIRFlex N-500™*. Retrieved December 28, 2021, from https://www.buchi.com/en/products/instruments/nirflex-n-500.

CAMAG. (2021). *DBS-MS 500 HCT*. Retrieved July 8, 2021, from https://dbs.camag.com/product/camag-dbs-ms-500-hct#downloads.

Capiau, S., Veenhof, H., Koster, R. A., Bergqvist, Y., Boettcher, M., Halmingh, O., Keevil, B. G., Koch, B. C. P., Linden, R., Pistos, C., Stolk, L. M., Touw, D. J., Stove, C. P., & Alffenaar, J. W. C. (2019). Official International Association for Therapeutic Drug Monitoring and Clinical Toxicology Guideline: Development and validation of dried blood spot-based methods for therapeutic drug monitoring. *Therapeutic Drug Monitoring*, *41*, 409–430.

Capiau, S., Wilk, L. S., Aalders, M. C., & Stove, C. P. (2016a). A novel, nondestructive, dried blood spot-based hematocrit prediction method using noncontact diffuse reflectance spectroscopy. *Analytical Chemistry*, *88*, 6538–6546.

Capiau, S., Wilk, L. S., De Kesel, P. M. M., Aalders, M. C. G., & Stove, C. P. (2018). Correction for the hematocrit bias in dried blood spot analysis using a nondestructive, single-wavelength reflectance-based hematocrit prediction method. *Analytical Chemistry*, *90*, 1795–1804.

Capiau, S. A., Alffenaar, J., & Stove, C. P. (2016b). Alternative sampling strategies for therapeutic drug monitoring. *Clinical Challenges in Therapeutic Drug Monitoring*, *2016*, 279–336.

Cox, H. D., & Eichner, D. (2017). Mass spectrometry method to measure membrane proteins in dried blood spots for the detection of blood doping practices in sport. *Analytical Chemistry*, *89*, 10029–10036.

Crawford, E., Gordon, J., Wu, J. T., Musselman, B., Liu, R., & Yu, S. (2011). Direct analysis in real time coupled with dried spot sampling for bioanalysis in a drug-discovery setting. *Bioanalysis*, *3*, 1217–1226.

Dantonio, P. D., Stevens, G., Hagar, A., Ludvigson, D., Green, D., Hannon, H., & Vogt, R. F. (2014). Comparative evaluation of newborn bloodspot specimen cards by experienced laboratory personnel and by an optical scanning instrument. *Molecular Genetics and Metabolism*, *113*, 62–66.

De Kesel, P. M., Capiau, S., Lambert, W. E., & Stove, C. P. (2014a). Current strategies for coping with the hematocrit problem in dried blood spot analysis. *Bioanalysis*, *6*, 1871–1874.

De Kesel, P. M., Sadones, N., Capiau, S., Lambert, W. E., & Stove, C. P. (2013). Hemato-critical issues in quantitative analysis of dried blood spots: Challenges and solutions. *Bioanalysis*, *5*, 2023–2041.

De Kesel, P. M. M., Lambert, W. E., & Stove, C. P. (2014b). Why dried blood spots are an ideal tool for CYP1A2 phenotyping. *Clinical Pharmacokinetics*, *53*, 763–771.

De Kesel, P. M. M., Lambert, W. E., & Stove, C. P. (2015). Does volumetric absorptive microsampling eliminate the hematocrit bias for caffeine and paraxanthine in dried blood samples? A comparative study. *Analytica Chimica Acta*, *881*, 65–73.

Deglon, J., Thomas, A., Cataldo, A., Mangin, P., & Staub, C. (2009). On-line desorption of dried blood spot: A novel approach for the direct LC/MS analysis of mu-whole blood samples. *Journal of Pharmaceutical and Biomedical Analysis*, *49*, 1034–1039.

Deglon, J., Thomas, A., Daali, Y., Lauer, E., Samer, C., Desmeules, J., Dayer, P., Mangin, P., & Staub, C. (2011). Automated system for on-line desorption of dried blood spots applied to LC/MS/MS pharmacokinetic study of flurbiprofen and its metabolite. *Journal of Pharmaceutical and Biomedical Analysis, 54*, 359–367.

Delahaye, L., Heughebaert, L., Luhr, C., Lambrecht, S., & Stove, C. P. (2021a). Near-infrared-based hematocrit prediction of dried blood spots: An in-depth evaluation. *Clinica Chimica Acta; International Journal of Clinical Chemistry, 523*, 239–246.

Delahaye, L., Veenhof, H., Koch, B. C. P., Alffenaar, J. W. C., Linden, R., & Stove, C. (2021b). Alternative sampling devices to collect dried blood microsamples: State-of-the-art. *Therapeutic Drug Monitoring, 43*, 310–321.

Denniff, P., & Spooner, N. (2014). Volumetric absorptive microsampling: A dried sample collection technique for quantitative bioanalysis. *Analytical Chemistry, 86*, 8489–8495.

Deprez, S., Paniagua-Gonzalez, L., Velghe, S., & Stove, C. P. (2019). Evaluation of the performance and hematocrit independence of the HemaPEN as a volumetric dried blood spot collection device. *Analytical Chemistry, 91*, 14467–14475.

Deprez, S., & Stove, C. (2022). Application of a fully automated dried blood spot method for therapeutic drug monitoring of immunosuppressants: Another step towards implementation of dried blood spot analysis. *Archives of Pathology and Laboratory Medicine* (in press). https://doi.org/10.5858/arpa.2021-0533-OA.

Deprez, S., & Stove, C. P. (2021). Fully automated dried blood spot extraction coupled to liquid chromatography-tandem mass spectrometry for therapeutic drug monitoring of immunosuppressants. *Journal of Chromatography A, 1653*, 462430.

Dib, J., Tretzel, L., Piper, T., Lagojda, A., Kuehne, D., Schanzer, W., & Thevis, M. (2017). Screening for adiponectin receptor agonists and their metabolites in urine and dried blood spots. *Clinical Mass Spectrometry, 6*, 13–20.

Duthaler, U., Berger, B., Erb, S., Battegay, M., Letang, E., Gaugler, S., Krahenbuhl, S., & Haschke, M. (2017). Automated high throughput analysis of antiretroviral drugs in dried blood spots. *Journal of Mass Spectrometry: JMS, 52*, 534–542.

Duthaler, U., Berger, B., Erb, S., Battegay, M., Letang, E., Gaugler, S., Natamatungiro, A., Mnzava, D., Donzelli, M., Krahenbuhl, S., & Haschke, M. (2018). Using dried blood spots to facilitate therapeutic drug monitoring of antiretroviral drugs in resource-poor regions. *The Journal of Antimicrobial Chemotherapy, 73*, 2729–2737.

Duthaler, U., Suenderhauf, C., Gaugler, S., Vetter, B., Krahenbuhl, S., & Hammann, F. (2019a). Development and validation of an LC-MS/MS method for the analysis of ivermectin in plasma, whole blood, and dried blood spots using a fully automatic extraction system. *Journal of Pharmaceutical and Biomedical Analysis, 172*, 18–25.

Duthaler, U., Suenderhauf, C., Karlsson, M. O., Hussner, J., Schwabedissen, H. M. Z., Krahenbuhl, S., & Hammann, F. (2019b). Population pharmacokinetics of oral ivermectin in venous plasma and dried blood spots in healthy volunteers. *British Journal of Clinical Pharmacology*, *85*, 626–633.

Edelbroek, P. M., Van der Heijden, J., & Stolk, L. M. L. (2009). Dried blood spot methods in therapeutic drug monitoring: Methods, assays, and pitfalls. *Therapeutic Drug Monitoring*, *31*, 327–336.

Erasmus MC. (2021). *Apotheek Laboratorium*. Retrieved December 28, 2021, from https://www.erasmusmc.nl/nl-nl/patientenzorg/laboratoriumspecialismen/apotheek.

EUR-LEX. (2017). *Regulation (EU) 2017/746 of the European Parliament and of the Council of 5 April 2017 on in vitro diagnostic medical devices and repealing Directive 98/79/EC and Commission Decision 2010/227/EU*. Retrieved December 28, 2021, from https://eur-lex.europa.eu/legal-content/EN/TXT/?uri=CELEX:32017R0746.

European Medicines Agency. (2011). *Guideline on bioanalytical method validation*. Retrieved December 28, 2021, from http://www.ema.europa.eu/docs/en_GB/document_library/Scientific_guideline/2011/08/WC500109686.pdf.

Fan, L., Katty, W., Kavetskaia, H., & Wu, H. (2014). Automation in dried blood spot sample collection, processing, and analysis for quantitative bioanalysis in pharmaceutical industry. In LI, W. and Lee, M. (ed.) *Dried blood spots: Applications and techniques*. John Wiley & Sons, Inc. 229–234.

Fatch, R., Luginbuhl, M., Cheng, D. M., Gaugler, S., Emenyonu, N. I., Ngabirano, C., Adong, J., Muyindike, W. R., Samet, J. H., Bryant, K., & Hahn, J. A. (2021). Comparison of automated determination of phosphatidylethanol (PEth) in dried blood spots (DBS) with previous manual processing and testing. *Alcohol (Fayetteville, N.Y.)*, *98*, 51–54.

Fingerhut, R., Silva Polanco, M. L., Silva Arevalo Gde, J. & Swiderska, M. A. 2014. First experience with a fully automated extraction system for simultaneous on-line direct tandem mass spectrometric analysis of amino acids and (acyl-)carnitines in a newborn screening setting. *Rapid Communications in Mass Spectrometry: RCM*, 28, 965–973.

Fontaine, E., & Saez, C. (2021). Analysis of SARS-CoV-2 antibodies from dried blood spot samples with the Roche Elecsys Immunochemistry method. *Practical Laboratory Medicine*, *25*, e00234.

Francke, M. I., Van Domburg, B., van de Velde, D., Hesselink, D. A., & De Winter, B. C. M. (2022). The use of freeze-dried blood samples affects the results of a dried blood spot analysis. *Clinical Biochemistry*, *104*, 70–73.

Frey, B. S., Damon, D. E., & Badu-Tawiah, A. K. (2020). Emerging trends in paper spray mass spectrometry: Microsampling, storage, direct analysis, and applications. *Mass Spectrometry Reviews*, *39*, 336–370.

Ganz, N., Singrasa, M., Nicolas, L., Gutierrez, M., Dingemanse, J., Dobelin, W., & Glinski, M. (2012). Development and validation of a fully automated online human dried blood spot analysis of bosentan and its metabolites using the Sample

Card And Prep DBS System. *Journal of Chromatography B, Analytical Technologies in the Biomedical and Life Sciences, 885-886*, 50–60.

Gaugler, S., Al-Mazroua, M. K., Issa, S. Y., Rykl, J., Grill, M., Qanair, A. & Cebolla, V. L. (2019a). Fully automated forensic routine dried blood spot screening for workplace testing. *Journal of Analytical Toxicology*, 43, 212–220.

Gaugler, S., Rykl, J., Wegner, I., Von Daniken, T., Fingerhut, R., & Schlotterbeck, G. (2018). Extended and fully automated newborn screening method for mass spectrometry detection. *International Journal of Neonatal Screening, 4*, 2.

Gaugler, S., Sottas, P. E., Blum, K., & Luginbuhl, M. (2021). Fully automated dried blood spot sample handling and extraction for serological testing of SARS-CoV-2 antibodies. *Drug Testing and Analysis, 13*, 223–226.

Gaugler, S. R.J., Cebolla, V.L. (2018). Fully automated drug screening of dried blood spots using online LC-MS/MS analysis. *Journal of Applied Bioanalysis*, 4, 7–15.

Gaugler, S. R. J.; Cebolla, V.L. 2019b. Validation of an automated extraction procedure for amino acids and acylcarnitines for use with tandem mass spectrometry for newborn screening. *Endocrinology, Diabetes and Metabolism Journal*, 3, 1–9.

GERSTEL. (2021). *Dried Blood Spot Autosampler (DBSA)*. Retrieved December 28, 2021, from https://www.gerstel.com/en/2464.htm.

Gruner, N., Stambouli, O., & Ross, R. S. (2015). Dried blood spots—Preparing and processing for use in immunoassays and in molecular techniques. *Jove-Journal of Visualized Experiments, 97*, 52619.

Heinig, K., Wirz, T., Bucheli, F. & Gajate-Perez, A. (2011). Determination of oseltamivir (Tamiflu (R)) and oseltamivir carboxylate in dried blood spots using offline or online extraction. *Bioanalysis*, 3, 421–437.

Heinig, K., Wirz, T. & Gajate-Perez, A. (2010). Sensitive determination of a drug candidate in dried blood spots using a TLC-MS interface integrated into a column-switching LC-MS/MS system. *Bioanalysis*, 2, 1873–1882.

Hempen, C. M., Maarten Koster, E. H., & Ooms, J. A. (2015). Hematocrit-independent recovery of immunosuppressants from DBS using heated flow-through desorption. *Bioanalysis, 7*, 2019–2029.

Henion, J., Oliveria, R. V., Li, F., Foley, T. P., & Pomponio, R. J. (2013). Dried blood spots: The future. In ZANE, P. and Emmons, G. (ed.) *Microsampling in pharmaceutical bioanalysis*. Future Science Ltd. 48–67.

Horber, S., Peter, A., Lehmann, R., & Hoene, M. (2021). Evaluation of the first immunosuppressive drug assay available on a fully automated LC-MS/MS-based clinical analyzer suggests a new era in laboratory medicine. *Clinical Chemistry and Laboratory Medicine, 59*, 913–920.

Hudson Robotics. (2021). *DBS Dry Blood Spot Processor*. Retrieved December 28, 2021, from http://hudsonrobotics.com/images/Library_Files/Brochures/DBS_Brochure.pdf.

Instech. (2021). *Automated Blood Sampler (ABS2)*. Retrieved December 28, 2021, from https://www.instechlabs.com/products/automated-blood-sampler/abs2.

Jing, J., Shan, Y. H., Liu, Z., Yan, H., Xiang, P., Chen, P. J. & Xu, X. (2022). Automated online dried blood spot sample preparation and detection of anabolic steroid esters for sports drug testing. *Drug Testing and Analysis.*

Lange, T., Thomas, A., Walpurgis, K., & Thevis, M. (2020). Fully automated dried blood spot sample preparation enables the detection of lower molecular mass peptide and non-peptide doping agents by means of LC-HRMS. *Analytical and Bioanalytical Chemistry, 412*, 3765–3777.

Leuthold, L. A., Heudi, O., Deglon, J., Raccuglia, M., Augsburger, M., Picard, F., Kretz, O., & Thomas, A. (2015). New microfluidic-based sampling procedure for overcoming the hematocrit problem associated with dried blood spot analysis. *Analytical Chemistry, 87*, 2068–2071.

Li, Y., Henion, J., Abbott, R., & Wang, P. (2012). Semi-automated direct elution of dried blood spots for the quantitative determination of guanfacine in human blood. *Bioanalysis, 4*, 1445–1456.

Lin, Y., Chen, J. H., He, R., Tang, B., Jiang, L., & Zhang, X. (2020). A fully validated high-throughput liquid chromatography tandem mass spectrometry method for automatic extraction and quantitative determination of endogenous nutritional biomarkers in dried blood spot samples. *Journal of Chromatography A, 1622*, 461092.

Luginbuhl, M., Angelova, S., Gaugler, S., Langin, A., & Weinmann, W. (2020a). Automated high-throughput analysis of tramadol and O-desmethyltramadol in dried blood spots. *Drug Testing and Analysis, 12*, 1126–1134.

Luginbühl, M., Fischer, Y., & Gaugler, S. (2020b). Fully automated optical hematocrit measurement from dried blood spots. *Journal of Analytical Toxicology*. 46, 187–193.

Luginbuhl, M., & Gaugler, S. (2020). The application of fully automated dried blood spot analysis for liquid chromatography-tandem mass spectrometry using the CAMAG DBS-MS 500 autosampler. *Clinical Biochemistry, 82*, 33–39.

Luginbuhl, M., & Gaugler, S. (2021). Addressing new possibilities and new challenges: Automated nondestructive hematocrit normalization for dried blood spots. *Therapeutic Drug Monitoring, 43*, 346–350.

Luginbuhl, M., Gaugler, S., & Weinmann, W. (2019). Fully automated determination of phosphatidylethanol 16:0/18:1 and 16:0/18:2 in dried blood spots. *Journal of Analytical Toxicology, 43*, 489–496.

Luginbuhl, M., Stoth, F., Schrock, A., Gaugler, S., & Weinmann, W. (2021a). Quantitative determination of phosphatidylethanol in dried blood spots for monitoring alcohol abstinence. *Nature Protocols, 16*, 283–308.

Luginbuhl, M., Stoth, F., Weinmann, W., & Gaugler, S. (2021b). Fully automated correction for the hematocrit bias of non-volumetric dried blood spot phosphatidylethanol analysis. *Alcohol (Fayetteville, N.Y.), 94*, 17–23.

Luginbuhl, M., Young, R. S. E., Stoeth, F., Weinmann, W., Blanksby, S. J. & Gaugler, S. (2021c). Variation in the relative isomer abundance of synthetic and biologically derived phosphatidylethanols and its consequences for reliable quantification. *Journal of Analytical Toxicology*, 45, 76–83.

Manicke, N. E., Abu-Rabie, P., Spooner, N., Ouyang, Z., & Cooks, R. G. (2011). Quantitative analysis of therapeutic drugs in dried blood spot samples by paper spray mass spectrometry: An avenue to therapeutic drug monitoring. *Journal of the American Society for Mass Spectrometry*, *22*, 1501–1507.

Martial, L. C., Aarnoutse, R. E., Schreuder, M. F., Henriet, S. S., Bruggemann, R. J., & Joore, M. A. (2016). Cost evaluation of dried blood spot home sampling as compared to conventional sampling for therapeutic drug monitoring in children. *PloS One*, *11*, e0167433.

Martial, L. C., Van den Hombergh, E., Tump, C., Halmingh, O., Burger, D. M., Van Maarseveen, E. M., Bruggemann, R. J., & Aarnoutse, R. E. (2018). Manual punch versus automated flow-through sample desorption for dried blood spot LC-MS/MS analysis of voriconazole. *Journal of Chromatography B-Analytical Technologies in the Biomedical and Life Sciences*, *1089*, 16–23.

Matuszewski, B. K., Constanzer, M. L., & Chavez-Eng, C. M. (2003). Strategies for the assessment of matrix effect in quantitative bioanalytical methods based on HPLC-MS/MS. *Analytical Chemistry*, *75*, 3019–3030.

Miller, J. H., Poston, P. A., & Karnes, H. T. 2011. Direct analysis of dried blood spots by in-line desorption combined with high-resolution chromatography and mass spectrometry for quantification of maple syrup urine disease biomarkers leucine and isoleucine. *Analytical and Bioanalytical Chemistry*, *400*, 237–244.

Munkplast AB. (2021). *YuMi collaborative robot—Robot preparation of Breath Explor samples*. Retrieved December 28, 2021, from https://breathexplor.com/resources/video-robot-preparation-of-breath-explor-samples/.

Neoteryx. (2015). *Neoteryx and Brechbühler partner to automate the mitra microsampling workflow*. Retrieved December 28, 2021, from https://www.neoteryx.com/microsampling-news/neoteryx-and-brechbühler-partner-to-automate-the-mitra-microsampling-workflow.

Neoteryx. (2021). *An economical DBS card automation alternative*. Retrieved December 28, 2021, from https://www.neoteryx.com/microsampling-blog/a-novel-dbs-card-automation-alternative.

Oliveira, R. V., Henion, J., & Wickremsinhe, E. (2014a). Fully-automated approach for online dried blood spot extraction and bioanalysis by two-dimensional-liquid chromatography coupled with high-resolution quadrupole time-of-flight mass spectrometry. *Analytical Chemistry*, *86*, 1246–1253.

Oliveira, R. V., Henion, J., & Wickremsinhe, E. R. (2014b). Automated direct extraction and analysis of dried blood spots employing on-line SPE high-resolution accurate mass bioanalysis. *Bioanalysis*, *6*, 2027–2041.

Oliveira, R. V., Henion, J. & Wickremsinhe, E. R. 2014c. Automated high-capacity on-line extraction and bioanalysis of dried blood spot samples using liquid chromatography/high-resolution accurate mass spectrometry. *Rapid Communications in Mass Spectrometry*, 28, 2415–2426.

Ooms, J. A., Knegt, L., & Koster, E. H. (2011). Exploration of a new concept for automated dried blood spot analysis using flow-through desorption and online SPE-MS/MS. *Bioanalysis*, *3*, 2311–2320.

Oostendorp, M., El Amrani, M., Diemel, E. C., Hekman, D., & Van Maarseveen, E. M. (2016). Measurement of hematocrit in dried blood spots using near-infrared spectroscopy: Robust, fast, and nondestructive. *Clinical Chemistry*, *62*, 1534–1536.

Perkin Elmer. (2004). *Wallac DBS Puncher*. Retrieved December 28, 2021, from https://www.perkinelmer.com/content/relatedmaterials/brochures/fly_dbspuncher.pdf.

Perkin Elmer. (2022). *Panthera-PuncherTM 9 instrument*. Retrieved March 9, 2022, from https://www.perkinelmer.com/nl/product/panthera-puncher-9-2081-0010.

Prolab Instruments GMBH. (2021). *Sample Prep And Card system DBS*. Retrieved December 28, 2021, from https://www.prolab.ch/wp-content/uploads/SCAP-DBS-A4-6page-web.pdf.

Protti, M., Mandrioli, R., & Mercolini, L. (2019). Tutorial: Volumetric absorptive microsampling (VAMS). *Analytica Chimica Acta*, *1046*, 32–47.

Rijksinstituut Voor Ziekte-En Invaliditeitsverzekering (RIZIV). (2020). *Opsporingstesten COVID-19 voorschrijven, reserveren en resultaten ontvangen: Eenvoudig en toegankelijk online*. Retrieved December 28, 2021, from https://www.riziv.fgov.be/nl/nieuws/Paginas/voorschrijven-reserveren-resultaten-ontvangen-eenvoudig-toegankelijk-online.aspx.

Robijns, K., Koster, R. A., & Touw, D. J. (2014). Therapeutic drug monitoring by dried blood spot: Progress to date and future directions. *Clinical Pharmacokinetics*, *53*, 1053.

Rudge, J., & Kushon, S. (2021). Volumetric absorptive microsampling: Its use in COVID-19 research and testing. *Bioanalysis*, *13*, 1851–1863.

Ryona, I., & Henion, J. (2016). A book-type dried plasma spot card for automated flow-through elution coupled with online SPE-LC-MS/MS bioanalysis of opioids and stimulants in blood. *Analytical Chemistry*, *88*, 11229–11237.

Sadones, N., Capiau, S., De Kesel, P. M. M., Lambert, W. E., & Stove, C. P. (2014). Spot them in the spot: Analysis of abused substances using dried blood spots. *Bioanalysis*, *6*, 2211–2227.

Saint-Marcoux, F. (2021). *Measurement of tacrolimus in dried blood spots: A fully automated sample preparation and LCMS method*. 14th EBF symposium (Barcelona), November 24–26.

Sarstedt. (2021). *Tempus600*. Retrieved December 28, 2021, from https://www.tempus600.com/.

Shimadzu. (2021). *CLAM-2030*. Retrieved December 28, 2021, from https://www.shimadzu.com/an/products/liquid-chromatograph-mass-spectrometry/lc-ms-option/clam-2030/index.html.

Shimadzu and Novilytic. (2021). *Noviplex^{TM} cards*. Retrieved December 28, 2021, from https://www.shimadzu.eu/noviplex-cards.

Spark Holland. (2021). *DBS Autosampler*. Retrieved December 28, 2021, from https://www.sparkholland.com/portfolio-item/dbs-autosampler/.

Stichting Kwaliteitsbewaking Medische Laboratoriumdiagnostiek. (2021). *Immunosuppressiva microsampling*. Retrieved December 28, 2021, from https://www.skml.nl/rondzendingen/overzicht/rondzending?id=251.

Stove, C. P., Ingels, A. S. M. E., De Kesel, P. M. M., & Lambert, W. E. (2012). Dried blood spots in toxicology: From the cradle to the grave? *Critical Reviews in Toxicology*, *42*, 230–243.

Tasso INC. (2021). *Tasso-M20*. Retrieved December 28, 2021, from https://www.tassoinc.com/tasso-m20.

Taylor, R. R., Hoffman, K. L., Schniedewind, B., Clavijo, C., Galinkin, J. L. & Christians, U. 2013. Comparison of the quantification of acetaminophen in plasma, cerebrospinal fluid and dried blood spots using high-performance liquid chromatography-tandem mass spectrometry. *Journal of Pharmaceutical and Biomedical Analysis*, 83, 1–9.

Thermofischer Scientific. (2019). *SPOT ON The faster path to MS analysis—Verispray PaperSpray Ion Source*. Retrieved December 28, 2021, from https://assets.thermofisher.com/TFS-Assets/CMD/brochures/br-65405-verispray-paperspray-ion-source-br65405-en.pdf.

Thermofischer Scientific. (2021a). *Clinical Mass Spectrometry Solutions by Thermo Fisher Scientific*. Retrieved December 28, 2021, from https://www.thermofisher.com/clinical-mass-spectrometry/eu/en/solutions.html.

Thermofischer Scientific. (2021b). *Transcend DSX-1 System: Automated dried spot analysis*. Retrieved December 28, 2021, from https://www.thermofisher.com/document-connect/document-connect.html?url=https%3A%2F%2Fassets.thermofisher.com%2FTFS-Assets%2FCMD%2Fbrochures%2Fbr-000093-transcend-dsx-1-system-br000093-en.pdf.

Thevis, M., Kuuranne, T., Dib, J., Thomas, A., & Geyer, H. (2020). Do dried blood spots (DBS) have the potential to support result management processes in routine sports drug testing? *Drug Testing and Analysis*, *12*, 704–710.

Thomas, A., Deglon, J., Steimer, T., Mangin, P., Daali, Y., & Staub, C. (2010). On-line desorption of dried blood spots coupled to hydrophilic interaction/reversed-phase LC/MS/MS system for the simultaneous analysis of drugs and their polar metabolites. *Journal of Separation Science*, *33*, 873–879.

TOMTEC. (2021). *Tomtec DBS*. Retrieved December 28, 2021, from http://www.tomtec.com/dried-blood-spots.html.

Tretzel, L., Thomas, A., Piper, T., Hedeland, M., Geyer, H., Schanzer, W., & Thevis, M. (2016). Fully automated determination of nicotine and its major metabolites in whole blood by means of a DBS online-SPE LC-HR-MS/MS approach for sports drug testing. *Journal of Pharmaceutical and Biomedical Analysis, 123*, 132–140.

U.S. Department of Health and Human Services Food and Drug Administration. (2018). Bioanalytical method validation. Guidance for industry.

UMC Groningen. (2021). *Wat is medicatiemeting via vingerprik—Dried blood spot (DBS)?* Retrieved December 28, 2021, from https://www.umcg.nl/NL/UMCG/AFDELINGEN/ZIEKENHUISAPOTHEEK/PATIENTEN/VINGER-PRIKKEN/Paginas/default.aspx.

UMC Leiden. (2021). *Medicatiemeting thuis via een vingerprik*. Retrieved December 28, 2021, from https://www.lumc.nl/org/transplantatie-centrum/patienten/praktische-informatie/2007342/.

UMC Maastricht. (2018). *Tacrolimus bloedspot*. Retrieved December 28, 2021, from https://diagnostiekenadvies.mumc.nl/analyses/tacrolimus-bloedspot.

UMC Utrecht. (2021). *Thuis bloedspot afnemen*. Retrieved December 28, 2021, from https://assets-eu-01.kc-usercontent.com/546dd520-97db-01b7-154d-79bb6d950a2d/027b5e3d-c965-469e-8dbc-b4c5a2ecf127/Bloedspot-prikken_LoTX.pdf.

Van Berkel, G. J., & Kertesz, V. (2009). Application of a liquid extraction based sealing surface sampling probe for mass spectrometric analysis of dried blood spots and mouse whole-body thin tissue sections. *Analytical Chemistry, 81*, 9146–9152.

Van de Velde, D., Van der Graaf, J. L., Boussaidi, M., Huisman, R., Hesselink, D. A., Russcher, H., Kooij-Egas, A. C., Van Maarseveen, E., & De Winter, B. C. M. (2021). Development and validation of hematocrit level measurement in dried blood spots using near-infrared spectroscopy. *Therapeutic Drug Monitoring, 43*, 351–357.

van den Broek, I., Fu, Q., Kushon, S., Kowalski, M. P., Millis, K., Percy, A., Holewinski, R. J., Venkatraman, V., & Van Eyk, J. E. (2017). Application of volumetric absorptive microsampling for robust, high-throughput mass spectrometric quantification of circulating protein biomarkers. *Clinical Mass Spectrometry, 4-5*, 25–33.

Van Uytfanghe, K., Heughebaert, L., & Stove, C. P. (2021). Self-sampling at home using volumetric absorptive microsampling: Coupling analytical evaluation to volunteers' perception in the context of a large scale study. *Clinical Chemistry and Laboratory Medicine, 59*, e185–e187.

Veenhof, H., Koster, R. A., Brinkman, R., Senturk, E., Bakker, S. J. L., Berger, S. P., Akkerman, O. W., Touw, D. J., & Alffenaar, J. C. (2019). Performance of a web-based application measuring spot quality in dried blood spot sampling. *Clinical Chemistry and Laboratory Medicine, 57*, 1846–1853.

Velghe, S., Capiau, S., & Stove, C. P. (2016). Opening the toolbox of alternative sampling strategies in clinical routine: A key-role for (LC-)MS/MS. *Trac-Trends in Analytical Chemistry, 84*, 61–73.

Velghe, S., Delahaye, L., Ogwang, R., Hotterbeekx, A., Colebunders, R., Mandro, M., Idro, R. & Stove, C. P. 2020. Dried blood microsampling-based therapeutic drug monitoring of antiepileptic drugs in children with nodding syndrome and epilepsy in Uganda and the democratic Republic of the Congo. *Therapeutic Drug Monitoring*, 42, 481–490.

Velghe, S., Delahaye, L., & Stove, C. P. (2019a). Is the hematocrit still an issue in quantitative dried blood spot analysis? *Journal of Pharmaceutical and Biomedical Analysis, 163*, 188–196.

Velghe, S., Deprez, S., & Stove, C. P. (2019b). Fully automated therapeutic drug monitoring of anti-epileptic drugs making use of dried blood spots. *Journal of Chromatography A, 1601*, 95–103.

Velghe, S., & Stove, C. P. (2018a). Evaluation of the capitainer-B microfluidic device as a new hematocrit-independent alternative for dried blood spot collection. *Analytical Chemistry, 90*, 12893–12899.

Velghe, S., & Stove, C. P. (2018b). Volumetric absorptive microsampling as an alternative tool for therapeutic drug monitoring of first-generation anti-epileptic drugs. *Analytical and Bioanalytical Chemistry, 410*, 2331–2341.

Verplaetse, R., & Henion, J. (2016a). Hematocrit-independent quantitation of stimulants in dried blood spots: Pipet versus microfluidic-based volumetric sampling coupled with automated flow-through desorption and online solid phase extraction-LC-MS/MS bioanalysis. *Analytical Chemistry, 88*, 6789–6796.

Verplaetse, R., & Henion, J. (2016b). Quantitative determination of opioids in whole blood using fully automated dried blood spot desorption coupled to on-line SPE-LC-MS/MS. *Drug Testing and Analysis, 8*, 30–38.

VeruTech AB. (2019). *The DiLab® AccuSampler®*. Retrieved December 28, 2021, from http://www.verutech.com/accusampler_std.html.

Villar, L. M., Cruz, H. M., Deodato, R. M., Miguel, J. C., Da Silva, E. F., Flores, G. L., & Lewis-Ximenez, L. L. (2019). Usefulness of automated assays for detecting hepatitis B and C markers in dried blood spot samples. *BMC Research Notes, 12* (523).

Wagner, M., Tonoli, D., Varesio, E., & Hopfgartner, G. (2016). The use of mass spectrometry to analyze dried blood spots. *Mass Spectrometry Reviews, 35*, 361–438.

Wilhelm, A. J., Den Burger, J. C. G., & Swart, E. L. (2014). Therapeutic drug monitoring by dried blood spot: Progress to date and future directions. *Clinical Pharmacokinetics, 53*, 961–973.

Wiseman, J. M., Evans, C. A., Bowen, C. L., & Kennedy, J. H. (2010). Direct analysis of dried blood spots utilizing desorption electrospray ionization (DESI) mass spectrometry. *The Analyst, 135*, 720–725.

World Anti-Doping Agency. (2021). *Dried Blood Spots (DBS) for doping control: Requirements and procedures for collection, transport, analytical testing and storage*. Retrieved December 28, 2021, from https://www.wada-ama.org/sites/default/files/item_6_3_3_attach_1_finalversion_td2021dbs_final.pdf.

Yannell, K. E., Kesely, K. R., Chien, H. D., Kissinger, C. B., & Cooks, R. G. (2017). Comparison of paper spray mass spectrometry analysis of dried blood spots from devices used for in-field collection of clinical samples. *Analytical and Bioanalytical Chemistry*, *409*, 121–131.

Yeung, L., Ichetovkin, M., Yang, Y., & Dellatore, S. (2019). *Implementation of Tecan® automation with Neoteryx Mitra® volumetric abosrptive microsampling (VAMS) for reagent preparation for an Anti-Drug Antibody (ADA) assay*. Washington, DC: Society for Laboratory Automation and Screening (SLAS).

Yuan, L., Zhang, D. X., Aubry, A. F., & Arnold, M. E. (2012). Automated dried blood spots standard and QC sample preparation using a robotic liquid handler. *Bioanalysis*, *4*, 2795–2804.

6

Over 50 Years of Population-Based Dried Blood Spot Sampling of Newborns; Assuring Quality Testing and Lessons Learned

Amy M. Gaviglio[1], Kristina Mercer[2], Konstantinos Petritis[2], Carla D. Cuthbert[2] and Suzanne K. Cordovado[2]

[1] *4ES Corporation, San Antonio, TX, USA*
[2] *Centers for Disease Control and Prevention, Atlanta, GA, USA*

6.1 Overview of Population-Based Newborn Screening

Newborn screening (NBS) was named one of the 10 greatest public health achievements of the first decade of the twenty-first century by the US Centers for Disease Control and Prevention (CDC) (Centers for Disease Control and Prevention, 2011). It is the largest population-based screening program and is provided to all US newborns by state- and territory-sponsored public health programs to identify those babies at risk for a panel of treatable diseases. State-mandated NBS initially began in 1963 in Massachusetts for a single disease, phenylketonuria (PKU)—a genetic disease that, if left untreated, causes permanent intellectual disability, seizures, and behavioral issues (Guthrie and Susi, 1963). The rationale for screening all newborns for PKU is straightforward; when a newborn is identified prior to the onset of symptoms, the infant can be put on a special low phenylalanine diet, thereby preventing the severe intellectual impairment associated with PKU. Following the first successful implementation of statewide screening in Oregon and first statewide mandated screening in Massachusetts, Dr. Robert Guthrie, developer of the dried blood spot (DBS) PKU test, along with other stakeholders pushed state legislators to mandate NBS for PKU. As a result of this advocacy, all babies born in the United States were being screened for PKU by 1979 (Paul and Brosco, 2013; Polk et al., 2003–2008).

Additional diseases were added to NBS panels over time, many based on population-level screening criteria devised by Wilson and Jungner (1968). These

Patient Centric Blood Sampling and Quantitative Bioanalysis, First Edition.
Edited by Neil Spooner, Emily Ehrenfeld, Joe Siple, and Mike S. Lee.

criteria take into account factors, such as: severity of disease; whether a natural history is known; availability and suitability of a screening test; efficacy of clinical treatment; inability to clinically detect the disease at birth; and a balanced comparison between the cost of population-based testing and the costs associated with care of individuals with untreated disease. For more than 57 years, NBS for diseases, such as PKU, has allowed public health screening programs to successfully identify babies at risk for diseases that would result in disability and/or death if not treated early and before the onset of signs and symptoms.

Before 2005, in the absence of a federally recommended NBS panel, the number of diseases screened was inconsistent across all states. In 2003, 46 states were screening for six diseases, whereas a few were screening for more than 20 (March-of-Dimes, 2004). To help facilitate a more harmonized approach across state programs, as well as provide guidance on how to systematically evaluate diseases for inclusion in NBS, the Maternal and Child Health Bureau of the Health Resources and Services Administration (HRSA), and the American College of Medical Genetics (ACMG) published a report that sought to establish a uniform screening panel of diseases (American College of Medical Genetics Newborn Screening Expert Group, 2006). This panel of diseases is now known as the Recommended Uniform Screening Panel, or RUSP, and continues to be administered by HRSA (Health Resources & Services Administration, 2020). When a new disease is proposed for inclusion on the RUSP, the Health and Human Services' established Advisory Committee on Heritable Disorders in Newborns and Children performs an extensive evidence-based review and, if the committee recommends inclusion, it is then sent to the US Secretary of Health and Human Services, who ultimately decides whether or not to include the disease on the RUSP. As of August 2023, the RUSP consists of 37 core and 26 secondary diseases. Secondary diseases are those that are identified unintentionally when screening for or confirming a core disease and do not necessarily meet all the criteria for population-level screening. While the diseases on the RUSP are recommended for NBS, it is important to note that the adoption and implementation of any new disease are still decided individually by each state. Thus, while the RUSP has helped ease disparate screening across state programs, it does not mandate screening of the recommended diseases, so discrepancies still exist.

While the RUSP is a US-centric mechanism for adding diseases to screening panels, NBS programs worldwide often look to this work to inform their own decisions. However, just like each state and territory in the United States must individually implement screening, this is also the case globally, where each jurisdiction makes their own decisions on infrastructure and which diseases will be screened. The International Society of Neonatal Screening helps guide NBS programs and is currently made up of members from over 70 countries. In general, the premise of NBS is universal across the world with programs utilizing a DBS matrix to screen newborns for treatable diseases, though the timing of blood spot collection, diseases screened for, and clinical follow-up paradigms differ.

Almost all the core and secondary diseases that are part of NBS are detected using newborn blood collected at 24–48 hr of life. The blood is typically collected by a heel stick and placed on absorbent filter paper to dry for a minimum of 3 hours, as described in the Clinical and Laboratory Standard Institute's (CLSI) NBS01 guidance document (De Jesus et al., 2021a). This matrix is called a DBS and is subsequently shipped to the state's public health laboratory or their contract designee for testing (De Jesus et al., 2021a). Two NBS tests, hearing screening and pulse oximetry screening for critical congenital heart disease, do not use the DBS for testing. Rather, these screens are point-of-care tests that are performed and completed in the birthing facility with public health-based oversight to ensure compliance, timely diagnosis, and appropriate intervention.

Interest in DBS testing and other microsampling collection devices has been growing both in the public health and diagnostic community as measured by the number of publications per year that contain the term "dried blood spot" alone or in combination with the term "newborn screening" (Figure 6.1). This figure demonstrates the increased publication activity surrounding DBS over the past

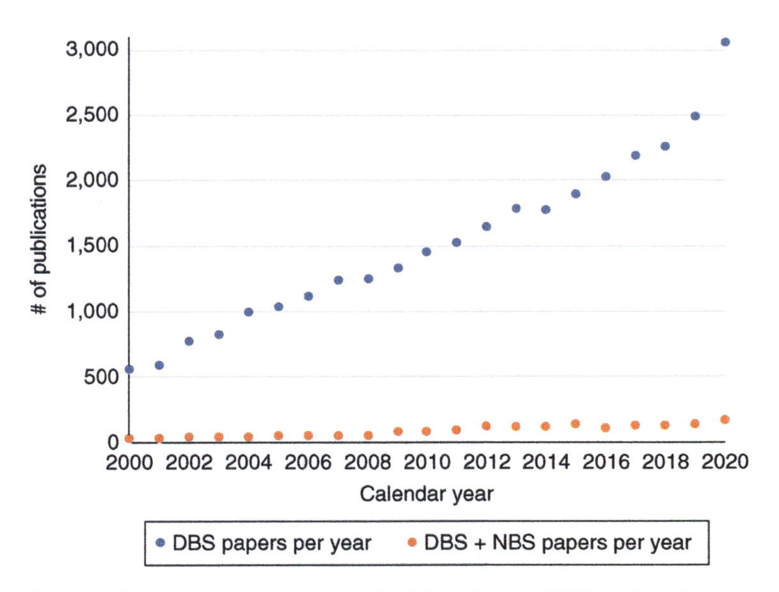

Figure 6.1 Publications featuring dried blood spots (DBS) and newborns screening (NBS). Blue dots: Number of publications (research and review journal articles only) per year from 2000 to 2020 that contain the term "dried blood spot." Orange dots: Number of publications per year from 2000 to 2020 that contained both terms "dried blood spot" and "newborn screening."

20 years (Science Direct, 2020). It is noteworthy that there has been a dramatic increase in publications in the last 10 years on DBS outside of NBS, indicating that there is increasing interest in using DBS in other fields of science and for other types of testing beyond NBS. In the last 57 years, NBS has not only evolved in the number and types of diseases screened but also in the technologies used to accomplish the testing. The first NBS test began with the simple inhibition assay for PKU as developed by Dr. Robert Guthrie (1961) and has since transitioned to incorporating some of the most contemporary, cutting-edge biochemical and molecular technologies. As of 2020, approximately 13,000 babies in the United States are identified each year through public health NBS programs, many of whom are saved from significant morbidity or mortality (Sontag et al., 2020). While there has been vast progression in the technologies utilized in NBS, primarily in response to the addition of new diseases to the RUSP, one thing has remained constant: the use of the DBS matrix for all RUSP disorders other than critical congenital heart disease and hearing loss (Grosse et al., 2017). Indeed, the DBS matrix continues to survive the test of time because of its broad-spectrum utility. Considerations for use of this matrix across a number of assays and platforms as well as numerous lessons learned throughout the history of NBS are further elucidated below.

6.2 Public Perceptions of NBS

NBS has generally been accepted by the public, largely due to its success in significantly reducing morbidity and mortality in newborns (Miller et al., 2015). Communication efforts around NBS have focused on the large potential benefits of screening and early diagnosis over relatively small, identified risks. Typically, the risks attributed to NBS center around the general risks associated with the capillary blood draw and the potential harms that may arise with false positive or ambiguous screening results (Gurian et al., 2006).

As NBS is often a newborn's first blood draw, parents may be concerned about pain experienced by the newborn. A number of studies have examined pain responses in neonates comparing capillary blood draws (such as a heel lance) versus venipuncture. Conflicting outcomes have been reported from these studies, with some studies indicating that venipuncture results in less pain and others coming to the opposite conclusion (Mohamed et al., 2019; Shah and Ohlsson, 2011). To minimize pain, a number of strategies have been suggested, including oral dextrose, skin-to-skin contact, and swaddling (Morrow et al., 2010). Regardless of collection technique or pain mitigation strategy used, any pain experienced for the purpose of NBS is usually quick and temporary.

Public perceptions surrounding an expanded NBS scope (e.g. including later-onset diseases and/or whole genome sequencing) along with the ability for programs to store and reuse residual DBS specimens are varied (DeLuca, 2018; Rothwell et al., 2012). Whereas some individuals laud the potential of residual DBS samples to advance NBS programs and public health research, others have expressed concerns around the lack of consent and potential for misuse (Couzin-Frankel, 2009). This controversy was best summed up by Jennifer Couzin-Frankel in her 2009 article titled *Newborn blood collections: Science Gold Mine, Ethical Minefield*.

Understanding that the public and families are the most important stakeholder in NBS, programs will need to continue to engage and communicate with them to ensure trust and transparency in the process. The increasing complexity of NBS programs—across the spectrum of pre-analytical, analytical, and post-analytical realms—necessitates this renewed focus on education and awareness.

6.3 Characteristics of the DBS Matrix and Its Utility in NBS

The DBS matrix has been adopted for use by NBS programs worldwide owing to several factors that make this approach, over other blood sampling methods, ideal for wide-scale, population-based use. Lim (2018) outlines the benefits of DBS over more classic sampling methods like venipuncture and indicates that the following aspects contribute to the utility of DBS for population and global health: reduced workforce requirements, smaller blood volume needed, direct and indirect application options, simplified transport, shipment, and disposal, simplified biobanking for retrospective analyses, and easily mitigated risks of contamination and exposure. Some of these factors are further explicated in the following section.

The convenience, flexibility, and stability of blood collected and dried on filter paper have been shown over the years to be an ideal collection method for detecting infectious disease, biomarkers, environmental pollutants, drug adherence, and alterations in DNA or RNA sequences (McClendon-Weary et al., 2020). The first use of this collection method dates back over 100 years ago to 1913 when Ivar Christian Bang used cellulose paper to collect blood to identify risk factors for hyperglycemia through analysis of blood glucose levels (Schmidt, 1986). For the following 50 years, DBS samples were used mostly as collection and storage media for the detection of infectious diseases, such as syphilis, measles, and poliovirus, using serology testing (Chapman, 1924; Guo, 1938; Hannon and Therell, 2014; Zimmermann, 1939). DBS aliquots for these tests were taken using copper tubes and scissors. The idea of using something as simple as an ordinary ticket hole puncher to produce standardized dried blood aliquots for testing was introduced in 1958 by Pellegrino and Brener (1958).

In 1961, Doctors Robert Guthrie and Ada Susi expanded on previous methods, collecting blood from a newborn heel puncture onto Schleicher & Schuell # 903 (now referred to as Grade 903, formerly Whatman 903) filter paper to screen for a rare genetic disease, PKU. The collection of blood on filter paper from a heel prick provided a relatively painless and minimally invasive way to collect and transport blood from newborns. The use of newborn DBS for population-based screening was revolutionary, enhancing public health efforts worldwide, and continues to be used for the detection of an increasing number of congenital diseases and conditions (Bhattacharya et al., 2014; Malsagova et al., 2020).

Based on standards described in the CLSI Standard NBS01Ed7: 2021, filter paper for the purpose of NBS should be made of 100% pure cotton linters with no wet-strength additives, have a pH between 5.7 and 7.5 and a maximum ash content of 0.1% (De Jesus et al., 2021a). Grade 903 and PerkinElmer 226 (previously referred to as Ahlstrom 226) are high-quality, cellulose-based filter papers and are the only two that have been approved by the US Food and Drug Administration (FDA) as Class II Medical Devices for the collection of blood (45C.F.R.§862.1675, 2022) and that meet the specifications for NBS as per CLSI guidelines (De Jesus et al., 2021a). As part of the registration of these two filter papers by the FDA as class II collection devices, it is important to note that the filter paper has been given an expiration date and should not be used past that date. In addition, it is important that there be uniform spreading of blood on the filter paper to ensure an equal distribution of analytes throughout the DBS. A study by Mei et al., which evaluated both Grade 903 and PerkinElmer 226 filter paper, found that the interference of filter paper fibers on blood spreading accounted for less than 2% of variability in blood homogeneity (Mei et al., 2001). Thus, when these two filter papers are used according to a standardized protocol, they both meet the qualifications for NBS and can effectively be used for the collection of newborn blood that can then be tested using a variety of laboratory methods (Chace and De Jesus, 2014; Urv and Parisi, 2017).

6.3.1 Recovery of Biochemical and Molecular Analytes from DBS for NBS

The utility of the DBS matrix for NBS depends largely on qualities of the filter paper onto which the blood is applied. The filter paper must have the capacity to capture and stabilize whole blood biochemical and genomic analytes used to detect NBS diseases. In addition, the laboratory must be able to recover sufficient and measurable quantities of the different NBS analytes to ensure a robust and reliable test. Mei et al. assessed the recovery of 26 NBS biochemical analytes using multiple analyte-enriched blood spotted onto Grade 903 and PerkinElmer 226 filter paper (Mei et al., 2010). Recovery was measured for amino acids and acylcarnitines by tandem mass spectrometry (MS/MS) and for immunoreactive trypsinogen (IRT), thyroid-stimulating hormone (TSH), 17-α-hydroxyprogesterone (17OHP), and thyroxine (T4) by immunoassay. The authors observed a linear dose response for both types of filter paper when mean assay values were plotted against the expected concentrations of analyte. Based on their results, they concluded that blood, when properly collected on the filter paper to generate DBS, yielded consistent biochemical analyte measurements for both FDA-approved collection devices (Mei et al., 2010).

Retention, stability, and recovery of DNA from filter paper are also critical to ensure reliable results for molecular NBS assays, including real-time polymerase chain reaction (PCR), genotyping assays, and gene sequencing. The discovery of DNA immobilization and stability on cellulose dates back to 1978 (Bagioni et al., 1978). Twenty years later, in 1998, Su et al. investigated the efficiency of DNA recovery from secondary fibril-associated cellulose (SF-cellulose) fibers contained in Grade 903 filter paper. The authors reported the ability to recover 70–90% of DNA from a sample containing 0.5 μg of genomic DNA. Furthermore, they recovered DNA ranging in size from 100 bp to 50 kb, suggesting that DNA was stable on the cellulose matrix and high-molecular-weight DNA could be recovered (Su & Comeau, 1999).

Over the past few decades, many methods have been developed to extract genomic DNA from filter paper using manual or automated extraction techniques. Depending on the method used, the resulting DNA ranges in quality from crude to more purified and has been successfully employed for a variety of molecular NBS assays, including simple genotyping or real-time PCR tests (Lang et al., 2012; McCabe et al., 1987; Saavedra-Matiz et al., 2013), to next-generation sequencing (Baker et al., 2009, 2016; Hendrix et al., 2016, 2020), while achieving the throughput required of population-based screening (Saavedra-Matiz et al., 2013). In addition, research studies have shown that DNA extracted from DBS works well in methylome and transcriptome studies (Reust et al., 2018; Staunstrup et al., 2016).

6.3.2 Evaluation of Lot-to-Lot Variability in NBS Collection Devices

Since DBS samples are created from approximately 3.8 million US newborns every year, it is important that quality control (QC) measures are in place to ensure that consistent test results are obtained over time. To examine potential lot-to-lot variability in filter paper, Mei et al. spotted blood enriched with ^{125}I-L-thyroxine to measure serum volume in a 3.2 mm disc punch (Mei et al., 2010). DBS were tested from eight lots of Grade 903 produced over a period of 10 years and six different lots of PerkinElmer 226 produced over 6 years. Lot-to-lot differences were assessed based on the equivalency of serum volumes. Remarkably, the mean serum volumes were almost the same for the Grade 903 and PerkinElmer 226 filter papers at 1.474 and 1.476 mL, respectively. Additionally, all mean serum volumes fell within the standardized acceptable volume range of 1.37–1.71 mL, and the differences in mean serum volumes between lots were not statistically significant (Mei et al., 2010). These findings are further supported by a study of Grade 903 filter paper that examined within and between lot of variation for the detection of small molecules by liquid chromatography with tandem mass spectrometry (LC-MS/MS). Data from multiple analytical runs revealed that the largest deviation in results between filter paper lots fell well below the predetermined acceptance criteria for quantitative measurement of the test molecules (Luckwell et al., 2013).

6.3.3 Effect of Hematocrit on DBS Homogeneity, Data Analysis, and Results

Hematocrit, the proportion of red blood cells to total volume of whole blood, differs based on sex, age, and general health status, with a normal range of 28–67% (O'Mara et al., 2011). For full-term newborns, the range in percent hematocrit is 42–65% (Hall et al., 2015). Because percent hematocrit affects the viscosity of blood, equal volumes of blood with varying hematocrits will spread differently on filter paper. Several studies have evaluated the effect of hematocrit on homogeneity and analyte measurements. DBS made using blood with a hematocrit of 30% contained 47% more serum volume per punch than blood with a hematocrit of 70%. Another study by Hall et al. compared serum volumes in DBS from spots made using blood with hematocrits of 40, 45, 50, 55, 60, and 65%. Their results support those of Mei et al, revealing that percent hematocrit of spotted blood is inversely proportional to the amount of serum in a DBS punch and that percent hematocrit is positively correlated with blood volume per DBS punch (Hall et al., 2015). Adam et al. also evaluated the effect of hematocrit and blood spot volume on quantitative measurements of phenylalanine by comparing DBS reference materials with differing hematocrits and blood spot volumes. The analyte recovery was only slightly dependent on percentage hematocrit. Additionally, recovery was 10.9% higher for the 100 ml versus the 35-mL blood spot when testing center punches of DBS made using blood with equivalent phenylalanine concentrations. The authors noted they expect similar results when measuring other serum-based NBS analytes (Adam et al., 2000). Thus, both hematocrit and blood volume should be controlled when preparing DBS reference materials to ensure even distribution of analytes across DBS and to avoid variation in analytical results, both within and between DBS specimens.

Of course, hematocrit levels within individual newborns cannot be controlled and routinely measured, so instead, other indicators are used to help assess whether a DBS is acceptable for screening. For example, discrepancies in DBS volumes, above or below the 75–100 μL target, can be identified by evaluating the degree to which the blood fills the printed circle on the filter paper. Differences in hematocrit can also be identified by assessing the borders of the DBS. DBS made from lower-hematocrit blood will have smoother edges compared to those made from higher-hematocrit blood, which tend to have irregular edges (Hall et al., 2015).

6.3.4 DBS Specimen Collection Transport and Safe Handling

In addition to the elements mentioned above, the accuracy of DBS testing depends on standardized collection and acceptability parameters. Since proper DBS collection is so critical to NBS, much effort has gone into generating clear standards on how to generate and safely handle newborn DBS specimens (De Jesus et al., 2021a). To facilitate education and training of specimen collectors, Schleicher and Schuell, the original manufacturer of Grade 903 filter paper published a guidance document known as the Simple Spot Check that provides images of DBS to illustrate valid and invalid DBS. Variations of this document continue to be utilized by programs throughout the world in an effort to help assure quality specimen collection.

Because it is expected that standard size punches taken from a DBS specimen contain a certain volume of blood (calculated to be 3.42 μL at 55% hematocrit for a 3.2-mm punch per CLSI standards), deviation from this expected volume as a result of poor-quality DBS can result in inaccurate results or the inability to appropriately perform testing. Indeed, analysis of the impact of quality on NBS results has found that poor-quality specimens result in more heterogeneous results both between and within the DBS in the sample (George and Moat, 2016). Despite standardized training efforts and guidelines, unacceptable DBS specimens are still routinely submitted to NBS programs (De Jesus et al., 2021a). In many cases, these specimens represent the desire of the collector to submit a specimen at the expense of proper collection and handling techniques.

Table 6.1 provides examples of the most commonly observed unacceptable specimen types from experienced and trained collectors of newborn DBS specimens. However, these do not represent the full gamut of causes of unacceptable specimens received by NBS programs over the years. Other, less common, though perhaps more intriguing causes of unacceptable specimens have included the use of a phlebotomist's own blood to supplement specimens that might otherwise be unacceptable (e.g. adding additional blood to a specimen that is insufficiently saturated with a newborn's blood) as well as simply coloring in the filter paper circles with a red pen or marker.

Potential issues with DBS specimens are not limited to the act of blood collection. Proper drying, handling, transport, and storage of the specimens are also pivotal to the utility of the DBS matrix. Specimens must be dried horizontally before being packaged for transport as drying samples vertically can result in a blood

Table 6.1 Observed Causes of Unacceptable DBS Specimens.

Observed specimen issue	Possible causes	Image
Layered blood	Excess blood applied; usually as a result of multiple drops of blood placed in circle or placing blood on both sides of the filter paper	
Serum rings	Excessive squeezing or milking of the heel; filter paper coming in contact with alcohol; application of blood with a capillary tube containing anticoagulant	
Smeared blood with poor saturation	Touching the heel to the filter paper	
Filter paper abraded or torn	Application of blood using a capillary tube or other application device	 * Angled to better illustrate the abrasion.
Poor saturation	Not allowing filter paper to completely fill circle or before it has soaked through to the other side; allowing filter paper to come into contact with hands, lotion, or powders	
Blood on overlay biohazard cover	Biohazard flap closed over blood before it has dried	

Contamination	Allowing filter paper to come in contact with hands, or other substances such as alcohol, formula, water, lotion, powders, etc.	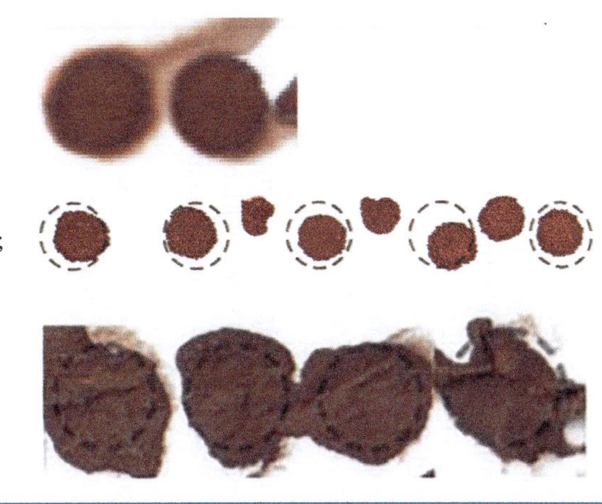
Not enough blood	Removing filter paper before blood has completely filled circle; not allowing an ample blood drop to form; inadequate collection procedure	
Wrinkled or creased filter paper	Filter paper becomes creased or torn when blood is still wet	

gradient or pooling of blood that leads to heterogeneity within each DBS. Additionally, specimens should be air-dried for at least 3 hours prior to being packaged to avoid some of the issues mentioned above, specifically, the wicking of blood onto the biohazard overlay flap. In some cases, collection and/or shipping staff have tried to expedite the drying process by placing specimens under a heat lamp or on a windowsill. These tactics will result in either unacceptable specimens or false results, particularly in heat-labile enzymes, such as biotinidase (BTD), galactose-1-phosphate uridyltransferase (GALT), acid α-glucosidase (GAA), and α-l-iduronidase (IDUA), among others (March-of-Dimes, 2004).

A nice feature of DBS samples is that they can be shipped at ambient temperature, which is cost saving compared to shipping of blood tubes, which weigh more and may have shipping requirements, such as ice packs for temperature control and special packaging. Thus, DBS are a more convenient way of shipping a blood specimen and do not require an elaborate infrastructure for shipping. In fact, standard couriers and mail services are routinely used for transport of DBS specimens. It is noteworthy that while standard couriers can be used, NBS DBS specimens must have timely delivery to ensure rapid identification of critical conditions; any delivery system for newborn DBS samples must be able to guarantee overnight delivery. Expeditious transit of specimens from the collection facility to the screening laboratory is necessary to facilitate accurate testing results and ensure that affected newborns are identified, diagnosed, and treated prior to symptom onset.

It is also important that care is taken when handling the DBS samples, as the time from spotting to processing will be less than the time required for inactivation of the most common bloodborne pathogens. All human blood, whether liquid or dried, should be handled according to standard practices to reduce transmission of hepatitis C virus (HCV), hepatitis B virus (HBV), human immunodeficiency virus (HIV), and other bloodborne pathogens (March-of-Dimes, 2004). While it remains unclear if filter paper promotes the inactivation of common bloodborne pathogens, several studies have documented the time frame of virus inactivation in dried blood. HCV remains infectious in dried blood for at least 16 hr, but for no more than 4 days; HIV in a dried state remains infectious for 3–7 days; and HBV in dried blood can be transmitted for up to 1 week and possibly longer (Bond et al., 1981; Kamili et al., 2007; Resnick et al., 1986).

6.3.5 DBS Analyte Stability and Storage

Several studies have examined the impact of temperature and humidity on the stability of metabolic, endocrine, and genomic analytes in DBS on Grade 903 and PerkinElmer 226 filter paper as shown in Table 6.2. The effect of time in storage under varying temperature and humidity conditions has also been evaluated.

Table 6.2 Impact of Temperature, Humidity, Desiccant, and Time on Analytes in DBS Samples.

Analyte type	Analysis method	Storage time	Storage temperature	Storage humidity/ use of desiccant	Analytes effected	Outcome[a]
21 amino acids[b]	LC/ESI-MS	6 days	Temp. cycling: −20 <–> 40	Controlled with desiccant	Serine, threonine, glutamine, methionine, histidine, ornithine	11.5–33% degradation
		30 days	25 °C	38%	Histidine	38.2% degradation
		30 days	40 °C	38%	Histidine and lysine	68.1 and 39.7% degradation
		30 days	25 °C	75%	Glutamic acid and aspartic acid	43 and 144% concentration increase
		30 days	40 °C	75%	Asparagine, tryptophan, methionine, glutamine	70–90% degradation
		30 days	40 °C	75%	Glutamic acid and aspartic acid	41–112% concentration increase
244 polar metabolites[c]	DIMS	4 weeks	Room temp	N/P[d]	Creatine, L-glutamine, glucose, L-carnitine	Stable
33 inborn disorder markers[e]	Fluoroimmunoassay, various enzymatic methods, MS/MS	31–35 days	37 °C	<30%	Galactose-1-phospate uridly-transferase and biotinidase	>60% degradation
		31–35 days	37 °C	>90%	Galactose-1-phospate uridly-transferase, biotinidase, succinylacetone, decenoylcarniting, tetradecenoylcarnitine	>50% degradation in 1 week

(Continued)

Table 6.2 (Continued)

Analyte type	Analysis method	Storage time	Storage temperature	Storage humidity/ use of desiccant	Analytes effected	Outcome[a]
		31–35 days	37 °C	>90%	Biotinidase, succinylacetone, arginine, acetylcarnitine, decenoylcarnitine, tetradecenoylcarnitine, malonylcarnitine	>90% degradation
5 lysosomal enzymes[f]	MS/MS	187 days	−20 and 4 °C	Controlled with desiccant	Acid a-glucocerebrosidase (ABG), acid a-galactosidase A (GLA), acid a-glucosidase (GAA), galactocerebrosidase (GALC), acid sphingomyelinase (ASM)	Stable
			25 °C	Controlled with desiccant	Acid a-glucocerebrosidase (ABG), acid a-galactosidase A (GLA), acid sphingomyelinase (ASM)	40–60% degradation

			37 °C	Controlled with desiccant	Acid a-glucocerebrosidase (ABG), acid a-galactosidase A (GLA), acid sphingomyelinase (ASM)	>60% degradation
5 steroids[g]	MS/MS	1 year	−70 °C	N/P	21-Deoxycortisol, 11-deoxycortisol, 4-androstenedione, 17-hydroxyprogesterone, cortisol	Stable
			−20 °C	N/P	21-Deoxycortisol, 11-deoxycortisol, 4-androstenedione, 17-hydroxyprogesterone, cortisol	Stable
			4 °C	N/P	21-Deoxycortisol, 11-deoxycortisol, 4-androstenedione, 17-hydroxyprogesterone, cortisol	Stable
			RT	N/P	Cortisol	Degradation at 6 months
			RT	N/P	4-Androstenedione	concentration increase at 4 weeks

(*Continued*)

Table 6.2 (Continued)

Analyte type	Analysis method	Storage time	Storage temperature	Storage humidity/ use of desiccant	Analytes effected	Outcome[a]
DNA[h]	Abbott m2000 real-time HIV-1 quantitative assay	3 years	−20 °C	N/P	HIV-1 DNA	All samples DNA amplifiable /stable
	Roche Amplicor HIV-1 DNA test	12 weeks	37 °C	77–87%	HIV-1 DNA	All samples DNA amplifiable/stable
DNA[i]	Roche Amplicor HIV-1 DNA test	1 year	37 °C	85%	HIV-1 DNA	78% of samples DNA amplifiable/stable
DNA[j]	AmpFLSTR Identifiler PCR amplification kit	10 years	RT	No desiccant	Human genomic DNA	All samples DNA amplifiable /stable

[a] At end of storage time unless otherwise noted.
[b] Han et al. (2018).
[c] Trifonova et al. (2019).
[d] N/P—information not provided.
[e] Adam et al. (2011).
[f] De Jesus et al. (2009).
[g] Grecso et al. (2020).
[h] Masciotra et al. (2012).
[i] Mitchell et al. (2008).
[j] Cordovado et al. (2009).

Successful use of DBS matrices for testing depends on how these environmental variables may or may not impact the stability of targeted analytes. In general, stability studies of various metabolites, particularly those utilized for NBS, have shown that analytes degrade at high temperature and humidity in a time-course-dependent manner. For NBS, it is particularly important to know the potential for analyte degradation when DBS specimens are exposed to extreme temperature and humidity during shipping from the birth facility to the screening laboratory. Fortunately, even under extreme conditions, there is minimal decline over the short period of time (typically 1–3 days) required for transport, allowing for the use of standard courier and shipping services as mentioned above. For specimens that require longer-term storage, including NBS specimens that require retesting, laboratory storage can be controlled and storage at lower temperatures with controlled humidity (e.g. using desiccant) is recommended. Table 6.2 further outlines the impact of temperature, humidity, desiccant, and time on analytes in DBS from select studies (Adam et al., 2011; Chace and Hannon, 2010; Cordovado et al., 2009; De Jesus et al., 2009; Golbahar et al., 2014; Grecso et al., 2020; Han et al., 2018; Masciotra et al., 2012; Mitchell et al., 2008; Trifonova et al., 2019). Furthermore, DNA in DBS samples is remarkably stable and can withstand high temperatures and humidity conditions, for short term, with little to no impact on the integrity of DNA recovery and stability. Analytes utilized for the purpose of NBS should be stable enough in a filter paper matrix to allow for accurate analysis after shipping from the birth facility to the public health laboratory. To account for possible degradation of analytes, many NBS programs consider specimens that are in transit for more than 2 weeks unacceptable.

Taken as a whole, these studies indicate that the relatively short-term storage and shipping conditions prior to analysis of DBS for NBS, regardless of season or geography, still allows for accurate analysis of the wide range of components examined during the NBS process.

6.3.6 Known Interferences with DBS use for NBS

Since NBS's inception more than five decades ago, the program has identified numerous interferences that can impact NBS assays using the DBS matrix. While some of these were expected, based on either the chemistry of the assay or analyte being measured, others were unanticipated and have only been discovered through screening millions of newborns.

In general, NBS analyte interferences fall into one of several categories: issues with specimen collection and handling; maternal conditions or treatments; or infant conditions or treatments. The Association of Public Health Laboratories maintains a dynamic and continuously updated list of observed analyte interferences, which can also be referenced (APHL, 2018, 2020). Understanding the array of potential interferences in DBS samples for NBS is critical to providing high-quality screening results and interpretative recommendations. Indeed, as NBS has continued to expand to screen for more and more rare diseases, so too, has the need to understand how results may be affected by any number of potential interfering elements.

Specimen collection and handling issues may occur in DBS samples collected from any newborn or from any collection sites; however, it is well known that newborns in the Neonatal Intensive Care Unit (NICU) are more likely to have infant or maternal factors that interfere with testing (CLSI, 2023). Because of this, many NBS programs have implemented serial screening protocols in this population in an effort to reduce the impact of known interferences on screening results, and subsequent burden on families and providers.

Tables 6.3–6.5 describe current known or possible analyte interferences within NBS. A thorough understanding of the diseases being screened for, the assays utilized, and the biomarker expression indicative of each disease is pivotal, as a single interfering factor may result in both false-positive and false-negative results depending on the above variables. The information in these tables is largely focused on known interferences impacting biochemical-based NBS assays. But, inhibitor factors in DBS specimens can also negatively affect measurements of genomic analytes in NBS. As first observed in 1988, Taq DNA polymerase, required for DNA amplification, is inhibited by compounds in blood (de Franchis et al., 1988). Later, in 1994, Akane et al. reported that inhibition of Taq polymerase was caused by heme complexes that result from the degradation of blood hemoglobin (Akane et al., 1994). Because PCR has become an indispensable method used in most molecular assays, the effect of inhibitors on Taq polymerase must be reduced or eliminated for optimal assay performance. Early solutions to address inhibition of this enzyme focused on identifying methods of extracting and purifying DNA from blood that would reduce the percent of inhibitors in the final eluate (Akane et al., 1994). More recently, mutants of Taq DNA polymerase that are resistant to blood inhibitors have become commercially available and allow for DNA amplification by PCR using even crude DNA extract from blood (Kermekchiev et al., 2009). These products have been so successful that some NBS laboratories have used DNA extracted from DBS punches that have not undergone any washes to remove heme with successful results (Baker et al., 2016; Hendrix et al., 2020).

6.3.7 History of NSQAP—40 Years of Quality Assurance

Today, NBS represents the largest and most successful population-based genetic screening effort in the world. Part of the success of NBS stems from a focus and commitment to high-quality testing that was put in place shortly after the advent

Table 6.3 Specimen Collection and Handling Interferences and Their Impact on NBS Analytes.

Issue	Analyte(s) affected	Effect on screening result	Effect on screening outcome
Blood collection shortly after birth	17-OHP TSH IRT GALC GLA	Elevation Elevation Elevation Reduction Reduction	False positive(s)
Blood collection with EDTA	17-OHP TSH IRT	Elevation Reduction Reduction	False positive False negative False negative
Blood collection with heparin	T-cell receptor excision circles Molecular assays utilizing PCR	Reduction	False positive
Use of antistatic spray (canned air) near screening collection devices	Acylcarnitine analysis, specifically C6, C8, C10, and C6DC	Elevation	False positive(s)
Use of benzocaine-containing cream on infant prior to collection	Phenylalanine	Elevation	False positive
Use of nipple-fissure cream with neopentanoate esters	C5	Elevation	False positive
Use of Sani-Cloth surface disinfectant wipes near screening collection devices	C3DC	Elevation	False positive
Use of sivelestat (with pivalic acid) to treat acute respiratory distress syndrome	C5	Elevation	False positive
Use of watermelon-containing creams or sprays around specimen collection devices	Citrulline	Elevation	False positive

Table 6.4 Maternal Conditions or Treatments that can Cause Interferences in NBS Analytes.

Issue	Analyte(s) affected	Effect on screening result	Effect on screening outcome
^{131}I (radioactive iodine) treatment during pregnancy	T4 TSH	Reduction Elevation	False positive or permanent CH (depending on gestational age during treatment)
3-MCC deficiency	C5-OH	Elevation	False positive
Antibiotics containing pivalic acid	C5	Elevation	False positive
Carnitine deficiency	Carnitine	Reduction	False positive
Carnitine supplementation	Carnitine Acylcarnitines	Elevated Elevated	False negative False positive
Congenital adrenal hyperplasia	17-OHP	Elevation	False positive
Dopamine treatment	TSH	Reduction	False negative
Fatty liver of pregnancy or HELLP syndrome	Even-chain acylcarnitines	Elevation	False positive
Essential oils (lavender, tea tree, cypress)	C8 and C10	Elevation	False positive[a]
Gastric bypass surgery	C3	Elevation	False positive
Glucocorticoid treatment	T4 TSH	Reduction Reduction	False negative
Hashimoto's thyroiditis	TSH	Elevation	False positive
Hyperthyroidism, treated with PTU or methimazole (thiamazole)	T4 TSH	Reduction Elevation	False positive
Hypothyroidism, untreated or treated with carbamazepine	T4 TSH	Reduction Elevation	False positive
Immunosuppressant drugs	TREC KREC	Reduction Reduction	False positive
PKU, uncontrolled	Phenylalanine	Elevated	False positive
Steroid administration	17-OHP	Reduction	False negative
Vitamin B_{12} deficiency, with or without pernicious anemia	C3	Elevated	False positive
Vegan diet	C3	Elevated	False positive

[a] Mikami-Saito et al. (2020).

Table 6.5 Infant Conditions or Treatments that can Cause Interferences in NBS Analytes.

Issue	Analyte(s) affected	Effect on screening result	Effect on screening outcome
Carnitine supplementation	Acylcarnitines Carnitine	Elevation Elevation	False positive False negative
Dopamine treatment	TSH	Reduction	False negative
Generalized critical illness	17-OHP T4 TSH IRT	Elevation Reduction Elevation Elevation	False positive(s)
Hypoxia	17-OHP TSH IRT	Elevation Elevation Elevation	False positive(s)
Iodine deficiency	T4 TSH	Reduction Elevation	False positive
Iodine exposure with povidone-iodine preps	T4 TSH	Reduction Elevation	False positive
Liver disease	Tyrosine Methionine Galactose Biotinidase Phenylalanine	Elevation Elevation Elevation Elevation Elevation	False positive(s)
Meconium ileus, intestinal perforation, and abdominal wall defects	IRT	Decreased	False negative
Parenteral nutrition administration	Amino acids	Elevated, especially if drawn from an arterial or venous line	False positive
Prematurity	Biotinidase TREC	Reduction Reduction	False positive(s)
Prematurity, specifically an immature hypothalamic–pituitary thyroid axis	T4 TSH	Reduction Reduction/delayed Elevation	False negative
Prematurity, specifically liver enzyme immaturity	Tyrosine Methionine Galactose	Elevation Elevation Elevation	False positive(s)

(Continued)

Table 6.5 (Continued)

Issue	Analyte(s) affected	Effect on screening result	Effect on screening outcome
Renal immaturity or renal disease	17OHP Multiple amino acids Multiple organic acids	Elevation	False positive(s)
Steroid administration	T4 TSH	Reduction Reduction	False negative
Thoracic surgery with thymectomy	TREC	Reduction	False positive
Transfusion, red blood cells	Biotinidase GALT Hemoglobins TREC T4	Elevation Elevation	False negative False negative

of screening for PKU. Recognizing the high stakes involved in accurate early identification of newborns at risk for significant morbidity and mortality, the CDC created a quality assurance (QA) program in 1978, which later became the Newborn Screening Quality Assurance Program (NSQAP). This program was developed to assist NBS programs in providing high-quality screening to their population, thereby accurately detecting at-risk newborns so they can quickly get the critical care they need. More specifically, NSQAP provides this assistance by offering certified DBS materials for proficiency testing (PT) and QC analysis. Testing of these materials by the NBS programs is used to assess the accuracy of their testing, minimize false-positive and -negative results and ensure high-quality performance is sustained. For more than 40 years, NSQAP has provided certified DBS materials that currently include 76 analytes. These analytes represent diseases currently being screened for in the United States, including all of the current core and secondary diseases on the RUSP, which is the most progressive and comprehensive list of newborn diseases recommended for inclusion in a population-based NBS program (Urv and Parisi, 2017).

All NBS laboratories in the United States must meet strict QA criteria in order to perform testing on human specimens as outlined in the Clinical Laboratory Improvement Amendments (CLIA) of 1988 (Centers for Medicare & Medicaid Services, 2022). CLIA criteria require that laboratories participate in a PT program that is designed to evaluate the quality of laboratory performance on a periodic basis using matrix appropriate specimens. PT examines the analytical performance of individual laboratories for specific assays and is used to monitor laboratories' continuing performance at one point in time. In addition, CLIA regulations require that laboratories have QC materials and processes to monitor the accuracy and precision of the complete testing process.

NSQAP is housed in the Newborn Screening and Molecular Biology Branch (NSMBB) of the Division of Laboratory Sciences in the National Center for Environmental Health at the CDC. NSMBB's QA program is free of charge, and participation is voluntary. CDC is the only comprehensive provider of PT and QC DBS materials for the purpose of NBS in the world. The Branch is ISO 17043 accredited as a PT provider. NBS laboratories rely on CDC for both PT challenges to comply with their clinical testing regulatory requirements and QC materials to monitor method performance over time and harmonize results and cutoffs among laboratories. In addition, NSMBB works to improve the quality and scope of laboratory services by providing technical assistance and consultation to NBS programs.

In 2009, NSMBB expanded their services and began offering funding to two state NBS programs to pilot-test NBS for severe combined immunodeficiency (SCID). After successful implementation in these two programs, NSQAP continued to fund eight more programs over the next 8 years, successfully ensuring that all state NBS programs had implemented SCID NBS as of 2018. This funding has been expanded to aid in implementation of other new diseases and, thus far, 12 initiatives have been funded to add NBS for Pompe disease, mucopolysaccharidosis, type I, X-linked adrenoleukodystrophy (ALD), and spinal muscular atrophy (SMA).

Responding to the ongoing expansion of NBS and associated increased needs, NSMBB underwent a reorganization in 2011, creating two new laboratories—the Biochemical and Mass Spectrometer Laboratory and the Molecular Quality Improvement Program. These programs were created to assist NBS programs with the ever-increasing demands of new biochemical and molecular testing technologies. These two laboratories support the NSQAP program in creating new QA materials, developing new or improved screening methods, providing hands-on training to laboratorians engaged in NBS, as well as offering routine and ad hoc technical assistance. A review of current QC and PT offerings by NSQAP can be found in Table 6.6.

Over the past few decades, NSQAP has expanded from one analyte to over 70 with almost 700 participants (Figure 6.2). NSQAP also performs filter paper evaluations, special consultations, and technical assistance to public health and private laboratories engaged in NBS. As a result, NSQAP enables participating programs to meet the QA requirement for verifying test accuracy and helps ensure

Table 6.6 Current QC and PT DBS Offerings from NSQAP as of 2023.

NBS disease	Analyte(s)	PT program available (Y/N)	QC materials available (Y/N)
2,4 Dienoyl-CoA reductase deficiency	C10:2 acylcarnitine	Y	Y
2-Methyl-3-hydroxybutyric aciduria	C5-OH acylcarnitine	Y	Y
2-Methylbutyrylglycinuria	C5 acylcarnitine	Y	Y
3-Hydroxy-3-methyglutaric aciduria	C5-OH acylcarnitine	Y	Y
3-Methylcrotonyl-CoA carboxylase deficiency	C5-OH acylcarnitine	Y	Y
3-Methylglutaconic aciduria	C5-OH acylcarnitine	Y	Y
Argininemia	Arginine	Y	Y
Argininosuccinic aciduria	Argininosuccinic acid	N	N
	Citrulline	Y	Y
Benign hyperphenylalaninemia	Phenylalanine	Y	Y
Beta-ketothiolase deficiency	C5:1 acylcarnitine	Y	Y
Biopterin defect in cofactor biosynthesis	Phenylalanine	Y	Y
Biopterin defect in cofactor regeneration	Phenylalanine	Y	Y
Biotinidase deficiency	Biotinidase activity	Y	N
Carnitine acylcarnitine translocase deficiency	C0 acylcarnitine	N	Y
	C16 acylcarnitine	N	Y
	C18 acylcarnitine	N	Y
Carnitine palmitoyltransferase type I deficiency	C0 acylcarnitine	N	Y
	C16 acylcarnitine	N	Y
	C18 acylcarnitine	N	Y
Carnitine palmitoyltransferase type II deficiency	C2 acylcarnitine	N	Y
	C16 acylcarnitine	N	Y
	C18:1 acylcarnitine	N	Y
Carnitine uptake defect/ carnitine transport defect	C0 acylcarnitine	Y	Y
	C2 acylcarnitine	Y	Y
Citrullinemia type I	Citrulline	Y	Y

Table 6.6 (Continued)

NBS disease	Analyte(s)	PT program available (Y/N)	QC materials available (Y/N)
Citrullinemia type II	Citrulline	Y	Y
Classic galactosemia	Total galactose	Y	Y
	GALT activity	Y	Y
Classic phenylketonuria	Phenylalanine	Y	Y
Congenital adrenal hyperplasia	17-OHP	Y	Y
	4-Androstenedione	Y	Y
	Cortisol	Y	Y
	11-Deoxycortisol	Y	Y
	21-Deoxycortisol	Y	Y
Cystic fibrosis	Immunoreactive trypsinogen	Y	Y
	CFTR DNA (71 variants)	Y	N
Fabry disease	Alpha-Gal A	N	Y
Galacto epimerase deficiency	GALE	N	N
Galactokinase deficiency	GALK	N	N
Guanidinoacetate methyltransferase deficiency	Creatine	Y	Y
	Guanidinoacetic acid	Y	Y
Gaucher disease	GBA	N	Y
Glucose-6-phophate dehydrogenase deficiency	G6PD	Y	N
Glutaric acidemia type I	C5-DC acylcarnitine	Y	Y
Glutaric acidemia type II	C4 acylcarnitine	Y	Y
	C5 acylcarnitine	Y	Y
HIV	Anti-HIV-1 antibodies	Y	N
Holocarboxylase synthase deficiency	C5-OH acylcarnitine	Y	Y
Homocystinuria	Methionine	Y	Y
	Homocysteine	N	Y
Hypermethioninemia	Methionine	Y	Y
Isobutyrylglycinuria	C4 acylcarnitine	Y	Y
Isovaleric acidemia	C5 acylcarnitine	Y	Y

(*Continued*)

Table 6.6 (Continued)

NBS disease	Analyte(s)	PT program available (Y/N)	QC materials available (Y/N)
Krabbe disease	GALC	Y	Y
Long-chain L-3 hydroxyacyl-CoA dehydrogenase deficiency	C16-OH acylcarnitine	Y	Y
	C18-OH acylcarnitine	Y	Y
Malonic acidemia	Malonic Acid	N	Y
	C3-DC acylcarnitine	Y	Y
Maple syrup urine disease	Alloisoleucine	N	Y
	Isoleucine	N	Y
	Leucine	Y	Y
	Valine	Y	Y
Medium/short-chain L-3-hydroxyacyl-CoA dehydrogenase deficiency	C4-OH acylcarnitine	Y	Y
Medium-chain acyl-CoA dehydrogenase deficiency	C6 acylcarnitine	Y	Y
	C8 acylcarnitine	Y	Y
	C10 acylcarnitine	Y	Y
	C10:1 acylcarnitine	Y	Y
Medium-chain ketoacyl-CoA thiolase deficiency	C6-OH acylcarnitine	N	N
	C8 acylcarnitine	Y	Y
Methylmalonic acidemia (cobalamin disorders)	C3 acylcarnitine	Y	Y
	Methylmalonic acid	N	Y
	Methylcitric acid	N	Y
Methylmalonic acidemia (methylmalonyl-CoA mutase)	C3 acylcarnitine	Y	Y
	Methylmalonic acid	N	Y
	Methylcitric acid	N	Y
Methylmalonic acidemia with homocystinuria	C3 acylcarnitine	Y	Y
	Homocysteine	N	Y
Mucopolysaccharidosis, Type I	IDUA	Y	Y
Mucopolysaccharidosis, Type II	I2S	N	Y
Niemann-Pick A/B	ASM	N	Y
Pompe disease	GAA	N	Y
Primary congenital hypothyroidism	TSH	Y	Y
	T4	Y	Y

Table 6.6 (Continued)

NBS disease	Analyte(s)	PT program available (Y/N)	QC materials available (Y/N)
Propionic acidemia	C3 acylcarnitine	Y	Y
S, beta-Thalassemia	Various hemoglobins	Y	N
S,C Disease		Y	N
S,S Disease (Sickle cell anemia)		Y	N
Severe combined immunodeficiency	TREC	Y	N
T cell related lymphocyte deficiencies		Y	N
Short-chain acyl-CoA dehydrogenase deficiency	C4 acylcarnitine	Y	Y
Spinal Muscular Atrophy	SMN1 exon 7 deletion	Y	N
Toxoplasmosis gondii	Anti-toxoplasma gondii	Y	N
	Immunoglobulins M and G	Y	N
Trifunctional protein deficiency	C16-OH acylcarnitine	Y	Y
	C18-OH acylcarnitine	Y	Y
Tyrosinemia type I	SUAC	Y	Y
	Tyrosine	Y	Y
Tyrosinemia type II	Tyrosine	Y	Y
Tyrosinemia type III	Tyrosine	Y	Y
Various other hemoglobinopathies	Various hemoglobins	Y	Y
Very long-chain acyl-CoA dehydrogenase deficiency	C12 acylcarnitine	Y	Y
	C14:1 acylcarnitine	Y	Y
	C14 acylcarnitine	Y	Y
X-linked adrenoleukodystrophy	C20-LPC	N	Y
	C22-LPC	N	Y
	C24-LPC	Y	Y
	C26-LPC	Y	Y

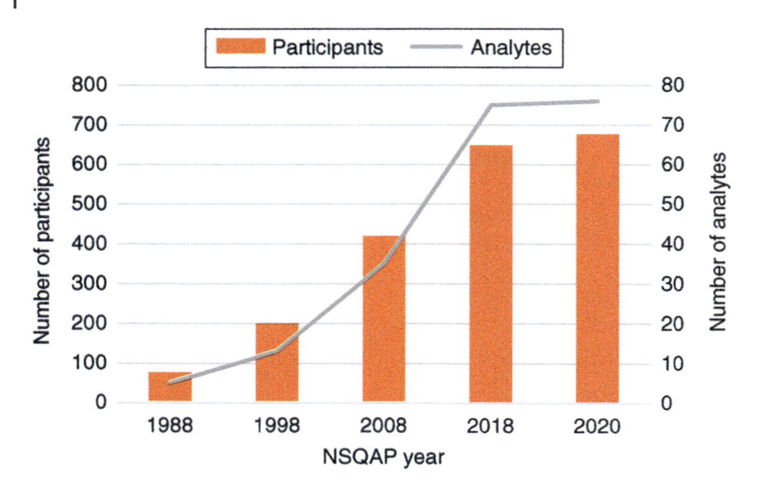

Figure 6.2 Expansion of the Newborn Screening Quality Assurance Program (NSQAP) including number of analytes and participants from its inception (1978) to December 2020.

the quality and accuracy of screening tests for almost 3.8 million babies born each year in the United States. A timeline of NSQAP highlights is illustrated in Figure 6.3.

6.4 Methods Used in NBS

6.4.1 Origins of NBS and Expansion of Biochemical Testing

As described earlier, the first NBS test employed in the United States aimed to detect PKU. Dr. Guthrie, in his development of the phenylalanine analytical test, used a phenylalanine analogue, β-2-thienylalanine, shown to inhibit bacterial growth through its incorporation into proteins and their subsequent inactivation (Cohen and Munier, 1956; Munier and Cohen, 1959). Dr. Guthrie discovered that, in the case of the bacterium *Bacillus subtilis*, this growth inhibition can be specifically prevented by the addition of phenylalanine. This discovery lead to the development of a bacterial inhibition assay (BIA) using an agar plate containing β-2-thienylalanine and inoculated with *Bacillus subtilis* on which punches from DBS specimens were applied (Guthrie and Susi, 1963). The extent of the bacterial growth around the DBS punch could be measured and correlated with blood phenylalanine concentrations using calibrators containing known amounts of phenylalanine. The PKU BIA assay met a number of important criteria for a population-wide screening assay. It was simple, had higher clinical sensitivity

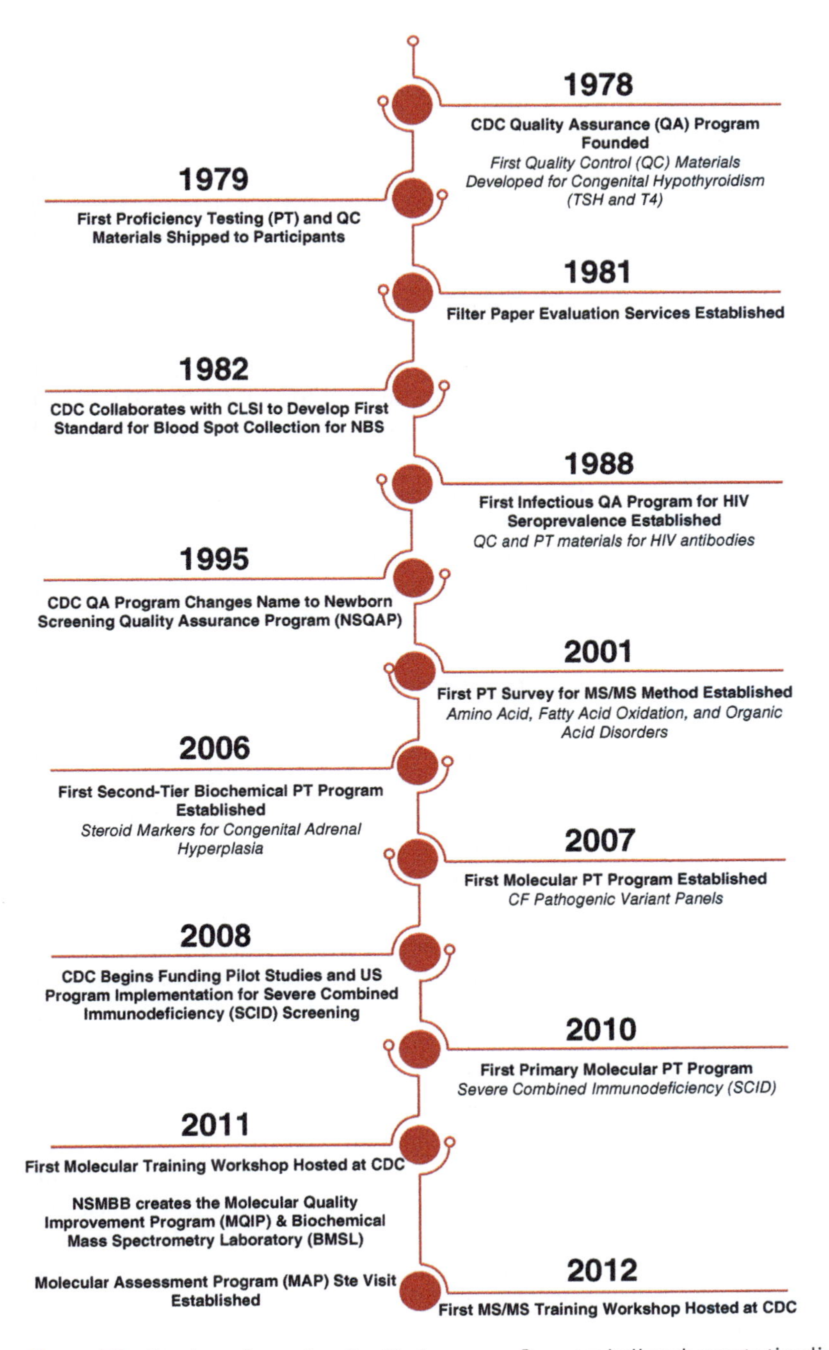

Figure 6.3 Newborn Screening Quality Assurance Program hallmark events timeline.

than the previously utilized ferric chloride urine test (Allen and Wilson, 1964; Munier and Cohen, 1959; Phenylketonuria-screening, 1968), had reasonable throughput, and could be easily adopted by public health labs that were, at the time, employing mainly microbiologists with ample experience with this type of assay. Dr. Guthrie expanded the BIA concept to the detection of other metabolic diseases, such as maple syrup urine disease (MSUD) (Levy and Hammersen, 1978), galactosemia (GALT) (Levy and Hammersen, 1978), and homocystinuria (Accinni et al., 2003; Yap and Naughten, 1998). However, these tests were not as widely adopted as screening for PKU, likely due to a number of factors, including disease incidence, assay performance, and a lack of coordinated advocacy.

Approximately 10 years after PKU screening had been established in the United States, several NBS programs started screening for congenital hypothyroidism (CH) using radioactively labeled antibodies to measure thyroxine (T4) in DBS specimens. CH is the most common treatable cause of intellectual disability with a birth prevalence of 1 in 3,000–4,000 (Buyukgebiz, 2013; Olney et al., 2010), and its screening during the newborn period has been shown to have considerable clinical and economic benefits. Hence, its addition to NBS panels is often considered an important early development in NBS. Today, most programs screen for TSH as the primary biomarker for CH (Kilberg et al., 2018; LaFranchi, 2010).

In the 1980s, a number of enzymatic (Gerasimova et al., 1989) and fluorometric assays (Heard et al., 1984) were developed that eventually replaced BIAs. These more modern assays solved the limitations of BIAs that included slower processing times and higher costs, the qualitative nature of BIA analysis, imprecise ability to measure the bacterial growth zones, and inaccurate test results due to other biological factors that can inhibit bacterial growth (Chace and Hannon, 2016). During the same period, novel screening assays were developed for additional diseases, such as biotinidase deficiency (BIOT) (Heard et al., 1984), hemoglobinopathies (Benson and Therrell, 2010; Black, 1984), and congenital adrenal hyperplasia (Levy, 1998). At that point, screening for a large panel of diseases was limited by the fact that each disease required a separate analytical assay, a separate DBS punch, and separate extraction buffers. Several researchers of the time, including Dr. Guthrie, soon realized that the "one test—one disease" model would be the limiting factor for significant expansion of NBS; however, early efforts to multiplex several bacterial or chromatography assays failed (Tarini et al., 2006).

The "one test—many diseases" or "multiplexing" concept did not materialize until the 1990s with the introduction of tandem mass spectrometry (MS/MS) for the screening of inborn errors of metabolism, specifically through the measurement of amino acids and acylcarnitines. The adoption of MS/MS by NBS programs led to a significant expansion of NBS test panels from just a few to several dozen diseases. Indeed, from 1995 to 2005, on average, states increased the number of diseases tested approximately five times (from about 5–25 diseases) (Tarini et al., 2006).

Thanks to the high specificity of MS/MS, expanding NBS to >20 diseases resulted in a total false-positive rate of only 0.33%, which was similar to the false-positive rate obtained using the former PKU enzymatic assay alone (0.23%) (Schulze et al., 2003). Since the introduction of MS/MS for the screening of PKU in 1993 (Chace et al., 1993), the MS/MS screening panel has continued to expand significantly and now includes biomarkers for multiple amino acid, fatty acid oxidation, and organic acid diseases; lysophosphatidylcholines for the screening of ALD (Haynes and De Jesus, 2012; Hubbard et al., 2009), as well as nucleosides for the screening of SCID caused by adenosine deaminase deficiency (ADA-SCID) (la Marca et al., 2014; Young et al., 2020). Furthermore, the clinical sensitivity and specificity for some diseases have significantly improved by the addition of more specific biomarkers, such as succinylacetone for tyrosinemia, type I (Turgeon et al., 2008), and argininosuccinic acid for argininosuccinic aciduria (De Biase et al., 2015).

At the onset of MS/MS applications in NBS, one particular challenge arose in the extraction of biomarkers from DBS samples that differ significantly in hydrophobicity (from arginine to cerotyl-carnitine and other lipids). Because the extraction buffer composition needs to be designed to allow for extraction of both hydrophilic and hydrophobic compounds, this compromise can lead to a sensitivity decrease for some biomarkers due to lower biomarker recoveries. Today, this issue has been mitigated by advances in MS instrumentation, which is now much more sensitive than earlier models.

More recently, lysosomal storage disorders (LSDs) have been added to NBS panels, with programs using different technologies for their detection. Chamoles et al. (2001) were the first to apply a fluorometric assay to measure enzymatic activity for IDUA (a biomarker for mucopolysaccharidoses, type I) in DBS. A few years later, in 2004, Gelb and co-workers developed an MS/MS-based enzymatic method to screen for Krabbe disease, and soon after, multiplexed this assay to include several other LSD biomarkers (Gelb et al., 2019; Li et al., 2004a, 2004b).

Multiplexing of fluorometric-based enzymatic assays has been made possible using digital microfluidics (DMF). DMF is a well-established technology that can manipulate microdroplets on printed circuit boards using electrical impulses (Choi et al., 2012). Several NBS laboratories have adopted DMF for the screening of LSD diseases (Washburn and Millington, 2020) due to the ease of the assay and space-savings as compared to MS/MS. Residual enzymatic activity in DBS with either fluorometric, DMF, or MS/MS assays is measured indirectly by using synthetic reagents that can be used as substrates for the enzymes of interest, and their products can be analyzed by the respective technologies (Acker and Auld, 2014).

The expansion of NBS panels from 1 to over 50 diseases inherently means an increased number of false-positive results—with some disease additions contributing more to this metric than others. To mitigate this increase in false-positive rates, second-tier screening assays have been utilized as a way to improve the specificity of a NBS assay that produces a high number of false positives, without compromising the corresponding sensitivity. Second-tier screening involves further analysis using the same DBS specimen, after an out-of-range first-tier result. The second-tier analysis is done using an assay that either measures a more specific biomarker or separates the biomarker analyzed during first-tier screening from any interferences. One of the most widespread approaches in second-tier screening is using LC to separate the biomarkers of interest before detection by MS/MS (Chace and Hannon, 2010). Second-tier biochemical NBS has been successfully used to significantly reduce the number of false positives in amino acid disorders (Matern et al., 2007; Sinclair et al., 2016), organic acid disorders (La Marca et al., 2007; Matern et al., 2007), LSDs (Peck et al., 2020), and ALD (Turgeon et al., 2015).

Currently, more than 80% of core and secondary RUSP diseases can be screened by MS/MS, and this platform will likely continue to allow for further expansion of NBS using DBS testing. Already, in addition to the biomarkers and diseases mentioned above, MS/MS has been used to show the possibility of screening for numerous other diseases that may be on the horizon for addition to NBS panels (Gelb et al., 2019; Stapleton et al., 2020). Examples of candidate disease biomarkers include: bile acids for cerebrotendinous xanthomatosis (DeBarber et al., 2018) and Niemann-Pick C disease (Jiang et al., 2016; Mazzacuva et al., 2016); sulfatides for metachromatic leukodystrophy (MLD) (Spacil et al., 2016); amino acids and organic acids for the screening of guanidinoacetate methyltransferase (Asef et al., 2016) and ornithine transcarbamylase deficiency (Janzen et al., 2014); adenosine and deoxyadenosine for the screening of adenosine deaminase deficient SCID (Young et al., 2020); and bilirubin for the screening of hyperbilirubinemia (Gong et al., 2018). Many of these markers have been multiplexed with the common amino acid, acylcarnitine and succinylacetone analytes already used in routine NBS (Asef et al., 2016).

Two promising areas for the development of biochemical-based NBS include larger multiplexed assay options as well as the use of high-resolution tandem mass spectrometry (HRMS) for metabolomic screening in newborns (Pickens and Petritis, 2020). In HRMS, increased instrument resolution allows for several isobaric interferences to be resolved and the use of a higher number of internal standards used. Furthermore, the use of HRMS allows targeted and untargeted assays to be combined (Pickens and Petritis, 2020), providing the information NBS laboratories are used to through tandem mass spectrometry (i.e. targeted quantitation of analyte of interest) while allowing many more analytes to be detected under the full scan mode. The additional metabolic profiling obtained through untargeted metabolomics could further increase the sensitivity and specificity of those assays in the future. It is important to note that the cost of HRMS has been decreasing over the years, while becoming much more user-friendly than they used to be which has led to the adoption of HRMS in several public health laboratories for environmental and food analysis but not yet for NBS. Adoption of HRMS instrumentation for NBS in the future is possible if those platforms end up providing significant advantages over the current triple quadrupole platforms.

In regard to expanded multiplexed platforms, Hong et al. (2020) recently published a high-throughput, 18-plex MS/MS assay for the simultaneous screening of 15 LSDs as well as MLD, ALD, GALT, and BIOT. The ability to multiplex several disorders in one assay is necessary to be able to further expand the NBS panels due to the limited availability of newborn blood. Multiplexing further simplify testing as it decreases the amount of detection platforms that need to be used, as well as laboratory space and personnel required. In addition to allowing more disorders to be tested in the future, residual blood can be used for second-tier assays when necessary to increase the positive predictive value of NBS assays.

This approach highlights that diseases currently screened by conventional fluorometric/colorimetric "one disease—one test" assays could be multiplexed with future mass spectrometry assays. The ability to multiplex testing of relevant analytes for NBS has allowed the amount of blood collected by NBS programs to remain relatively stable over time (typically around 500 μL total) even as the number of diseases being screened has greatly increased (De Jesus et al., 2021b). Keeping the amount of blood collected from newborns to a minimum is an important component of NBS programs (especially for those newborns born prematurely), and efforts to multiplex have increased in order to maintain the ability to test more with the same amount of blood.

Targeted (Scolamiero et al., 2015; Yoon, 2015) and untargeted (Denes et al., 2012; Knottnerus et al., 2020) metabolomics have also been investigated for NBS of DBS samples. These approaches have shown higher specificity, thanks to their ability to separate isobars as well as identification of diseases through multivariate analysis of metabolomic profiles. Pickens and Petritis (2020) combined targeted and untargeted metabolomics in one assay in order to take advantage of benefits from both approaches. The high-resolution mass spectrometers needed for this type of analysis are already in use in several public health laboratories to support food safety and environmental-related applications. In addition, HRMS have, over time, become more user-friendly and less expensive. Thus, a shift toward HRMS from quadrupole-based MS/MS may be possible in the future. This technology may offer higher clinical specificity/sensitivity by eliminating interferences, increasing the ability to measure biomarkers under flow injection analysis (FIA) conditions (e.g. homocysteine in place of methionine) and allowing for much higher multiplexing of biomarkers of interest.

6.4.2 Origins of Molecular DBS Testing and Expansion in NBS

In 1987, McCabe et al. showed that DNA could reproducibly be extracted and isolated in sufficient quantities (0.5 μg) from a 50-μl DBS for use in molecular assays, such as Southern blot analysis (McCabe et al., 1987). Not long after this study, a seminal paper presenting the technique of PCR amplification was published, which catapulted the use of DBS samples for the molecular study of diseases for the next decade and beyond (Jinks et al., 1989). No longer was it necessary to extract large quantities of DNA from a DBS specimen containing 50 μL of blood. The method was modified to allow for extraction of DNA from a 3.2-mm DBS punch containing as little as 3 μL of blood. The resulting DNA obtained from this extraction was used to analyze genomic DNA for the variant causative of sickle cell anemia using allele-specific oligonucleotide (ASO) probes (Jinks et al., 1989).

In 1994, Washington State's NBS Program adopted the first molecular test that was used for the purpose of confirming a sickle cell disease positive screen. This second-tier molecular test was performed on DNA extracted from a DBS punch when the first-tier test, isoelectric focusing (IEF), indicated the presence of a variant hemoglobin S protein. The molecular test identified the pathogenic variant, which caused the S sickle protein and was an independent verification of the first-tier assay (Glass, 2015). This assay was the first routine use of a DBS-based molecular assay within a NBS program. Shortly thereafter, also in 1994, the Wisconsin NBS Program adopted a molecular assay designed to identify the most common pathogenic variant (p.Phe508del; c.1521_1523delCTT; legacy F508del) found in patients with cystic fibrosis (CF). This assessment was adopted as a second-tier molecular screening test to enhance the sensitivity of the first-tier IRT assay. The addition of a second-tier molecular test allowed IRT cutoffs to be lowered while still maintaining specificity. A screen positive result was defined, then, as having both elevated IRT and at least one identified F508del variant (Farrell et al., 1994; Gregg et al., 1997).

While the F508del variant is the most common CF-causing variant worldwide, there are many other known pathogenic variants in the *CFTR* gene that result in CF as seen in the CFTR1 online database (Tsui, 1989) and CFTR2 website (CFTR2, 2011). To account for this, the Massachusetts Newborn Screening Program implemented a similar approach to Wisconsin, but instead of limiting detection to the F508del pathogenic variant, they included a panel of 27 *CFTR* variants. This expansion of the variant panel enhanced the sensitivity of the CF second-tier molecular assay from about 85% (with the single F508del variant) to over 95% (with the expanded variant panel), and thus, detected the vast majority of CF-affected infants in Massachusetts (Comeau et al., 2004). The multiple-*CFTR* variant panel resulted in an increase from 50 to 75% of affected cases where both CF-causing variants were detected through NBS, which facilitated more rapid referral and intervention.

Using data from these early adopting states, Grosse and co-workers determined that evidence supported the benefit of early identification of CF by NBS (Grosse et al., 2004), and by the end of 2009, all 50 states in the United States were screening their newborn populations for CF, using at least an immunoassay to detect elevated IRT. Of these 50 state programs, 36 programs were also using a second-tier DNA-based test to detect, at a minimum, the F508del pathogenic variant, and in many cases, a panel of variants that were most often inclusive of the ACMG recommended 23 variants (Richards et al., 2002). Of note, the ACMG 23 pathogenic variant panel was developed as a recommendation for prenatal carrier screening, which was predominately recommended in the White population at that time, and not for full population NBS. Today, all 50 state programs use a second- and/or third-tier molecular test for CF. Since there are >2000 variants in the *CFTR* gene, including pathogenic variants, benign variants, and variants of unknown significance (1989), screening programs that use a molecular second-tier testing strategy with a pathogenic variant panel must define a screen positive as elevated IRT and at least one pathogenic variant, despite CF being an autosomal recessive disease. Thus, screen positives using this algorithm include newborns with CF, as well as newborns that are likely CF carriers and do not have overt disease.

Using this algorithm, clinicians must then perform a diagnostic sweat test to rule out carriers from those with classic CF or CF-related metabolic syndrome (CRMS). In CF, the sweat test is considered the "gold standard" for diagnosis as it measures the amount of chloride in the sweat. Individuals with CF have higher chloride levels in their sweat than unaffected individuals. Thus, this diagnostic test must be performed in order to definitively determine whether a child is a carrier, has classic CF, or CRMS.

An alternative approach, taken initially by the California NBS program, is a three-tiered strategy, which begins with identifying newborns with elevated IRT, followed by the evaluation of a panel of known CF pathogenic variants, and then gene sequencing using Sanger sequencing to identify additional nonpanel variants in newborns where only one pathogenic variant was found. The program noted several benefits to this approach, including detection of fewer carrier newborns and fewer diagnostic sweat tests (Kammesheidt et al., 2006; Prach et al., 2013). However, since the clinical consequence is not known for all *CFTR* variants, the addition of third-tier gene sequencing has resulted in more babies being identified with CRMS (Kharrazi et al., 2015). CRMS is defined as a patient with an elevated IRT and two *CFTR* variants, but with a normal or equivocal sweat test (Cystic Fibrosis et al., 2009). Most of these babies do not end up having symptoms of classic CF; however, a percentage have gone on to develop the disease and, typically, all CRMS cases are followed by a CF care center (Barben et al., 2020). Recently, the NY NBS program adopted a similar approach to

California, but instead of using Sanger sequencing for the third-tier *CFTR* sequencing, they are using next-generation sequencing (Kay, 2019). Since implementing third-tier sequencing, NY has seen an 84.3% reduction in their CF screen positive results requiring referral to sweat test; however, they have observed a threefold increase in CRMS diagnoses (Kay, 2020).

While CF was the first disease where multiple programs utilized molecular testing as part of the screening algorithm, it was not the first disease for which gene sequencing was used in routine screening. In 2004, the Texas (TX) Newborn Screening Program introduced third-tier hemoglobin B (*HBB*) gene sequencing with the goal of adding clarity to the hemoglobinopathy newborn screen. The algorithm for hemoglobinopathy screening in TX begins with a first-tier biochemical test, IEF, followed by high performance liquid chromatography (HPLC) confirmation to identify variant hemoglobin proteins. When a variant hemoglobin protein is identified, it is then confirmed using a molecular-based, restriction fragment length polymorphism (RFLP) test to identify the common *HBB* variants as well as two beta thalassemia pathogenic variants. If this genotyping test does not identify any pathogenic variants, the *HBB* gene is then sequenced using Sanger sequencing to detect off-panel pathogenic variants that are specific to the genomes of certain ancestral groups (Thein, 2013).

In 2010, the Secretary of Health and Human services added SCID to the RUSP, which was a significant milestone for NBS molecular testing because SCID was the first NBS disease to utilize a first-tier molecular test (2018). Unlike the molecular second-tier tests already in use, the screening test for SCID did not identify pathogenic variants in genomic DNA. Since SCID is a group of diseases that can be caused by many different genes, it is not feasible to test for a panel of pathogenic variants or perform gene sequencing as a first-tier assay. Instead, the T-cell receptor excision circle (TREC), is used as a biomarker to measure T-cell lymphopenia, a hallmark of SCID. TREC is an extraneous DNA sequence that is excised from the genome during immune system maturation and when present indicates an active and maturing immune system. During normal T-cell differentiation in the thymus in the neonatal period, there is a rearrangement of the T-cell receptor genes leading to the excision of the V, D, and J gene segments. The excised DNA circularizes, forming extrachromosomal TRECs, where the signal joint (i.e. where the two ends join to form the circle) is a unique sequence not found in the human genome. One specific TREC, the δRec-ψJα TREC, is produced from approximately 70% of all T cells and is the target for the NBS assay (Douek et al., 2000; Verschuren et al., 1997). The presence and amount of TRECs can be identified and measured in DBS using real-time quantitative PCR using a probe that anneals specifically to the signal joint sequence. This assay serves as an excellent surrogate marker for the number of naïve T-cells, since TRECs will not be present or are in very low copy numbers in dysfunctional immune systems (Chan and Puck, 2005).

Until the addition of SCID screening to the RUSP, molecular testing had only been used as a second-tier assay and NBS laboratories did not have the infrastructure in place to accommodate the large-scale molecular testing required of a first-tier assay. Once this infrastructure was developed, it opened the door for additional molecular testing at all levels, including first-, second-, and third-tier applications. In 2018, a second condition that utilized a first-tier molecular test, SMA, was added to the RUSP. Like SCID, SMA can be detected using a real-time quantitative assay. However, rather than detecting a biomarker such a TREC, the NBS assay for SMA is designed to detect the presence or absence of a genomic sequence, specifically exon 7 of the *SMN1* gene, which accounts for approximately 95–98% of SMA cases (Kolb et al., 2017). Real-time PCR assays have been designed to detect the exon 7 region of the genome and then multiplexed into the SCID real-time PCR assay, making the adoption of SMA NBS relatively easy and cost-efficient from the laboratory perspective (Gutierrez-Mateo et al., 2019; Kraszewski et al., 2018). Additionally, since only one region in the gene or target is being assayed, neither of these assays requires the complex bioinformatics and interpretation associated with gene sequencing. Some NBS programs have incorporated a second-tier test for SMA that uses digital PCR to measure the number of *SMN2* gene copies present. Babies with higher copy numbers of the *SMN2* gene (i.e. $\geq$4) have better outcomes despite having a homozygous deletion of the *SMN1* exon 7, since the protein produced by *SMN2* produces some functional protein (Park et al., 2020; Vidal-Folch et al., 2018).

Today, there are a number of molecular tests ongoing in US NBS laboratories or their contract designees ranging from single-variant genotyping to next-generation sequencing (Table 6.7). These molecular assays have been employed as first-tier, second-tier, and third-tier assays. Those assays employed as second- or third-tier assays are often used to add specificity, clarify ambiguity, or add supplemental information to assist with the clinical diagnosis to ensure timely and equitable follow-up.

6.4.3 How the Expansion of Genomics may Impact NBS

In the era of genomic medicine, sequencing genomes, exomes, and panels of genes have become a staple tool of the clinical diagnostic community. To explore the utility of genome and exome sequencing in NBS diseases, the National Institutes of Health funded four investigators in 2013 to explore the implications, challenges, and opportunities associated with using genomic sequence information in the newborn period to identify the underlying genetic causes in sick children, as well as augment traditional NBS, which is largely done on a healthy population (National Human Genome Research Institute, 2020). This program was called the Newborn Sequencing in Genomic Medicine and Public Health

Table 6.7 Newborn Screening Molecular Tests Performed in the United States.

Disorder	Gene	Primary molecular test	Second-tier molecular test	Third-tier molecular test
Severe combined immunodeficiency (SCID)	TREC; >39 genes	RT-PCR	Pilot: NGS Gene panel	
Spinal muscular atrophy (SMA)	SMN1 and SMN2	RT-PCR SMN1 E7	ddPCR SMN2 copy number	
Carnitine palmitoyltransferase (CPT1)	CPT1A		Genotyping population specific variant	
Congenital adrenal hyperplasia (CAH)—pilot	CYP21A2		Genotyping panel of variants	
Cystic fibrosis (CF)	CFTR		Genotyping panel of variants	Gene sequencing
Galactosemia	GALT		Genotyping panel of variants	
Glutaric acidemia	GCDH		Genotyping panel of variants	
Hemoglobinopathies (HB)	HBA		Deletion analysis (HBA)	
	HBB		Genotyping panel of variants (HBB)	Gene sequencing (HBB)
Isovaleric acidemia	IVD		Genotyping panel of variants	
Krabbe disease	GALC		Deletion analysis Gene sequencing	
Maple syrup urine disease (MSUD)	BCKDHA		Genotyping population specific variant	
Medium-chain acyl-CoA dehydrogenase (MCAD)	ACADM		Genotyping panel of variants	
3-Methylcrotonyl-CoA carboxylase deficiency (3-MCC)	MCCC2		Genotyping panel of variants	
Mucopolysaccharidosis I (MPS 1)[a]	IDUA		Gene sequencing	Gene sequencing
Pompe disease[a]	GAA		Gene sequencing	Gene sequencing

Table 6.7 (Continued)

Disorder	Gene	Primary molecular test	Second-tier molecular test	Third-tier molecular test
Propionic acidemia	PCCB		Genotyping panel of variants	
Very long-chain Acyl-CoA dehydrogenase (VLCAD)	AACADVL		Gene sequencing	
X-linked adrenoleukodystrophy (X-ALD)[a]	ABCD1		Gene sequencing	Gene sequencing

[a] Some newborn screening laboratories use gene sequencing as second-tier assay and some use it as a third-tier assay for specific disorders.

(NSIGHT) program. At the time NSIGHT began, it was still unknown if genome or exome sequencing from a DBS sample was feasible. At the completion of the NSIGHT studies, it was clear this technology can readily be utilized with DNA extracted from a DBS specimen. Dr. Jennifer Puck, one of the NSIGHT investigators, noted in her talk at the NSIGHT Steering Committee Meeting "Is Sequencing Helpful in Population Screening for Metabolic Diseases?" that exome sequencing would not be a good replacement for MS at this point, but rather, it would be a useful tool in cases where there is no MS screen or as a second-tier to help reduce false positives. All NSIGHT investigators agreed that sequencing does not replicate NBS program results because sequencing has a lower sensitivity and specificity. They did agree that sequencing was useful in gathering information about prognosis, specific therapies, and solving complicated cases in newborns with unclear screening results (Adhikari et al., 2020).

The NSIGHT program also created the Ethics and Policy Advisory Board of the NSIGHT Consortium to weigh in on the ethical, legal, and social implications of implementing genome sequencing in the newborn period. The Board determined that targeted or whole exome/genome sequencing in symptomatic newborns was useful for guiding diagnosis or treatment when paired with a newborn's symptoms; however, it fell short as a screening tool because the complexity and uncertainty associated the data in an asymptomatic population were limited in its utility without symptoms to guide physicians (Johnston et al., 2018). They did determine that there was utility in using targeted sequencing for population-based screening for newborn diseases that are part of the RUSP (Urv and Parisi, 2017) as is being done in several US NBS laboratories today. It is anticipated that NBS laboratories will continue to expand their use of molecular testing and, in particular, the incorporation of NGS. While NGS is often used to sequence genes or gene segments, it can also be used to identify a large panel of preselected variants. In CF NBS, NGS is being used by different laboratories in both ways. Some NBS labs use NGS for CF screening as a second-tier assay where only preselected variants are visible, whereas other labs use NGS as a third-tier gene sequencing assay where they will find any variants in the coding regions of the gene. In addition, the New York State Department of Health Newborn Screening Program piloted a targeted NGS gene panel to identify pathogenic variants in the wide variety of genes that may cause SCID ($N = 39$ genes) (Stevens, 2017). Additionally, the Utah Department of Health Newborn Screening Program has begun using exome sequencing with an *a priori* restriction to disease-causing genes that are part of their second- or third-tier algorithms, thus eliminating the need to develop different gene panels or conduct single-gene sequencing assays for different diseases (Ruiz-Schultz, 2018; Ruiz-Schultz et al., 2021). As the field of genomics continues to expand at this exponential rate, the tools being developed will need to be evaluated for their utility in NBS, population-based health, and in follow-up clinical diagnosis.

6.4.4 Expansion of DBS Utility, Including Direct Patient Use

The success of DBS utilization within NBS programs suggested that this matrix could be similarly employed to accomplish testing that might otherwise be difficult due to the targeted population, resource limitations, geography, or the need for serial monitoring over an extended period. Today, a countless number of DBS applications are in use, including immunologic studies, nutritional evaluations, antibody assessments, and therapeutic drug monitoring. Emerging uses of DBS include toxicokinetic and pharmacokinetic studies, metabolic profiling, miRNA quantification, forensic toxicology, and understanding of environmental contamination. Addressing the realized and potential other uses of DBS samples is not within the scope of this chapter, but several reviews discuss these applications in much more detail (Freeman et al., 2018; Gupta and Mahajan, 2018; Lakshmy et al., 2014; Snowden et al., 2020).

It is also important to note that numerous NBS programs throughout the United States have expanded their use of the DBS matrix to provide phenylalanine monitoring services to patients affected with PKU. Strict monitoring and control of phenylalanine levels is necessary in all PKU patients, especially during pregnancy when poorly controlled maternal PKU is considered a teratogen (Murphy, 2015). To accomplish this monitoring, patient or caregiver-collected DBS specimens have been used. Bringing sample collection and monitoring into the home allows for the continuous assessment of phenylalanine levels and dietary control while reducing family and clinic burden (Clinic, 2005; Gregory et al., 2007; Moat et al., 2020). To accomplish at-home monitoring using DBS samples, PKU patients and their caregivers are educated about the need for frequent, regular monitoring, and they receive training similar to medical staff on how to properly collect a DBS specimen. While samples are obtained via heel stick in infants during routine NBS, in children and adults, samples are obtained via fingerstick. Samples are dried as usual (at least 3 hr) and mailed to the public health laboratory or clinic for testing.

The use of PKU monitoring by some NBS programs illustrates that at-home DBS collection and shipping can be accomplished through proper education and training initiatives. Building upon this work, numerous uses of this matrix for at-home sampling have been piloted and implemented (Al-Uzri et al., 2017; Karp et al., 2020; Klak et al., 2019; Prinsenberg et al., 2020; Roberts et al., 2016; Willemsen et al., 2018). The utility of patient-driven collection and the DBS matrix is especially appealing in resource-limited settings where trained phlebotomists, geographical barriers, or a scarcity of laboratory and clinical resources make blood sampling difficult or impossible. With the ever-increasing number of analytes shown to be measurable in DBS samples, the potential for this simple matrix continues to offer solutions for a range of surveillance and clinical needs.

6.5 Conclusion

NBS programs highlight the successful use of DBS specimens in a widespread population-based screening system. Throughout its over 50-year history, NBS programs in partnership with NSQAP have evolved in their understanding of the DBS matrix and the need for high-quality and ongoing QA. NBS programs rely on NSQAP to provide reference materials, PT programs, QC and PT reports, funding, and technical assistance, all of which have become necessary components in helping NBS programs deliver accurate results and minimizing false-positive and false-negative outcomes.

As NBS panels have expanded and testing has become more complex, so has the need for a more extensive understanding of pre-analytical components, a need for a wider array of QA materials, along with more modernized data analytics. While NBS has been successful in its efforts to reduce morbidity and mortality in affected newborns, its continuous growth and effectiveness will depend on the ability of programs and supporting organizations to work together to innovatively adapt their ability to interpret and address new and old challenges alike. While NBS programs can be found wordwide (Loeber et al., 2021; Martinez-Morillo et al., 2016), there is no single overarching governing body, and each jurisdiction makes their own decisions on the infrastructure of their program and the disorders screened. There have been some attempts at global harmonization and oversight, primarily through the consistent use of the WHO's commissioned population screening report by Wilson and Jungner (Wilson and Jungner, 1968) and through the Clinical and Laboratory Standards Institute's (CLSI) suite of NBS-specific guidelines. CLSI is a global organization that creates consensus-driven documents to help guide NBS programs throughout the world. As NBS continues to expand in quantity of diseases and complexity of technologies and interpretation, global collaboration will become more pivotal to the future of NBS worldwide.

Remarkably, the relatively simple DBS matrix has stood the test of time within NBS and continues to keep pace with emerging technologies. With the progression of this successful public health program showing no signs of slowing down, the DBS matrix, first utilized over a century ago, is certain to continue to impact thousands of lives each year through its ability to serve as a foundation of population-based screening for treatable congenital diseases.

Acknowledgments

The findings and conclusions in this report are those of the authors and do not necessarily represent the views of the Centers for Disease Control and Prevention. Use of trade names and commercial sources is for identification only and does not constitute endorsement by the US Department of Health and Human Services, or US Centers for Disease Control and Prevention. The authors specially thank to Elizabeth McCown and Joanne Mei, PhD, for contributions and critical review.

Conflicts of Interest

The authors declare no conflict of interest.

References

45C.F.R.§862.1675 (2022). *Blood specimen collection device.*

Accinni, R., Campolo, J., Parolini, M., De Maria, R., Caruso, R., Maiorana, A., Galluzzo, C., Bartesaghi, S., Melotti, D., & Parodi, O. (2003). Newborn screening of homocystinuria: Quantitative analysis of total homocyst(e)ine on dried blood spot by liquid chromatography with fluorimetric detection. *Journal of Chromatography. B, Analytical Technologies in the Biomedical and Life Sciences*, *785*, 219–226.

Acker, M., & Auld, D. (2014). Considerations for the design and reporting of enzyme assays in high-throughput screening applications. *Perspectives in Science*, *1*, 56–73.

Adam, B. W., Alexander, J. R., Smith, S. J., Chace, D. H., Loeber, J. G., Elvers, L. H., & Hannon, W. H. (2000). Recoveries of phenylalanine from two sets of dried-blood-spot reference materials: Prediction from hematocrit, spot volume, and paper matrix. *Clinical Chemistry*, *46*, 126–128.

Adam, B. W., Hall, E. M., Sternberg, M., Lim, T. H., Flores, S. R., O'Brien, S., Simms, D., Li, L. X., De Jesus, V. R., & Hannon, W. H. (2011). The stability of markers in dried-blood spots for recommended newborn screening disorders in the United States. *Clinical Biochemistry*, *44*, 1445–1450.

Adhikari, A. N., Gallagher, R. C., Wang, Y., Currier, R. J., Amatuni, G., Bassaganyas, L., Chen, F., Kundu, K., Kvale, M., Mooney, S. D., Nussbaum, R. L., Randi, S. S., Sanford, J., Shieh, J. T., Srinivasan, R., Sunderam, U., Tang, H., Vaka, D., Zou, Y., Koenig, B. A., Kwok, P. Y., Risch, N., Puck, J. M., & Brenner, S. E. (2020). The role of exome sequencing in newborn screening for inborn errors of metabolism. *Nature Medicine*, *26*, 1392–1397.

Akane, A., Matsubara, K., Nakamura, H., Takahashi, S., & Kimura, K. (1994). Identification of the heme compound copurified with deoxyribonucleic acid (DNA) from bloodstains, a major inhibitor of polymerase chain reaction (PCR) amplification. *Journal of Forensic Sciences*, *39*, 362–372.

Allen, R. J., & Wilson, J. L. (1964). Urinary phenylpyruvic acid in phenylketonuria. *JAMA*, *188*, 720–724.

Al-Uzri, A., Freeman, K. A., Wade, J., Clark, K., Bleyle, L. A., Munar, M., & Koop, D. R. (2017). Longitudinal study on the use of dried blood spots for home monitoring in children after kidney transplantation *Pediatric Transplantation*, 21.

American College of Medical Genetics Newborn Screening Expert Group. (2006). Newborn screening: Toward a uniform screening panel and system—executive summary. *Pediatrics*, *117*, S296–S307.

APHL. (2018). *Expanding the reach of SCID testing: A report on the severe combined immunodeficiency newborn screening implementation experience*. Association of Public Health Laboratories. https://www.aphl.org/aboutAPHL/publications/Documents/NBS-2018Nov-SCID-Implementation-Report.pdf.

APHL. (2020). *Newborn screening analyte interference list*. Retrieved November 27, 2020, from https://www.aphl.org/programs/newborn_screening/Pages/NBS%20Interference%20List.aspx.

Asef, C. K., Khaksarfard, K. M., & De Jesus, V. R. (2016). Non-derivatized assay for the simultaneous detection of amino acids, acylcarnitines, succinylacetone, creatine, and guanidinoacetic acid in dried blood spots by tandem mass spectrometry. *International Journal of Neonatal Screening*, *2*, 13.

Bagioni, S., Sisto, R., Ferraro, A., Caiafa, P., & Turano, C. (1978). A new method for the preparation of DNA—cellulose. *Analytical Biochemistry*, *89*, 616–619.

Baker, M. W., Atkins, A. E., Cordovado, S. K., Hendrix, M., Earley, M. C., & Farrell, P. M. (2016). Improving newborn screening for cystic fibrosis using next-generation sequencing technology: A technical feasibility study. *Genetics in Medicine*, *18*, 231–238.

Baker, M. W., Grossman, W. J., Laessig, R. H., Hoffman, G. L., Brokopp, C. D., Kurtycz, D. F., Cogley, M. F., Litsheim, T. J., Katcher, M. L., & Routes, J. M. (2009). Development of a routine newborn screening protocol for severe combined immunodeficiency. *The Journal of Allergy and Clinical Immunology*, *124*, 522–527.

Barben, J., Castellani, C., Munck, A., Davies, J. C., De Winter-De Groot, K. M., Gartner, S., Kashirskaya, N., Linnane, B., Mayell, S. J., Mccolley, S., Ooi, C. Y., Proesmans, M., Ren, C. L., Salinas, D., Sands, D., Sermet-Gaudelus, I., Sommerburg, O., Southern, K. W., & European CF Society Neonatal Screening Working Group. (2020). Updated guidance on the management of children with cystic fibrosis transmembrane conductance regulator-related metabolic syndrome/cystic fibrosis screen positive, inconclusive diagnosis (CRMS/CFSPID). *Journal of Cystic Fibrosis*, *20*, 810–819.

Benson, J. M., & Therrell, B. L. Jr. (2010). History and current status of newborn screening for hemoglobinopathies. *Seminars in Perinatology*, *34*, 134–144.

Bhattacharya, K., Wotton, T., & Wiley, V. (2014). The evolution of blood-spot newborn screening. *Translational Pediatrics*, *3*, 63–70.

Black, J. (1984). Isoelectric focusing in agarose gel for detection and identification of hemoglobin variants. *Hemoglobin*, *8*, 117–127.

Bond, W. W., Favero, M. S., Petersen, N. J., Gravelle, C. R., Ebert, J. W., & Maynard, J. E. (1981). Survival of hepatitis B virus after drying and storage for one week. *Lancet*, *1*, 550–551.

Buyukgebiz, A. (2013). Newborn screening for congenital hypothyroidism. *Journal of Clinical Research in Pediatric Endocrinology*, *5*(Suppl 1), 8–12.

Centers for Disease Control and Prevention. (2011). Ten great public health achievements—United States, 2001–2010. *MMWR. Morbidity and Mortality Weekly Report*, *60*, 619–623.

Centers for Medicare & Medicaid Services. (2022). *Clinical Laboratory Improvement Amendments (CLIA)*.

CFTR2. (2011). *The clinical and functional translation of CFTR (CFTR2)*. US CF Foundation, Johns Hopkins University, and The Hospital for Sick Children. http://cftr2.org.

Chace, D., & De Jesus, V. (2014). Applications of dried blood spots in newborn and metabolic screening. In W. Li & M. Lee (Eds.), *Dried blood spots. Applications and techniques*. Hobokan: Wiley.

Chace, D., & Hannon, W. (2016). Technological journey from colorimetric to tandem mass spectrometric measurements in the diagnostic investigation for phenylketonuria. *Journal of Inborn Errors of Metabolism and Screening*, *4*, 1–11.

Chace, D. H., & Hannon, W. H. (2010). Impact of second-tier testing on the effectiveness of newborn screening. *Clinical Chemistry*, *56*, 1653–1655.

Chace, D. H., Millington, D. S., Terada, N., Kahler, S. G., Roe, C. R., & Hofman, L. F. (1993). Rapid diagnosis of phenylketonuria by quantitative analysis for phenylalanine and tyrosine in neonatal blood spots by tandem mass spectrometry. *Clinical Chemistry*, *39*, 66–71.

Chamoles, N. A., Blanco, M., & Gaggioli, D. (2001). Diagnosis of alpha-L-iduronidase deficiency in dried blood spots on filter paper: The possibility of newborn diagnosis. *Clinical Chemistry*, *47*, 780–781.

Chan, K., & Puck, J. M. (2005). Development of population-based newborn screening for severe combined immunodeficiency. *The Journal of Allergy and Clinical Immunology*, *115*, 391–398.

Chapman, O. (1924). The complement-fixation test for syphilis. Use of patient's whole blood dried on filter paper. *Archives of Dermatology and Syphilology*, *9*, 607–611.

Choi, K., Ng, A. H., Fobel, R. & Wheeler, A. R. (2012). Digital microfluidics. *Annual Review of Analytical Chemistry (Palo Alto, California)*, 5, 413–40.

Clinic, P. (2005). *Monitoring blood phenylalanine levels at home*. University of Washington.

CLSI. (2023). Newborn Screening Follow-up and Education. 3rd ed. CLSI guideline NBS02. Clinical and Laboratory Standards Institute.

Cohen, G. N., & Munier, R. (1956). Incorporation of structural analogues of amino acids in bacterial proteins. *Biochimica et Biophysica Acta*, *21*, 592–593.

Comeau, A. M., Parad, R. B., Dorkin, H. L., Dovey, M., Gerstle, R., Haver, K., Lapey, A., O'Sullivan, B. P., Waltz, D. A., Zwerdling, R. G., & Eaton, R. B. (2004). Population-based newborn screening for genetic disorders when multiple mutation DNA testing is incorporated: A cystic fibrosis newborn screening model demonstrating increased sensitivity but more carrier detections. *Pediatrics*, *113*, 1573–1581.

Cordovado, S. K., Earley, M. C., Hendrix, M., Driscoll-Dunn, R., Glass, M., Mueller, P. W., & Hannon, W. H. (2009). Assessment of DNA contamination from dried blood spots and determination of DNA yield and function using archival newborn dried blood spots. *Clinica Chimica Acta*, *402*, 107–113.

Couzin-Frankel, J. (2009). Newborn blood collections. Science gold mine, ethical minefield. *Science*, *324*, 166–168.

Cystic Fibrosis Foundation, Borowitz, D., Parad, R. B., Sharp, J. K., Sabadosa, K. A., Robinson, K. A., Rock, M. J., Farrell, P. M., Sontag, M. K., Rosenfeld, M., Davis, S. D., Marshall, B. C., & Accurso, F. J. (2009). Cystic Fibrosis Foundation practice guidelines for the management of infants with cystic fibrosis transmembrane conductance regulator-related metabolic syndrome during the first two years of life and beyond. *The Journal of Pediatrics*, *155*, S106–S116.

De Biase, I., Liu, A., Yuzyuk, T., Longo, N., & Pasquali, M. (2015). Quantitative amino acid analysis by liquid chromatography-tandem mass spectrometry: Implications for the diagnosis of argininosuccinic aciduria. *Clinica Chimica Acta*, *442*, 73–74.

de Franchis, R., Cross, N. C., Foulkes, N. S., & Cox, T. M. (1988). A potent inhibitor of Taq polymerase copurifies with human genomic DNA. *Nucleic Acids Research*, *16*, 10355.

De Jesus, V., Borrajo, G., Mei, J., Adams, W., Chunko, K., Das, S., Dorley, M., Dunbar, D., El Mouden, H., George, R. S., Held, P., Kelm, K., Macdonald, J., Mcroberts, C., Piper, K., Quashnock, J., Schier-Pugsley, J., Therrell, B. L., Turner, K., Wiencek, J.,

& Yahyaoui, R. (2021a). *Blood collection on filter paper for newborn screening programs* 6th ed. NBS, CLSI.

De Jesus, V., Borrajo, G., Mei, J. V., Adams, W., Chunko, K., Das, S., Dorley, M., Dunbar, D., El Mouden, H., George, R. S., Held, P., Kelm, K., Macdonald, J., Mcroberts, C., Piper, K., Quashnock, J., Schier-Pugsley, J., Therrell, B. L., Turner, K., Weiencek, J., & Yahyaoui, R. (2021b). *Dried blood spot specimen collection for newborn screening (NBS01)*. 7th ed. NBS, CLSI.

De Jesus, V. R., Zhang, X. K., Keutzer, J., Bodamer, O. A., Muhl, A., Orsini, J. J., Caggana, M., Vogt, R. F., & Hannon, W. H. (2009). Development and evaluation of quality control dried blood spot materials in newborn screening for lysosomal storage disorders. *Clinical Chemistry*, *55*, 158–164.

Debarber, A. E., Kalfon, L., Fedida, A., Fleisher Sheffer, V., Ben Haroush, S., Chasnyk, N., Shuster Biton, E., Mandel, H., Jeffries, K., Shinwell, E. S., & Falik-Zaccai, T. C. (2018). Newborn screening for cerebrotendinous xanthomatosis is the solution for early identification and treatment. *Journal of Lipid Research*, *59*, 2214–2222.

Deluca, J. M. (2018). Public attitudes toward expanded newborn screening. *Journal of Pediatric Nursing*, *38*, e19–e23.

Denes, J., Szabo, E., Robinette, S. L., Szatmari, I., Szonyi, L., Kreuder, J. G., Rauterberg, E. W., & Takats, Z. 2012. Metabonomics of newborn screening dried blood spot samples: A novel approach in the screening and diagnostics of inborn errors of metabolism. *Analytical Chemistry*, *84*, 10113–20.

Douek, D. C., Vescio, R. A., Betts, M. R., Brenchley, J. M., Hill, B. J., Zhang, L., Berenson, J. R., Collins, R. H., & Koup, R. A. (2000). Assessment of thymic output in adults after haematopoietic stem-cell transplantation and prediction of T-cell reconstitution. *Lancet*, *355*, 1875–1881.

Farrell, P. M., Aronson, R. A., Hoffman, G., & Laessig, R. H. (1994). Newborn screening for cystic fibrosis in Wisconsin: First application of population-based molecular genetics testing. *Wisconsin Medical Journal*, *93*, 415–421.

Freeman, J. D., Rosman, L. M., Ratcliff, J. D., Strickland, P. T., Graham, D. R., & Silbergeld, E. K. (2018). State of the science in dried blood spots. *Clinical Chemistry*, *64*, 656–679.

Gelb, M. H., Lukacs, Z., Ranieri, E., & Schielen, P. (2019). Newborn screening for lysosomal storage disorders: Methodologies for measurement of enzymatic activities in dried blood spots. *International Journal of Neonatal Screening*, 5, 1.

George, R. S., & Moat, S. J. (2016). Effect of dried blood spot quality on newborn screening analyte concentrations and recommendations for minimum acceptance criteria for sample analysis. *Clinical Chemistry*, *62*, 466–475.

Gerasimova, N. S., Steklova, I. V., & Tuuminen, T. (1989). Fluorometric method for phenylalanine microplate assay adapted for phenylketonuria screening. *Clinical Chemistry*, *35*, 2112–2115.

Glass, M. (2015). *Molecular testing: Applications in screening newborns for hemoglobinopathies and galactosemia*. [Powerpoint Slides]. APHL. https://www.aphl.org/programs/newborn_screening/Documents/2015_Molecular-Workshop/Molecular-Testing-Applications-in-Screening-Newborns-for-Hemoglobinopathies-and-Galactosemia.pdf.

Golbahar, J., Altayab, D. D., & Carreon, E. (2014). Short-term stability of amino acids and acylcarnitines in the dried blood spots used to screen newborns for metabolic disorders. *Journal of Medical Screening, 21*, 5–9.

Gong, Z., Zheng, L., Wang, Y., Wu, Y., Tian, G., & Lv, Z. (2018). Quantification of bilirubin from dry blood spots using tandem mass spectrometry. *New Journal of Chemistry, 42*, 19701–19706.

Grecso, N., Zadori, A., Szecsi, I., Barath, A., Galla, Z., Bereczki, C., & Monostori, P. (2020). Storage stability of five steroids and in dried blood spots for newborn screening and retrospective diagnosis of congenital adrenal hyperplasia. *PLoS One, 15*, e0233724.

Gregg, R. G., Simantel, A., Farrell, P. M., Koscik, R., Kosorok, M. R., Laxova, A., Laessig, R., Hoffman, G., Hassemer, D., Mischler, E. H., & Splaingard, M. (1997). Newborn screening for cystic fibrosis in Wisconsin: Comparison of biochemical and molecular methods. *Pediatrics, 99*, 819–824.

Gregory, C. O., Yu, C., & Singh, R. H. (2007). Blood phenylalanine monitoring for dietary compliance among patients with phenylketonuria: Comparison of methods. *Genetics in Medicine, 9*, 761–765.

Grosse, S. D., Boyle, C. A., Botkin, J. R., Comeau, A. M., Kharrazi, M., Rosenfeld, M., Wilfond, B. S., & CDC. (2004). Newborn screening for cystic fibrosis: Evaluation of benefits and risks and recommendations for state newborn screening programs. *MMWR—Recommendations and Reports, 53*, 1–36.

Grosse, S. D., Riehle-Colarusso, T., Gaffney, M., Mason, C. A., Shapira, S. K., Sontag, M. K., Braun, K. V. N., & Iskander, J. (2017). CDC grand rounds: Newborn screening for hearing loss and critical congenital heart disease. *MMWR. Morbidity and Mortality Weekly Report, 66*, 888–890.

Guo, K. (1938). A blotting paper method for the floceulation reaction in syphilis. *Münchener Medizinische Wochenschrift (1950), 64*, 675–677.

Gupta, K., & Mahajan, R. (2018). Applications and diagnostic potential of dried blood spots. *International Journal of Applied & Basic Medical Research, 8*, 1–2.

Gurian, E. A., Kinnamon, D. D., Henry, J. J., & Waisbren, S. E. (2006). Expanded newborn screening for biochemical disorders: The effect of a false-positive result. *Pediatrics, 117*, 1915–1921.

Guthrie, R. (1961). Blood screening for phenylketonuria. *JAMA, 178*, 167.

Guthrie, R., & Susi, A. (1963). A simple phenylalanine method for detecting phenylketonuria in large populations of newborn infants. *Pediatrics, 32*, 338–343.

Gutierrez-Mateo, C., Timonen, A., Vaahtera, K., Jaakkola, M., Hougaard, D. M., Bybjerg-Grauholm, J., Baekvad-Hansen, M., Adamsen, D., Filippov, G., Dallaire, S.,

Goldfarb, D., Schoener, D., & Wu, R. (2019). Development of a multiplex real-time PCR assay for the newborn screening of SCID, SMA, and XLA. *International Journal of Neonatal Screening*, *5*, 39.

Hall, E. M., Flores, S. R., & De Jesus, V. R. (2015). Influence of hematocrit and total-spot volume on performance characteristics of dried blood spots for newborn screening. *International Journal of Neonatal Screening*, *1*, 69–78.

Han, J., Higgins, R., Lim, M. D., Lin, K., Yang, J., & Borchers, C. H. (2018). Short-term stabilities of 21 amino acids in dried blood spots. *Clinical Chemistry*, *64*, 400–402.

Hannon, W., & Therell, B. (2014). Overview of the history and applications of dried blood spot samples. In W. Li & M. Lee (Eds.), *Dried blood spots. Applications and techniques*. Hoboken: Wiley.

Haynes, C. A., & De Jesus, V. R. (2012). Improved analysis of C26:0-lysophosphatidylcholine in dried-blood spots via negative ion mode HPLC-ESI-MS/MS for X-linked adrenoleukodystrophy newborn screening. *Clinica Chimica Acta*, *413*, 1217–1221.

Health Resources & Services Administration. (2020). *Recommended uniform screening panel*. Retrieved December 3, 2020 from https://www.hrsa.gov/advisory-committees/heritable-disorders/rusp/index.html.

Heard, G. S., Secor Mcvoy, J. R., & Wolf, B. (1984). A screening method for biotinidase deficiency in newborns. *Clinical Chemistry*, *30*, 125–127.

Hendrix, M. M., Cuthbert, C. D., & Cordovado, S. K. (2020). Assessing the performance of dried-blood-spot DNA extraction methods in next generation sequencing. *International Journal of Neonatal Screening*, *6*, 36.

Hendrix, M. M., Foster, S. L., & Cordovado, S. K. (2016). Newborn Screening Quality Assurance Program for CFTR mutation detection and gene sequencing to identify cystic fibrosis. *Journal of Inborn Errors of Metabolism and Screening*, *4*. 1–11.

Hong, X., Sadilek, M., & Gelb, M. H. (2020). A highly multiplexed biochemical assay for analytes in dried blood spots: Application to newborn screening and diagnosis of lysosomal storage disorders and other inborn errors of metabolism. *Genetics in Medicine*, *22*, 1262–1268.

Hubbard, W. C., Moser, A. B., Liu, A. C., Jones, R. O., Steinberg, S. J., Lorey, F., Panny, S. R., Vogt, R. F., Jr., Macaya, D., Turgeon, C. T., Tortorelli, S., & Raymond, G. V. (2009). Newborn screening for X-linked adrenoleukodystrophy (X-ALD): Validation of a combined liquid chromatography-tandem mass spectrometric (LC-MS/MS) method. *Molecular Genetics and Metabolism*, *97*, 212–220.

Janzen, N., Terhardt, M., Sander, S., Demirkol, M., Gokcay, G., Peter, M., Lucke, T., Sander, J., & Das, A. M. (2014). Towards newborn screening for ornithine transcarbamylase deficiency: Fast non-chromatographic orotic acid quantification from dried blood spots by tandem mass spectrometry. *Clinica Chimica Acta*, *430*, 28–32.

Jiang, X., Sidhu, R., Mydock-Mcgrane, L., Hsu, F. F., Covey, D. F., Scherrer, D. E., Earley, B., Gale, S. E., Farhat, N. Y., Porter, F. D., Dietzen, D. J., Orsini, J. J., Berry-Kravis, E., Zhang, X., Reunert, J., Marquardt, T., Runz, H., Giugliani, R., Schaffer, J. E., & Ory, D. S. (2016). Development of a bile acid-based newborn screen for Niemann-Pick disease type C. *Science Translational Medicine*, *8*, 337ra63.

Jinks, D. C., Minter, M., Tarver, D. A., Vanderford, M., Hejtmancik, J. F., & Mccabe, E. R. (1989). Molecular genetic diagnosis of sickle cell disease using dried blood specimens on blotters used for newborn screening. *Human Genetics*, *81*, 363–366.

Johnston, J., Lantos, J. D., Goldenberg, A., Chen, F., Parens, E., Koenig, B. A., Members of the NSIGHT Ethics and Policy Advisory Board (2018). Sequencing newborns: A call for nuanced use of genomic technologies. *The Hastings Center Report*, *48*(Suppl 2), S2–S6.

Kamili, S., Krawczynski, K., Mccaustland, K., Li, X., & Alter, M. J. (2007). Infectivity of hepatitis C virus in plasma after drying and storing at room temperature. *Infection Control and Hospital Epidemiology*, *28*, 519–524.

Kammesheidt, A., Kharrazi, M., Graham, S., Young, S., Pearl, M., Dunlop, C., & Keiles, S. (2006). Comprehensive genetic analysis of the cystic fibrosis transmembrane conductance regulator from dried blood specimens—implications for newborn screening. *Genetics in Medicine*, *8*, 557–562.

Karp, D. G., Danh, K., Espinoza, N. F., Seftel, D., Robinson, P. V., & Tsai, C. T. (2020). A serological assay to detect SARS-CoV-2 antibodies in at-home collected finger-prick dried blood spots. *Scientific Reports*, *10*, 20188.

Kay, D. (2019). *Implementing next generation sequencing as a third-tier newborn screen for cystic fibrosis in New York State*. APHL.

Kay, D. (2020). *Improvements to the New York State IRT-DNA-SEQ cystic fibrosis newborn screening algorithm: Results from the first two years*. [Powerpoint Slides] APHL.

Kermekchiev, M. B., Kirilova, L. I., Vail, E. E., & Barnes, W. M. (2009). Mutants of Taq DNA polymerase resistant to PCR inhibitors allow DNA amplification from whole blood and crude soil samples. *Nucleic Acids Research*, *37*, e40.

Kharrazi, M., Yang, J., Bishop, T., Lessing, S., Young, S., Graham, S., Pearl, M., Chow, H., Ho, T., Currier, R., Gaffney, L., Feuchtbaum, L., & California Cystic Fibrosis Newborn Screening Consortium (2015). Newborn screening for cystic fibrosis in California. *Pediatrics*, *136*, 1062–1072.

Kilberg, M. J., Rasooly, I. R., Lafranchi, S. H., Bauer, A. J., & Hawkes, C. P. (2018). Newborn screening in the US may miss mild persistent hypothyroidism. *The Journal of Pediatrics*, *192*, 204–208.

Klak, A., Pauwels, S., & Vermeersch, P. (2019). Preanalytical considerations in therapeutic drug monitoring of immunosuppressants with dried blood spots. *Diagnosis (Berlin)*, *6*, 57–68.

Knottnerus, S. J. G., Pras-Raves, M. L., Van Der Ham, M., Ferdinandusse, S., Houtkooper, R. H., Schielen, P., Visser, G., Wijburg, F. A., & De Sain-Van Der Velden, M. G. M. (2020). Prediction of VLCAD deficiency phenotype by a metabolic fingerprint in newborn screening bloodspots. *Biochimica et Biophysica Acta—Molecular Basis of Disease*, *1866*, 165725.

Kolb, S. J., Coffey, C. S., Yankey, J. W., Krosschell, K., Arnold, W. D., Rutkove, S. B., Swoboda, K. J., Reyna, S. P., Sakonju, A., Darras, B. T., Shell, R., Kuntz, N., Castro, D., Parsons, J., Connolly, A. M., Chiriboga, C. A., Mcdonald, C., Burnette, W. B., Werner, K., Thangarajh, M., Shieh, P. B., Finanger, E., Cudkowicz, M. E., Mcgovern, M. M., Mcneil, D. E., Finkel, R., Iannaccone, S. T., Kaye, E., Kingsley, A., Renusch, S. R., Mcgovern, V. L., Wang, X., Zaworski, P. G., Prior, T. W., Burghes, A. H. M., Bartlett, A., Kissel, J. T., & NeuroNEXT Clinical Trial Network on behalf of the NN101 SMA Biomarker Investigators. (2017). Natural history of infantile-onset spinal muscular atrophy. *Annals of Neurology*, *82*, 883–891.

Kraszewski, J. N., Kay, D. M., Stevens, C. F., Koval, C., Haser, B., Ortiz, V., Albertorio, A., Cohen, L. L., Jain, R., Andrew, S. P., Young, S. D., Lamarca, N. M., De Vivo, D. C., Caggana, M., & Chung, W. K. (2018). Pilot study of population-based newborn screening for spinal muscular atrophy in New York state. *Genetics in Medicine*, *20*, 608–613.

La Marca, G., Giocaliere, E., Malvagia, S., Funghini, S., Ombrone, D., Della Bona, M. L., Canessa, C., Lippi, F., Romano, F., Guerrini, R., Resti, M., & Azzari, C. (2014). The inclusion of ADA-SCID in expanded newborn screening by tandem mass spectrometry. *Journal of Pharmaceutical and Biomedical Analysis*, *88*, 201–206.

La Marca, G., Malvagia, S., Pasquini, E., Innocenti, M., Donati, M. A., & Zammarchi, E. (2007). Rapid 2nd-tier test for measurement of 3-OH-propionic and methylmalonic acids on dried blood spots: Reducing the false-positive rate for propionylcarnitine during expanded newborn screening by liquid chromatography-tandem mass spectrometry. *Clinical Chemistry*, *53*, 1364–1369.

Lafranchi, S. H. (2010). Newborn screening strategies for congenital hypothyroidism: An update. *Journal of Inherited Metabolic Disease*, *33*, S225–S233.

Lakshmy, R., Tarik, M., & Abraham, R. A. (2014). Role of dried blood spots in health and disease diagnosis in older adults. *Bioanalysis*, *6*, 3121–3131.

Lang, P. O., Govind, S., Drame, M., & Aspinall, R. (2012). Comparison of manual and automated DNA purification for measuring TREC in dried blood spot (DBS) samples with qPCR. *Journal of Immunological Methods*, *384*, 118–127.

Levy, H. L. (1998). Newborn screening by tandem mass spectrometry: A new era. *Clinical Chemistry*, *44*, 2401–2402.

Levy, H. L., & Hammersen, G. (1978). Newborn screening for galactosemia and other galactose metabolic defects. *The Journal of Pediatrics*, *92*, 871–877.

Li, Y., Brockmann, K., Turecek, F., Scott, C. R., & Gelb, M. H. (2004a). Tandem mass spectrometry for the direct assay of enzymes in dried blood spots: Application to newborn screening for Krabbe disease. *Clinical Chemistry*, *50*, 638–640.

Li, Y., Scott, C. R., Chamoles, N. A., Ghavami, A., Pinto, B. M., Turecek, F., & Gelb, M. H. (2004b). Direct multiplex assay of lysosomal enzymes in dried blood spots for newborn screening. *Clinical Chemistry*, *50*, 1785–1796.

Lim, M. D. (2018). Dried blood spots for global health diagnostics and surveillance: Opportunities and challenges. *The American Journal of Tropical Medicine and Hygiene*, *99*, 256–265.

Loeber, J. G., Platis, D., Zetterstrom, R. H., Almashanu, S., Boemer, F., Bonham, J. R., Borde, P., Brincat, I., Cheillan, D., Dekkers, E., Dimitrov, D., Fingerhut, R., Franzson, L., Groselj, U., Hougaard, D., Knapkova, M., Kocova, M., Kotori, V., Kozich, V., Kremezna, A., Kurkijarvi, R., La Marca, G., Mikelsaar, R., Milenkovic, T., Mitkin, V., Moldovanu, F., Ceglarek, U., O'Grady, L., Oltarzewski, M., Pettersen, R. D., Ramadza, D., Salimbayeva, D., Samardzic, M., Shamsiddinova, M., Songailiene, J., Szatmari, I., Tabatadze, N., Tezel, B., Toromanovic, A., Tovmasyan, I., Usurelu, N., Vevere, P., Vilarinho, L., Vogazianos, M., Yahyaoui, R., Zeyda, M., & Schielen, P. (2021). Neonatal screening in Europe revisited: An ISNS perspective on the current state and developments since 2010. *International Journal of Neonatal Screening*, *7*, 15.

Luckwell, J., Denniff, P., Capper, S., Michael, P., Spooner, N., Mallender, P., Johnson, B., Clegg, S., Green, M., Ahmad, S., & Woodford, L. (2013). Assessment of the within- and between-lot variability of Whatman FTA((R)) DMPK and 903((R)) DBS papers and their suitability for the quantitative bioanalysis of small molecules. *Bioanalysis*, *5*, 2613–2630.

Malsagova, K., Kopylov, A., Stepanov, A., Butkova, T., Izotov, A., & Kaysheva, A. (2020). Dried blood spot in laboratory: Directions and prospects. *Diagnostics (Basel)*, *10*, 248.

March-of-Dimes. (2004). *March of Dimes statement on newborn screening report*. https://www.eurekalert.org/pub_releases/2004-09/modb-mod092004.php.

Martinez-Morillo, E., Prieto Garcia, B., & Alvarez Menendez, F. V. (2016). Challenges for worldwide harmonization of newborn screening programs. *Clinical Chemistry*, *62*, 689–698.

Masciotra, S., Khamadi, S., Bile, E., Puren, A., Fonjungo, P., Nguyen, S., Girma, M., Downing, R., Ramos, A., Subbarao, S., & Ellenberger, D. (2012). Evaluation of blood collection filter papers for HIV-1 DNA PCR. *Journal of Clinical Virology*, *55*, 101–106.

Matern, D., Tortorelli, S., Oglesbee, D., Gavrilov, D., & Rinaldo, P. (2007). Reduction of the false-positive rate in newborn screening by implementation of MS/MS-based second-tier tests: The Mayo Clinic experience (2004–2007). *Journal of Inherited Metabolic Disease*, *30*, 585–592.

Mazzacuva, F., Mills, P., Mills, K., Camuzeaux, S., Gissen, P., Nicoli, E. R., Wassif, C., Te Vruchte, D., Porter, F. D., Maekawa, M., Mano, N., Iida, T., Platt, F., & Clayton, P. T. (2016). Identification of novel bile acids as biomarkers for the early diagnosis of Niemann-Pick C disease. *FEBS Letters*, *590*, 1651–1662.

Mccabe, E. R., Huang, S. Z., Seltzer, W. K., & Law, M. L. (1987). DNA microextraction from dried blood spots on filter paper blotters: Potential applications to newborn screening. *Human Genetics*, *75*, 213–216.

Mcclendon-Weary, B., Putnick, D. L., Robinson, S., & Yeung, E. (2020). Little to give, much to gain-what can you do with a dried blood spot? *Current Environmental Health Reports*, *7*, 211–221.

Mei, J. V., Alexander, J. R., Adam, B. W., & Hannon, W. H. (2001). Use of filter paper for the collection and analysis of human whole blood specimens. *The Journal of Nutrition*, *131*, 1631S–1636S.

Mei, J. V., Zobel, S. D., Hall, E. M., De Jesus, V. R., Adam, B. W., & Hannon, W. H. (2010). Performance properties of filter paper devices for whole blood collection. *Bioanalysis*, *2*, 1397–1403.

Mikami-Saito, Y., Maekawa, M., Wada, Y., Kanno, T., Kurihara, A., Sato, Y., Yamamoto, T., Arai-Ichinoi, N., Kure, S. (2020). Essential oils can cause false-positive results of medium-chain acyl-CoA dehydrogenase deficiency. *Molecular Genetics and Metabolism Reports*, 25, 1–7.

Miller, F. A., Hayeems, R. Z., Bombard, Y., Cressman, C., Barg, C. J., Carroll, J. C., Wilson, B. J., Little, J., Allanson, J., Chakraborty, P., Giguere, Y., & Regier, D. A. (2015). Public perceptions of the benefits and risks of newborn screening. *Pediatrics*, *136*, e413–e423.

Mitchell, C., Jennings, C., Brambilla, D., Aldrovandi, G., Amedee, A. M., Beck, I., Bremer, J. W., Coombs, R., Decker, D., Fiscus, S., Fitzgibbon, J., Luzuriaga, K., Moye, J., Palumbo, P., Reichelderfer, P., Somasundaran, M., Stevens, W., Frenkel, L., & ACTN Dried Blood Spot Working Group of the Infant Maternal Pediatric Adolescent. (2008). Diminished human immunodeficiency virus type 1 DNA yield from dried blood spots after storage in a humid incubator at 37 degrees C compared to −20 degrees C. *Journal of Clinical Microbiology*, *46*, 2945–2949.

Moat, S. J., Schulenburg-Brand, D., Lemonde, H., Bonham, J. R., Weykamp, C. W., Mei, J. V., Shortland, G. S., & Carling, R. S. (2020). Performance of laboratory tests used to measure blood phenylalanine for the monitoring of patients with phenylketonuria. *Journal of Inherited Metabolic Disease*, *43*, 179–188.

Mohamed, S., Elhamid Dabash, S., Mohamed Rashad, H., & Moselhi, E. (2019). Comparison of pain response to vein puncture versus heel lance among preterm infants undergoing blood sampling. *Egyptian Nursing Journal*, *16*, 155–161.

Morrow, C., Hidinger, A., & Wilkinson-Faulk, D. (2010). Reducing neonatal pain during routine heel lance procedures. *MCN: American Journal of Maternal Child Nursing*, *35*, 346–354; quiz 354-356.

Munier, R., & Cohen, G. N. (1959). Incorporation of structural analogues of amino acids into bacterial proteins during their synthesis in vivo. *Biochimica et Biophysica Acta, 31*, 378–391.

Murphy, E. (2015). Pregnancy in women with inherited metabolic disease. *Obstetric Medicine, 8*, 61–67.

National Human Genome Research Institute. (2020). *Newborn Sequencing in Genomic Medicine and Public Health (NSIGHT).* https://www.genome.gov/Funded-Programs-Projects/Newborn-Sequencing-in-Genomic-Medicine-and-Public-Health-NSIGHT.

Olney, R. S., Grosse, S. D., & Vogt, R. F., Jr. (2010). Prevalence of congenital hypothyroidism—current trends and future directions: Workshop summary. *Pediatrics, 125*(Suppl 2), S31–S36.

O'Mara, M., Hudson-Curtis, B., Olson, K., Yueh, Y., Dunn, J., & Spooner, N. (2011). The effect of hematocrit and punch location on assay bias during quantitative bioanalysis of dried blood spot samples. *Bioanalysis, 3*, 2335–2347.

Park, S., Lee, H., Shin, S., Lee, S. T., Lee, K. A., & Choi, J. R. (2020). Analytical validation of the droplet digital PCR assay for diagnosis of spinal muscular atrophy. *Clinica Chimica Acta, 510*, 787–789.

Paul, D., & Brosco, J. (2013). *The PKU paradox: A short history of a genetic disease.* Baltimore: Johns Hopkins University Press.

Peck, D. S., Lacey, J. M., White, A. L., Pino, G., Studinski, A. L., Fisher, R., Ahmad, A., Spencer, L., Viall, S., Shallow, N., Siemon, A., Hamm, J. A., Murray, B. K., Jones, K. L., Gavrilov, D., Oglesbee, D., Raymond, K., Matern, D., Rinaldo, P., & Tortorelli, S. (2020). Incorporation of second-tier biomarker testing improves the specificity of newborn screening for mucopolysaccharidosis type I. *International Journal of Neonatal Screening, 6*, 10.

Pellegrino, J., & Brener, Z. (1958). Reação de fixação de complemento com sangue dessecado no diagnóstico do calazar canino. *Revista Brasileira Malariologia Doenças Tropicais Publicações Avulsas, 10*, 39–44.

Phenylketonuria-Screening. (1968). Screening tests for phenylketonuria. *British Medical Journal, 1*, 656–657.

Pickens, C. A., & Petritis, K. (2020). High resolution mass spectrometry newborn screening applications for quantitative analysis of amino acids and acylcarnitines from dried blood spots. *Analytica Chimica Acta, 1120*, 85–96.

Polk, B., Hoggatt, P., Walker, L., Cannon, G., Bish, C., Han, N., Richardson, M., & Cooper, L. (2003–2008). *Newborn screening report 2003–2008.* Mississippi State Department of Health.

Prach, L., Koepke, R., Kharrazi, M., Keiles, S., Salinas, D. B., Reyes, M. C., Pian, M., Opsimos, H., Otsuka, K. N., Hardy, K. A., Milla, C. E., Zirbes, J. M., Chipps, B., O'Bra, S., Saeed, M. M., Sudhakar, R., Lehto, S., Nielson, D., Shay, G. F., Seastrand, M., Jhawar, S., Nickerson, B., Landon, C., Thompson, A., Nussbaum, E., Chin, T.,

Wojtczak, H., & California Cystic Fibrosis Newborn Screening Consortium. (2013). Novel CFTR variants identified during the first 3 years of cystic fibrosis newborn screening in California. *The Journal of Molecular Diagnostics*, *15*, 710–722.

Prinsenberg, T., Rebers, S., Boyd, A., Zuure, F., Prins, M., Van Der Valk, M., & Schinkel, J. (2020). Dried blood spot self-sampling at home is a feasible technique for hepatitis C RNA detection. *PLoS One*, *15*, e0231385.

Resnick, L., Veren, K., Salahuddin, S. Z., Tondreau, S., & Markham, P. D. (1986). Stability and inactivation of HTLV-III/LAV under clinical and laboratory environments. *JAMA*, *255*, 1887–1891.

Reust, M. J., Lee, M. H., Xiang, J., Zhang, W., Xu, D., Batson, T., Zhang, T., Downs, J. A., & Dupnik, K. M. (2018). Dried blood spot RNA transcriptomes correlate with transcriptomes derived from whole blood RNA. *The American Journal of Tropical Medicine and Hygiene*, *98*, 1541–1546.

Richards, C. S., Bradley, L. A., Amos, J., Allitto, B., Grody, W. W., Maddalena, A., Mcginnis, M. J., Prior, T. W., Popovich, B. W., Watson, M. S., & Palomaki, G. E. (2002). Standards and guidelines for CFTR mutation testing. *Genetics in Medicine*, *4*, 379–391.

Roberts, S. C., Seav, S. M., Mcdade, T. W., Dominick, S. A., Gorman, J. R., Whitcomb, B. W., & Su, H. I. (2016). Self-collected dried blood spots as a tool for measuring ovarian reserve in young female cancer survivors. *Human Reproduction*, *31*, 1570–1578.

Rothwell, E., Anderson, R., Goldenberg, A., Lewis, M. H., Stark, L., Burbank, M., Wong, B., & Botkin, J. R. (2012). Assessing public attitudes on the retention and use of residual newborn screening blood samples: A focus group study. *Social Science & Medicine*, *74*, 1305–1309.

Ruiz-Schultz, N. (2018). *Targeted second-tier confirmatory sequencing NBS pipeline*. [Powerpoint Slides] NewSTEPs. https://www.newsteps.org/sites/default/files/resources/download/newsteps_new_disorders_webinar_utah_august_2018_slides_kh.pdf.

Ruiz-Schultz, N., Sant, D., Norcross, S., Dansithong, W., Hart, K., Asay, B., Little, J., Chung, K., Oakeson, K., Young, E., Eilbeck, K., & Rohrwasser, A. (2021). Methods and feasibility study for exome sequencing as a universal second-tier test in newborn screening. *Genetics in Medicine*, *23*, 767–776.

Saavedra-Matiz, C. A., Isabelle, J. T., Biski, C. K., Duva, S. J., Sweeney, M. L., Parker, A. L., Young, A. J., Diantonio, L. L., Krein, L. M., Nichols, M. J., & Caggana, M. (2013). Cost-effective and scalable DNA extraction method from dried blood spots. *Clinical Chemistry*, *59*, 1045–1051.

Schmidt, V. (1986). Ivar Christian Bang (1869–1918), founder of modern clinical microchemistry. *Clinical Chemistry*, *32*, 213–215.

Schulze, A., Lindner, M., Kohlmuller, D., Olgemoller, K., Mayatepek, E., & Hoffmann, G. F. (2003). Expanded newborn screening for inborn errors of metabolism by electrospray ionization-tandem mass spectrometry: Results, outcome, and implications. *Pediatrics*, *111*, 1399–1406.

Science Direct. (2020). Bibliographic Database. *Sciene Direct*. https://www.sciencedirect.com/topics/computer-science/bibliographic-database.

Scolamiero, E., Cozzolino, C., Albano, L., Ansalone, A., Caterino, M., Corbo, G., Di Girolamo, M. G., Di Stefano, C., Durante, A., Franzese, G., Franzese, I., Gallo, G., Giliberti, P., Ingenito, L., Ippolito, G., Malamisura, B., Mazzeo, P., Norma, A., Ombrone, D., Parenti, G., Pellecchia, S., Pecce, R., Pierucci, I., Romanelli, R., Rossi, A., Siano, M., Stoduto, T., Villani, G. R., Andria, G., Salvatore, F., Frisso, G., & Ruoppolo, M. (2015). Targeted metabolomics in the expanded newborn screening for inborn errors of metabolism. *Molecular BioSystems*, *11*, 1525–1535.

Shah, V. S., & Ohlsson, A. 2011. Venepuncture versus heel lance for blood sampling in term neonates. *Cochrane Database of Systematic Reviews*, *18*, CD001452.

Sinclair, G., Ester, M., Horvath, G., Cvan Karnebeek, C., Stockler-Ipsirogu, S., & Vallance, H. (2016). Integrated multianalyte second-tier testing for newborn screening for MSUD, IVA, and GAMT deficiencies. *Journal of Inborn Errors of Metabolism and Screening*, *4*, 1–7.

Snowden, S. G., Korosi, A., De Rooij, S. R., & Koulman, A. (2020). Combining lipidomics and machine learning to measure clinical lipids in dried blood spots. *Metabolomics*, *16*, 83.

Sontag, M. K., Yusuf, C., Grosse, S. D., Edelman, S., Miller, J. I., Mckasson, S., Kellar-Guenther, Y., Gaffney, M., Hinton, C. F., Cuthbert, C., Singh, S., Ojodu, J., & Shapira, S. K. (2020). Infants with congenital disorders identified through newborn screening—United States, 2015–2017. *MMWR. Morbidity and Mortality Weekly Report*, *69*, 1265–1268.

Spacil, Z., Babu Kumar, A., Liao, H. C., Auray-Blais, C., Stark, S., Suhr, T. R., Scott, C. R., Turecek, F., & Gelb, M. H. (2016). Sulfatide analysis by mass spectrometry for screening of metachromatic leukodystrophy in dried blood and urine samples. *Clinical Chemistry*, *62*, 279–286.

Stapleton, M., Kubaski, F., Mason, R. W., Shintaku, H., Kobayashi, H., Yamaguchi, S., Taketani, T., Suzuki, Y., Orii, K., Orii, T., Fukao, T., & Tomatsu, S. (2020). Newborn screening for mucopolysaccharidoses: Measurement of glycosaminoglycans by LC-MS/MS. *Molecular Genetics and Metabolism Reports*, *22*, 100563.

Staunstrup, N. H., Starnawska, A., Nyegaard, M., Christiansen, L., Nielsen, A. L., Borglum, A., & Mors, O. (2016). Genome-wide DNA methylation profiling with MeDIP-seq using archived dried blood spots. *Clinical Epigenetics*, *8*, 81.

Stevens, C. (2017). *Next generation sequencing in the New York State*. Newborn Screening Molecular Lab.

Su, X., & Comeau, A. M. (1999). Cellulose as a matrix for nucleic acid purification. *Analytical Biochemistry*, *267*, 415–418.

Tarini, B. A., Christakis, D. A., & Welch, H. G. (2006). State newborn screening in the tandem mass spectrometry era: More tests, more false-positive results. *Pediatrics, 118*, 448–456.

Thein, S. L. (2013). The molecular basis of beta-thalassemia. *Cold Spring Harbor Perspectives in Medicine, 3*, a011700.

Trifonova, O. P., Maslov, D. L., Balashova, E. E., & Lokhov, P. G. (2019). Evaluation of dried blood spot sampling for clinical metabolomics: Effects of different papers and sample storage stability. *Metabolites, 9*, 277.

Tsui, L.-C. (1989). *Cystic Fibrosis Mutation Database (CFTR1)*. Cystic Fibrosis Centre at the Hospital for Sick Children in Toronto. http://www.genet.sickkids.on.ca/Home.html.

Turgeon, C., Magera, M. J., Allard, P., Tortorelli, S., Gavrilov, D., Oglesbee, D., Raymond, K., Rinaldo, P., & Matern, D. (2008). Combined newborn screening for succinylacetone, amino acids, and acylcarnitines in dried blood spots. *Clinical Chemistry, 54*, 657–664.

Turgeon, C. T., Moser, A. B., Morkrid, L., Magera, M. J., Gavrilov, D. K., Oglesbee, D., Raymond, K., Rinaldo, P., Matern, D., & Tortorelli, S. (2015). Streamlined determination of lysophosphatidylcholines in dried blood spots for newborn screening of X-linked adrenoleukodystrophy. *Molecular Genetics and Metabolism, 114*, 46–50.

Urv, T. K., & Parisi, M. A. (2017). Newborn screening: Beyond the spot. *Advances in Experimental Medicine and Biology, 1031*, 323–346.

Verschuren, M. C., Wolvers-Tettero, I. L., Breit, T. M., Noordzij, J., Van Wering, E. R., & Van Dongen, J. J. (1997). Preferential rearrangements of the T cell receptor-delta-deleting elements in human T cells. *Journal of Immunology, 158*, 1208–1216.

Vidal-Folch, N., Gavrilov, D., Raymond, K., Rinaldo, P., Tortorelli, S., Matern, D., & Oglesbee, D. (2018). Multiplex droplet digital PCR method applicable to newborn screening, carrier status, and assessment of spinal muscular atrophy. *Clinical Chemistry, 64*, 1753–1761.

Washburn, J., & Millington, D. S. (2020). Digital microfluidics in newborn screening for mucopolysaccharidoses: A progress report. *International Journal of Neonatal Screening, 6*, 78.

Willemsen, R. H., Burling, K., Barker, P., Ackland, F., Dias, R. P., Edge, J., Smith, A., Todd, J., Lopez, B., Mander, A. P., Guy, C., & Dunger, D. B. (2018). Frequent monitoring of C-peptide levels in newly diagnosed type 1 subjects using dried blood spots collected at home. *The Journal of Clinical Endocrinology and Metabolism, 103*, 3350–3358.

Wilson, J., & Jungner, G. (1968). Principles and practice of screening for disease. *Public health papers*. Geneva: World Health Organization.

Yap, S., & Naughten, E. (1998). Homocystinuria due to cystathionine beta-synthase deficiency in Ireland: 25 years' experience of a newborn screened and treated population with reference to clinical outcome and biochemical control. *Journal of Inherited Metabolic Disease*, *21*, 738–747.

Yoon, H. R. (2015). Screening newborns for metabolic disorders based on targeted metabolomics using tandem mass spectrometry. *Annals of Pediatric Endocrinology & Metabolism*, *20*, 119–124.

Young, B., Hendricks, J., Foreman, D., Pickens, C., Hovell, C., De Jesus, V., Haynes, C., & Petritis, K. (2020). Development of dried blood spot quality control materials for adenosine deaminase severe combined immunodeficiency and an LC-MS/MS method for their characterization. *Clinical Mass Spectrometry*, *17*, 4–11.

Zimmermann, E. (1939). The dried blood test for syphilis. A contribution to its simplification. *Münchener Medizinische Wochenschrift (1950)*, *86*, 1732–1733.

7

Considerations for Implementation of Microsampling in Pediatric Clinical Research and Patient Care

Ganesh S. Moorthy, Christina Vedar, and Athena F. Zuppa

Childrens Hospital of Philadelphia, Perelman School of Medicine, University of Pennsylvania, Philadelphia, PA, USA

7.1 Introduction

Microsampling has become an increasingly popular tool for its patient centric capability to obtain blood samples with minimal invasiveness. This type of sampling satisfies the wants and needs of pediatric patients, their parents/caregivers, and the clinical team caring for them. In recent years, there has been an increase in the use of microsampling in pediatric clinical care and pediatric clinical research studies (Abu-Rabie et al., 2019; Cordell et al., 2018; Dorofaeff et al., 2016; Moorthy et al., 2019a, 2020). Its ability to enable the collection of small volumes of blood (10–30 μL) is important within pediatrics, where blood collection is often limited and/or difficult to obtain. It allows for the ease of collections of multiple time points for therapeutic drug monitoring (TDM) and pharmacokinetic (PK) studies without an invasive phlebotomy or access to an invasive catheter. Sampling can be done in the clinical setting or at home, adding another layer of convenience. Although this technology is beneficial for pediatric patients, it also poses some challenges to consider. The novel method of microsampling, the samples from which laboratory personnel are generally not yet accustomed to analyzing, and the inconsistencies with sample collection (underloading or overloading of the devices) are just a few issues that have been seen in laboratory and clinical use. This chapter will discuss the benefits and challenges for pediatric clinical studies experienced by clinical staff, laboratory personnel, patients, and parents. In addition, the strategies to effectively implement these assays into pediatric clinical research and clinical care will be discussed.

Patient Centric Blood Sampling and Quantitative Bioanalysis, First Edition.
Edited by Neil Spooner, Emily Ehrenfeld, Joe Siple, and Mike S. Lee.

7.2 Considerations for Implementation

7.2.1 Benefits

The authors have noted that microsampling technologies have become more popular and demand continues to increase for pediatric clinical research. Since newer bioanalytical instruments with enhanced sensitivities are able to detect the required concentration ranges of drugs using a tiny sample volume (10–30 μL), collecting small volumes (<0.2 mL) from patients is feasible and can be used for analysis. Microsampling techniques are easy to use and patient-friendly. This section will discuss the advantages of microsampling that one should consider for implementation in pediatric clinical care and research.

Collecting smaller sample volumes in pediatrics is important for the well-being of the patients whose blood volumes are already low. It is reported that children have a total blood volume (TBV) of approximately 75–80 mL/kg, where a 10 kg child will have a TBV of about 750–800 mL (Howie, 2011). In 2011, the Bulletin of the World Health Organization highlighted existing guidelines or policies from nine different institutions for blood collection. It showed that blood collection should not exceed 1–5% of TBV over 24 hr or 10% over 8 weeks. Blood collection volumes for clinical care, as well as research studies, need to be considered before blood is drawn from the patient. Devices such as the Mitra® tip, or volumetric absorptive microsampling (VAMS®) devices (Neoteryx®), allow for the collection of small-volume samples (in whole blood or plasma) at 10, 20, or 30 μL volumes. With this device, a whole blood collection of more than 0.5 mL per sample typically collected for plasma assays is not necessary. This blood-conservation approach is significant when multiple time points are needed for pediatric PK studies or TDM. For example, a 10 kg patient with antibacterial treatment in the hospital often undergoes vancomycin and tobramycin TDM (0.5 mL per sample). The number of samples collected depends on several factors including dosing regimen and changes in clinical status. This patient may also need a daily complete blood count (CBC) and electrolyte panels (0.7–2 mL/day) over 7 days along with the TDM studies. The total amount of blood needed over the 7 days would be approximately 13–22 mL, depending on the weight of the patient, in order to perform these tests. This volume could be significantly reduced by 90% (100 vs. 10 μL) utilizing VAMS for vancomycin and tobramycin TDM.

Because collection of these small volumes is feasible, staying within the blood collection limits that are deemed safe for patients is achievable and will not place these children at increased risk of transfusion which is often a secondary effect of repeated high-volume phlebotomy. This becomes exceptionally important if a child who is undergoing clinical care requiring multiple blood tests is also enrolled in a clinical research study that requires blood (such as biomarker and PK studies). These small-volume collections also ensure that the blood collected will not go unused, as sometimes seen with plasma collections where the entire sample is not needed. With the microsampling approach, 50% of the blood collected is used for analysis. One out of two VAMS devices is used for analysis and the second device serves as a backup sample. In addition, when drawing blood from an indwelling catheter, the fluid that is first removed is often "wasted" as it is not comprises of just blood and may be contaminated by the fluid or medication that was previously administered. In pediatric care, especially with neonates, every drop of blood is precious and should not be unnecessarily collected or wasted.

Central line-associated bloodstream infection (CLABSI) is a common healthcare-associated issue, particularly harmful for critically ill patients in pediatric intensive care units (PICU; Woods-Hill et al., 2020; Wylie et al., 2010). Developing CLABSI can increase morbidity, mortality, length of stay, and costs. In a study done by the Children's Hospital of Philadelphia in the PICU, the rates for CLABSI per 1,000 central venous catheter (CVC) days were 1.95 in 2017 (Woods-Hill et al., 2020). Because patients in the PICU have daily clinical blood draws, along with blood draws for research studies, accessing the catheters multiple times a day increases the chance of infection. Utilizing VAMS devices in the PICU can potentially reduce the chance of infection, since the CVC is not required for microsampling collection. Instead, finger or heel sticks can eliminate the constant access to the catheter. VAMS devices are minimally invasive, which will decrease infection, therefore decreasing morbidity, mortality, length of stay, and associated costs.

With the utilization of microsampling technology, less invasive sample collection is possible and this can be performed in a geographical location that is more convenient to the patient and their carers. A simple finger or heel prick makes it easier for parents to collect samples from their children in the comfort of their homes. A trained professional (phlebotomist, nurse, or clinician) is not needed for these collections because of the user-friendliness of microsampling. Many children (and some adults too) have a fear of needles when needing to draw blood. Using VAMS or other microsampling devices can ease these young patients of their anxiety about needles and eliminate the need for venipuncture blood draws, which can cause the patient discomfort. Microsampling allows for someone the child trusts to collect samples in a place they feel safe, which can reduce stress for the child.

Another benefit to consider with VAMS and other microsampling collection approaches is that parents no longer have to bring their children to a clinic or laboratory for blood draws. They have the ability to collect samples in the comfort of their own homes when it is convenient. With microsampling, trips to and from any clinical setting can be eliminated. Taking time off work for the parent(s) or taking time out of school for the children is no longer a consideration. Children who are immunocompromised or technology-dependent, requiring special medical equipment (mechanical ventilators, cardiorespiratory monitors, oxygen supplementation, etc.), can stay at home. The ability to sample at home is beneficial for immunocompromised patients, where traveling and sitting in close proximity to other sick patients in a clinic or laboratory setting poses a potential risk for infection. In recent times, during the COVID-19 pandemic, sampling at home for immunocompromised patients is extremely important since this infectious disease is easily transmissible. Parents are no longer burdened with the need to get children ready to go, pack necessary essentials, and spend time in traffic with this convenient form of sample collection.

Because of technologies such as Facetime or Skype, training parents to collect samples using microsampling techniques is manageable and easy to achieve. Using these technologies has become more "normal" with the COVID-19 pandemic, making it easier for patients and parents to communicate with healthcare providers without needing to risk the chances of getting sick by being present in a doctor's office with other sick patients. Parents can learn the technique for sample collection, voice any concerns, and ask any questions face-to-face (virtually) without coming into the clinic or laboratory. There are also videos available for training that provide parents with easy to follow instructions on how to properly use microsampling. These online tutorials allow the viewer to go back and revisit the training video in case they have forgotten specific steps for collection at any time it is needed (via the Internet or download).

Once samples are collected, parents or caregivers can keep them at room temperature for drying and mail them straight to the laboratory for analysis. Dried microsamples can be shipped at ambient temperatures, depending on the stability of the drug of interest, eliminating the need for cold-chain shipping (cold packs or dry ice). This also reduces costs. It also allows for PK studies and TDM to be done according to the patient's schedule in the comfort of their homes.

7.2.1.1 Clinical Research

Several laboratories have successfully developed and validated bioanalytical methods utilizing VAMS devices for the quantitation of various drugs and metabolites (Abu-Rabie et al., 2019; Barco et al., 2017; Cordell et al., 2018; D'Urso et al., 2019; Friedl et al., 2019; Marahatta et al., 2016; Moorthy et al., 2019a, 2020; Velghe et al., 2020). Some of these assays have been implemented for pediatric clinical research studies (Abu-Rabie et al., 2019; Barco et al., 2017; Cordell et al., 2018; Delahaye et al., 2019; Moorthy et al., 2019a, 2020; Velghe et al., 2020). A simple schematic for the process of sample collection, extraction, and sample analysis for VAMS devices in a PK study for cefepime is shown in Figure 7.1 (derived from Moorthy et al., 2020). With more experience and usage of the VAMS devices, more insight is gained on how to effectively use and implement this

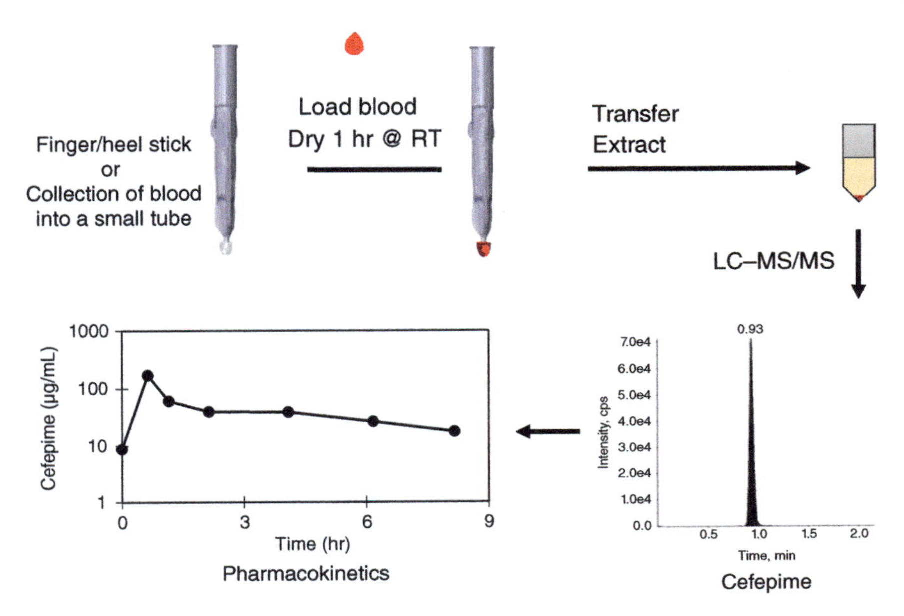

Figure 7.1 Simple schematic for the process of VAMS™ sample collection, extraction, and data analysis. Blood can be loaded onto the VAMS device from a finger, heel prick, or blood collection tube.

technology into pediatric clinical studies. These analytical and clinically validated methods in clinical research show promise for increased implementation in future clinical care.

7.2.1.2 Clinical Care

Microsampling is a valuable tool for pediatric clinical research. However, very few assays are currently in routine use for clinical care. Recently, a TDM VAMS assay was developed, validated, and implemented for the immunosuppressant tacrolimus (Vethe et al., 2019). Tacrolimus concentrations in capillary microsamples and liquid venous samples were investigated in stable renal recipients. Sampling throughout the 12-hr dose interval was examined on 2 separate days. Venous wet whole blood and capillary VAMS samples were obtained at each time point ($n = 2$ samples per time point). One set was delivered by courier directly to the laboratory for analysis the following day and the other was sent to the lab through standard mail. This study demonstrated that tacrolimus concentrations were reliably quantified throughout the dosing interval by using a VAMS assay for renal transplant recipients, and the results were not affected by shipping. Advances and familiarity with microsampling techniques in healthcare settings will increase the future implementation of this technology in clinical care.

7.2.2 Challenges

With all the great benefits of microsampling mentioned in the previous section, there are some challenges utilizing these devices for pediatric sample collection in the clinic, the laboratory, and at home. There are often concerns with training clinical staff, learning how to collect and analyze this new sample matrix (dried microsamples), and consistently collecting whole blood with the VAMS devices for analysis. This section will discuss some of the typical challenges encountered when implementing the microsampling technique in the pediatric care setting and some ways in which such challenges can be overcome.

A common challenge with microsampling is the concern with underloading and overloading of samples using VAMS devices in clinical sample collection. Figure 7.2 shows a visual representation of a blank tip (a), normally loaded tip (b), underloaded tip (c) and overloaded tip (d). Often, duplicate samples are collected at the same time point in case of the need for reanalysis. However, sometimes one or both of the samples are not loaded correctly (either underloaded or overloaded based on visual inspection), which makes analysis or reanalysis difficult to assess. Calculated values from these samples will either be under- or overestimated. In this case, the data should be interpreted with caution, and the principal investigator should be notified of any discrepancies with sample loading. Quantitative assays require exact matrix volume for accurate results. The challenges with underloading and overloading prevent accurate and precise results, rendering samples nonevaluable when analyzing TDM or PK data.

Under- and over-loading of VAMS devices are occasionally observed in clinical studies conducted in the clinic and at home. In our practice, under- and over-loaded clinical samples are documented during sample receipt. From our studies, there was an overall 82.1% success rate for correctly (normal) loaded samples in

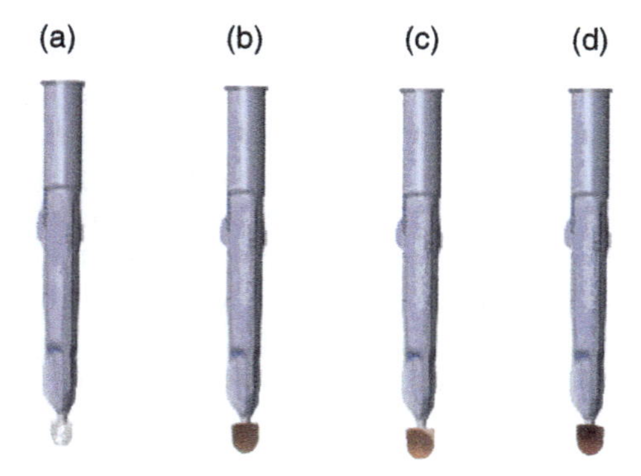

Figure 7.2 Visual representation of (a) blank VAMS device (no blood), (b) normally loaded VAMS device, (c) under-loaded VAMS device, and (d) over-loaded VAMS device.

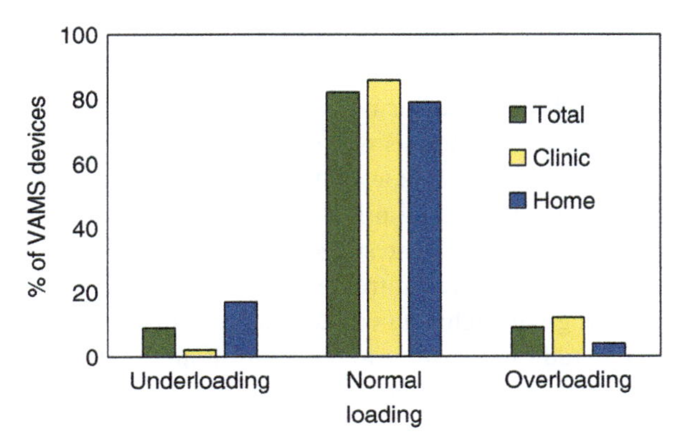

Figure 7.3 Comparison of the % difference in under -, normal, and overloading of VAMS devices in the clinic, at-home, and at both clinic and home (total).

these clinic and at-home studies (Figure 7.3). Underloaded samples are more common for at-home sampling (17.0 %), and overloading occurs more frequently for samples collected in the clinic (12.1%). Underloading samples at home most likely happens because patients or in the case of these studies, parents or caregivers, are not able to get enough blood from the finger stick alone and may be unaware of certain techniques to produce more blood flow. For the studies performed in the clinic, samples are collected (<100 μL) into a syringe or a small vial (often along with blood draws for other clinical tests) and then loaded onto the VAMS devices. There are also substudies conducted in the clinic to collect samples via direct finger or heel stick to see the relationship between samples collected from an existing arterial catheter to the capillary (finger/heel stick) samples at the same time point (unpublished data). Clinical staff may be overloading these samples by keeping the absorbent tip in the blood for too long, or completely submerging the absorbent tip plus a bit of the plastic holder into the blood instead of just touching the tip to the surface of the blood. Under- and overloading can be prevented with proper and thorough training for parents, caregivers, and clinical staff.

7.2.2.1 Clinical Research

Clinical research staff are a critical and essential part of the team when implementing any research study in a clinical setting. All members should be trained and knowledgeable of study conduct and microsampling procedures in their unit. However, training all clinical staff can be a challenging task. Shift changes in hospitals can be problematic, especially having some staff available during overnight hours. There are typically two types of training that can be conducted individually or in a group setting: (a) just-in-time (JIT) training and (b) structured or scheduled training. JIT training is an approach where training is provided when it is needed or being implemented by the trainee. Structured or scheduled training is the approach where training is provided at a scheduled time in advance before it is implemented. With JIT, parents at home or clinical staff in the hospital can be trained on the spot for VAMS sample collection. However, this becomes difficult during overnight shifts when the trainers are not available. For scheduled training, overnight staff may not be available during normal business hours, when microsampling training would typically occur. One suggestion for mitigating this challenge is to ask a small number of clinical staff in the unit to become champions and to serve as a resource during overnight and weekend shifts when research staff may not be available. TDM studies can run into overnight hours making it important that these staff members are properly trained for sample collection. A lack of training may cause inconsistency with sample loading, which will cause deviation in results during sample analysis. Although it is an easy technology to learn and use, it may be overwhelming, especially to a new nurse who has a busy schedule with several patients to care for during his or her shift. In addition to normal workload, staff need to be aware of how long samples are dried at room temperature (before transferring to lower temperatures if necessary), time of sample collection, and how to properly load the devices with blood without undeloading or overloading them.

Collecting multiple time points at home could cause some inconvenience to the patient, parent, or caregiver. In a study conducted in a children's focus group, children (ages 10–22) were asked questions regarding their opinions on sampling at home with VAMS devices versus sampling in the clinic (Wickremsinhe et al., 2019). The participants preferred collecting a single sample once a day, rather than collecting multiple samples throughout an entire day via finger stick. It can be a grueling task for a child to receive multiple finger or heel sticks throughout the day. A child will not want to be bothered with more than one finger or heel stick per day, especially if the first one was painful. There are ways to overcome the pain associated with the finger or heel stick such as using lidocaine (4%), an external cream, to numb the area of sampling prior to using the lancet for sample collection.

7.2.2.2 Clinical Care

In routine TDM, serum or plasma is the most commonly used biological matrix, and whole blood is used only for some therapies such as immunosuppressive and antimalarial drugs. The drug concentrations of VAMS samples are measured as whole blood values. If it is common in regular TDM to report drug levels and reference values in whole blood, there will likely be no difference between the reporting of venous samples or VAMS samples as previously shown for other drugs (Vethe et al., 2019). However, most drug concentrations in TDM are reported as serum or plasma values, and reference ranges are based on these serum or plasma levels. Hence, if the results are reported as whole blood VAMS values, another reference range should be defined, which may be confusing for clinical practice. Therefore, it is critical to develop approaches to convert the whole blood VAMS values into plasma values to be successfully implemented in clinical care (see Section 7.2.3).

7.2.3 Laboratory Challenges and Considerations

Laboratories have mastered the handling and analysis of plasma assays in their facilities. However, dried microsamples can be challenging for laboratory personnel to adopt since it is a new type of sample matrix. VAMS extractions can be a simple process for drug quantitation. However, the extraction efficiency varies depending on the physical and chemical properties of the analytes. Typically, analyte(s) extract differently from the dried microsamples compared to plasma samples and often involves more troubleshooting than a traditional plasma assay.

There have been several approaches evaluated to establish the correlation between plasma and dried microsamples. One approach is utilizing *in vitro* blood-to-plasma partitioning to establish correlation between plasma and whole blood VAMS concentrations. While plasma assays have been used as the "gold standard" for many years, introducing a new sample matrix raises questions about the relationship between whole blood VAMS and plasma concentrations. Previous studies have evaluated blood-to-plasma partitioning of several antiepileptic drugs (D'Urso et al., 2019). While the measured blood to plasma ratios were close to 1 for several compounds, there were cases where some of the antimalarial drugs were distributed more in red blood cells or plasma. An *in vitro* blood to plasma study can be performed to understand the distribution of the drug in whole blood (Figure 7.4).

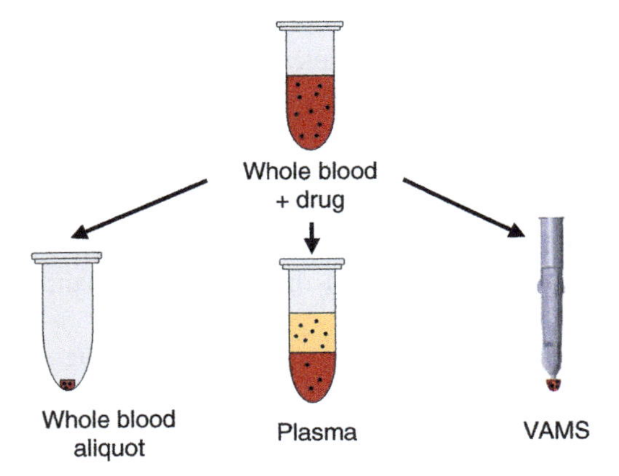

Figure 7.4 In vitro comparison of whole blood, whole blood VAMS, and plasma.

In a recent study, voriconazole had a VAMS to plasma ratio of 0.836–0.935 (Moorthy et al., 2019b), demonstrating that the drug is equally distributed between erythrocytes and plasma. However, cefepime had a VAMS to plasma ratio of 0.529–0.570 (Moorthy et al., 2020), demonstrating that more drug is associated with plasma compared to erythrocytes. *In vitro* VAMS to plasma ratios are useful to estimate plasma concentrations from whole blood VAMS. However, the results should be interpreted with caution as other parameters such as hematocrit and sample homogeneity could impact the whole blood VAMS—plasma correlation.

Assay method development, validation, and stability studies need to be performed prior to the start of a clinical study. This is important in order to clearly establish guidelines and instructions for clinical staff. In the Manual of Operations, sample collection, handling, and storage should be easy to understand in order to consistently achieve reliable and quality data. All staff need to be properly trained on loading samples, extraction, and data analysis, which can be a time-consuming process.

When performing the validation experiments, fresh blood needs to be collected from a healthy volunteer each day to prepare calibration standards (CS) and quality control (QC) samples to perform the analysis. It is unclear how long the quality of whole blood remains intact after storing at 4 °C for a given period of time. According to Ledvina et al., whole blood is said to be viable up to 5 days after collection in refrigerated conditions and is comparable to freshly collected whole blood (Ledvina et al., 2019). American Red Cross states that blood stability can be up to 21–35 days depending on the anticoagulant used (Cross, 2020). It is difficult to mimic the exact *in vivo* conditions in the laboratory for VAMS clinical studies where samples are collected from patients directly by finger or heel stick with no anticoagulant. The presence or lack of anticoagulant may affect drug concentrations in clinical samples. All VAMS samples collected will not have anticoagulant present since they will be collecting samples via finger sticks directly onto VAMS devices, unlike in the laboratory where CS and QC samples are first collected in EDTA or heparin tubes and then spiked with the drug of interest. Analyzing samples collected from the sampling site (via finger stick) and with anticoagulant (via venous blood draw) should be assessed and established for the patient's ease of sample collection at home (Cross, 2020). The effect of anticoagulant can also be explored as part of the validation of the bioanalytical method, for example by looking at the impact of alternative anticoagulants on the assay performance (Evans et al., 2015).

Variability in sample collection is often a significant issue in micro sampling. It is important that all team members can load microsamples consistently as it is essential that the exact volume is loaded for successful sample extraction and analysis. Samples that are not evenly loaded with blood can cause deviation from true drug concentrations. In a study performed by Denniff et al., they observed beginner interoperator variability on loading 10 μL VAMS devices. Accurate volume was measured by weighing the amount of blood absorbed divided by the blood density (Denniff & Spooner, 2014; Protti et al., 2019). They observed an 8.7–8.1% accuracy (bias) with a precision (%CV) ranged from 2.8 to 5.9% for interoperator variability (Denniff & Spooner, 2014). Although most of the operators were able to successfully load the VAMS devices without too much overloading, it is noted that one operator had a higher variance with loading samples. This demonstrates the importance of proper training for staff to successfully implement microsampling assays for clinical research and patient care.

7.2.4 Survey Results on Feasibility

A questionnaire was given to Research Assistants (RA) and Research Coordinators (RC) at our institution to evaluate the feasibility of microsampling compared to traditional plasma sampling. The survey participants have experience with traditional venipuncture whole blood and plasma collection as well as collecting VAMS samples. They were asked for their opinions on: (a) the ease of sample collection, (b) if the method is patient friendly, (c) if it benefits the patient's overall health and well-being, (d) if the method should continue or be implemented into routine clinical care, and (e) how often they experience issues, regarding both sampling collection processes (plasma versus VAMS). The questionnaire showed that VAMS sample collection is favorable over the traditional blood and plasma collections (Figure 7.5). It revealed that the RAs and RCs feel the need for better training materials and methods. If there were times dedicated to training where nurses can remain focused on the task at hand, a better outcome for accurately loading tips may be achieved. It would be helpful to have visual aids to help those collecting (whether at-home or in the clinic) to better understand the appearance of an under-loaded and over-loaded tip. The RAs and RCs feel as though it is difficult to know when a tip is fully loaded, especially if one does not have any prior

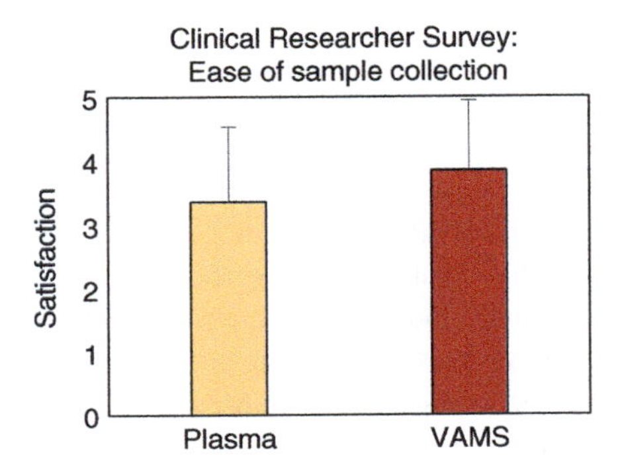

Figure 7.5 Summary of survey results for the preference of plasma versus VAMS devices from a questionnaire given to RAs and RCs at CHOP.

experience with VAMS and if there are no aids for guidance. Traditional blood collection can be difficult for RAs and RCs in research studies because it requires access to a patient's CVC. Because of this, blood collection for research studies is sometimes not feasible because a patient's CVC has already been accessed multiple times in the day for other clinical testing. Utilizing VAMS devices could reduce the number of times the CVC is accessed and increase the number of samples collected for clinical studies.

As suggested by the RAs and RCs on the clinical research team, dedicated time for training seminars could be held within the hospital community so clinical staff can learn more about microsampling, how they work, and explain the potential benefits they have for patients, as well as practice using the technologies. These seminars could be conducted virtually so that more clinical staff could attend offsite or refer back to them as needed. As clinical staff become more aware of how to use microsampling devices, they can be implemented into the hospital as part of standard of care involving routine sample collection.

7.3 Conclusion

This patient centric form of sample collection is a useful tool in many aspects of pediatric research and patient care. The at-home sampling capabilities allow for pediatric patients and their parents or caregivers to collect samples from home. With these technologies, collection of tiny volumes of blood is achievable, which benefits the patient by limiting excess blood collection. The low blood volumes can be used for clinical assays because modern instruments have the ability to detect low concentrations of drugs and metabolites. Microsampling has many benefits for the patient as described in this chapter including the collection of small volumes of blood, reducing the risk for infection, alleviating anxiety with venipuncture blood draws, and eliminating the need for a clinical visit. Despite some of the challenges discussed in this chapter, microsampling has the potential to be implemented into routine pediatric clinical care.

References

Abu-Rabie, P., Neupane, B., Spooner, N.Rudge J, Denniff P, Mulla H, Pandya H (2019). Validation of methods for determining pediatric midazolam using wet whole blood and volumetric absorptive microsampling. *Bioanalysis*, *11*(19), 1737–1754. 10.4155/bio-2019-0190. Epub 2019 Oct 16. PMID: 31617393.

Barco, S., Castagnola, E., Moscatelli, A., Rudge, J., Tripodi, G., & Cangemi, G. (2017). Volumetric adsorptive microsampling-liquid chromatography tandem mass spectrometry assay for the simultaneous quantification of four antibiotics in human blood: Method development, validation and comparison with dried blood spot. *Journal of Pharmaceutical and Biomedical Analysis, 145*, 704–710.

Cordell, R. L., Valkenburg, T. S. E., Pandya, H. C., Hawcutt, D. B., Semple, M. G., & Monks, P. S. (2018). Quantitation of salbutamol using micro-volume blood sampling—Applications to exacerbations of pediatric asthma. *The Journal of Asthma, 55*(11), 1205–1213.

Cross, A. R. (2020). *How can one donation help multiple people?* https://www.redcrossblood.org/donate-blood/how-to-donate/types-of-blood-donations/blood-components.html#:~:text=The%20transfusable%20components%20that%20canfrom%20donated%20blood%20before%20transfusion.

Delahaye, L., Dhont, E., De Cock, P., De Paepe, P., Stove, C. P. (2019). Volumetric absorptive microsampling as an alternative sampling strategy for the determination of paracetamol in blood and cerebrospinal fluid. *Analytical and Bioanalytical Chemistry*, 411(1), 181–191.

Denniff, P., & Spooner, N. (2014). Volumetric absorptive microsampling: A dried sample collection technique for quantitative bioanalysis. *Analytical Chemistry, 86*(16), 8489–8495.

Dorofaeff, T., Bandini, R. M., Lipman, J., Ballot, D. E., Roberts, J. A., & Parker, S. L. (2016). Uncertainty in antibiotic dosing in critically ill neonate and pediatric patients: Can microsampling provide the answers? *Clinical Therapeutics, 38*(9), 1961–1975.

D'Urso, A., Rudge, J., Patsalos, P. N., & de Grazia, U. (2019). Volumetric absorptive microsampling: A new sampling tool for therapeutic drug monitoring of antiepileptic drugs. *Therapeutic Drug Monitoring, 41*(5), 681–692.

Evans, C., Arnold, M., Bryan, P., Duggan J, James CA, Li W, Lowes S, Matassa L, Olah T, Timmerman P, Wang X, Wickremsinhe E, Williams J, Woolf E, Zane P. (2015). Implementing dried blood spot sampling for clinical pharmacokinetic determinations: Considerations from the IQ Consortium Microsampling Working Group. *The AAPS Journal, 17*(2), 292–300. 10.1208/s12248-014-9695-3. Epub 2014 Dec 9. PMID: 25488054; PMCID: PMC4365086.

Friedl, B., Kurlbaum, M., Kroiss, M., Fassnacht, M., & Scherf-Clavel, O. (2019). A method for the minimally invasive drug monitoring of mitotane by means of volumetric absorptive microsampling for a home-based therapeutic drug monitoring. *Analytical and Bioanalytical Chemistry, 411*(17), 3951–3962.

Howie, S. R. (2011). Blood sample volumes in child health research: Review of safe limits. *Bulletin of the World Health Organization 89*(1), 46–53.

Ledvina, A. R., Ewles, M., Pang, Y., & Cape, S. (2019). Whole blood stability in quantitative bioanalysis. *Bioanalysis, 11*(20), 1885–1897.

Marahatta, A., Megaraj, V., McGann, P. T., Ware, R. E., & Setchell, K. D. (2016). Stable-isotope dilution HPLC-electrospray ionization tandem mass spectrometry method for quantifying hydroxyurea in dried blood samples. *Clinical Chemistry*, *62*(12), 1593–1601.

Moorthy, G. S., Vedar, C., DiLiberto, M. A., & Zuppa, A. F. (2019a). A patient-centric liquid chromatography-tandem mass spectrometry microsampling assay for analysis of cannabinoids in human whole blood: Application to pediatric pharmacokinetic study. *Journal of Chromatography. B, Analytical Technologies in the Biomedical and Life Sciences*, *1130–1131*, 121828.

Moorthy, G. S., Vedar, C., Zane, N., Prodell, J. L., & Zuppa, A. F. (2019b). Development and validation of a volumetric absorptive microsampling assay for analysis of voriconazole and voriconazole N-oxide in human whole blood. *Journal of Chromatography B: Analytical Technologies in the Biomedical and Life Sciences*, *1105*, 67–75.

Moorthy, G. S., Vedar, C., Zane, N. R., Downes KJ, Prodell JL, DiLiberto MA, Zuppa AF. (2020). Development and validation of a volumetric absorptive microsampling-liquid chromatography mass spectrometry method for the analysis of cefepime in human whole blood: Application to pediatric pharmacokinetic study. *Journal of Pharmaceutical and Biomedical Analysis*, *179*, 113002. 10.1016/j.jpba.2019.113002. Epub 2019 Nov 20. PMID: 31785929; PMCID: PMC6943186.

Protti, M., Mandrioli, R., & Mercolini, L. (2019). Tutorial: Volumetric absorptive microsampling (VAMS). *Analytica Chimica Acta*, *1046*, 32–47.

Velghe, S., Delahaye, L., & Ogwang, R., Hotterbeekx A, Colebunders R, Mandro M, Idro R, Stove CP. (2020). Dried blood microsampling-based therapeutic drug monitoring of anti-epileptic drugs in children with nodding syndrome and epilepsy in Uganda and the Democratic Republic of the Congo. *Therapeutic Drug Monitoring*, *42*(3), 481–490.

Vethe, N. T., Gustavsen, M. T., Midtvedt, K., Lauritsen ME, Andersen AM, Åsberg A, Bergan S. (2019). Tacrolimus can be reliably measured with volumetric absorptive capillary microsampling throughout the dose interval in renal transplant recipients. *Therapeutic Drug Monitoring*, *41*(5), 607–614.

Wickremsinhe, E., Short, M., Talkington, B., & West, L. (2019). DIY blood sampling for pediatric clinical trials—The patients perspective. *Applied Clinical Trials*. http://www.appliedclinicaltrialsonline.com/diy-blood-sampling-pediatric-clinical-trials-patients-perspective?pageID=2.

Woods-Hill, C. Z., Srinivasan, L., Schriver, E., Haj-Hassan, T., Bezpalko, O., & Sammons, J. S. (2020). Novel risk factors for central-line associated bloodstream infections in critically ill children. *Infection Control and Hospital Epidemiology*, *41*(1), 67–72.

Wylie, M. C., Graham, D. A., Potter-Bynoe, G., Kleinman ME, Randolph AG, Costello JM, Sandora TJ. (2010). Risk factors for central line-associated bloodstream infection in pediatric intensive care units. *Infection Control and Hospital Epidemiology*, *31*(10), 1049–1056.

8

Simplification of Home Urine Sampling for Measurement of 2,8-Dihydroxyadenine in Patients with Adenine Phosphoribosyltransferase Deficiency

Unnur A. Thorsteinsdottir[1], *Hrafnhildur L. Runolfsdottir*[2], *Vidar O. Edvardsson*[3,4], *Runolfur Palsson*[2,4] *and Margret Thorsteinsdottir*[1,5]

[1] *Faculty of Pharmaceutical Sciences, School of Health Sciences, University of Iceland, Reykjavik, Iceland*
[2] *Internal Medicine Services, Landspitali–The National University Hospital of Iceland, Reykjavik, Iceland*
[3] *Children's Medical Center, Landspitali–The National University Hospital of Iceland, Reykjavik, Iceland*
[4] *Faculty of Medicine, School of Health Sciences, University of Iceland, Reykjavik, Iceland*
[5] *ArcticMass, Reykjavik, Iceland*

8.1 Introduction

The analysis of urine samples is an important part of clinical assessment, both for diagnostic purposes and therapeutic drug monitoring (TDM). At-home timed urine collections have through the years been used to evaluate the renal excretion of various biomarkers, and the collection of all urine voided in a 24-hr period is by far the best studied approach which for numerous metabolites is still considered the gold standard method. A correctly performed 24-hr urine collection accurately measures the quantity of a given biomarker excreted in that time period, and the results may be used for diagnosis and assessment of prognosis in numerous disorders (Corder et al., 2002; Fishel Bartal et al., 2022). However, despite the implementation of various strategies intended to assure adherence to 24-hr urine collection protocols, inaccuracy remains a major concern, thereby complicating the interpretation of test results (Côté et al., 2008).

The most commonly encountered problems associated with timed urine sampling include incomplete collection due to missed voids and collection beyond the recorded time period. In light of the above, simpler and more precise sampling methods are needed. Collection of the first-morning urine sample is preferred over a random urine specimen, as it is less likely to be affected by factors such as hydration and exercise. In recent years, several studies have shown an

Patient Centric Blood Sampling and Quantitative Bioanalysis, First Edition.
Edited by Neil Spooner, Emily Ehrenfeld, Joe Siple, and Mike S. Lee.

excellent correlation between first-morning void urine biomarker-to-creatinine (Cr) ratio and 24-hr urinary biomarker excretion. Hence, the assessment of solute or biomarker-to-Cr ratio in random urine samples has been increasingly used for the quantification of renal solute or biomarker excretion. A good example is the well-documented correlation between the first-morning void protein-to-Cr ratio and albumin-to-Cr ratio and the 24-hr urinary protein and albumin excretion, respectively (Jensen et al., 1997; Justesen et al., 2006; Price et al., 2005). This approach has in recent years almost eliminated the need for timed urine collections in the evaluation of proteinuria and albuminuria in clinical practice.

Liquid chromatography coupled to tandem mass spectrometry is a widely applied analytical technique for the measurement of a broad range of urinary analytes in clinical laboratories (van der Gugten, 2020; Wells, 2018), used for evaluation of response to therapies, various toxicology studies, diagnosis of inborn errors of metabolism, and for various other purposes (French, 2017; Grant and Rappold, 2018). Urine sampling is easy, the collection is noninvasive, and dilution can be used for sample preparation. Sample dilution is simple, inexpensive (Bi et al., 2020), can be used for analytes with different physicochemical properties, and is easily automated (Stone, 2017).

Repeated measurements of urinary 2,8-dihydroxyadenine (DHA) excretion are essential for monitoring of pharmacotherapy in patients with adenine phosphoribosyltransferase (APRT) deficiency, with the goal of preventing new urinary stone formation and progressive crystal nephropathy. The frequency of urinary DHA measurements required, and the problems associated with 24-hr urine collections described above are a matter of major concern. Thus, the use of single-void urine samples replacing timed urine collections for the purpose of TDM is an attractive option as they almost entirely eliminate sampling errors, are much less time-consuming and more convenient for patients. In this chapter, we present a case study of the correlation between DHA-to-Cr ratio in first-morning void urine samples and 24-hr urinary DHA excretion in patients with APRT deficiency. The quantification of urinary DHA concentration was carried out using a novel ultra-performance liquid chromatography coupled to tandem mass spectrometry (UPLC-MS/MS) assay developed by our group (Thorsteinsdottir et al., 2016). Urine Cr was measured using standard methodology in the clinical laboratory at our hospital.

8.1.1 Adenine Phosphoribosyltransferase Deficiency

APRT deficiency (OMIM, 2016) is a rare autosomal recessive disorder of adenine metabolism resulting in the generation and renal excretion of large amounts of poorly soluble DHA. Precipitation of DHA leads to stone formation in the urinary tract and crystal nephropathy which causes progressive chronic kidney disease, frequently resulting in end-stage kidney failure (Bollée et al., 2010; Runolfsdottir et al., 2016). APRT deficiency was first described in 1968 by Kelley and coworkers who reported partial deficiency of the enzyme in four asymptomatic subjects (Kelley et al., 1968). The first diagnosed symptomatic case of APRT deficiency was reported in 1974 in a patient from France (Cartier et al. 1974). The disorder has since been described in all ethnic groups, while the majority of reported cases have come from Japan, France, and Iceland (Edvardsson et al., 2012). APRT is a cytoplasmic enzyme that catalyzes the synthesis of 5′-adenosine monophosphate from adenine and 5-phosphoribosyl-1-pyrophosphate (PRPP), leading to effective recycling of adenine (Sahota et al., 2001) (Figure 8.1).

Radiolucent kidney stones are by far the most common clinical manifestation of APRT deficiency, described in at least 60% of cases. End-stage kidney disease, even without a past history of kidney stones, is the presenting feature in at least 15% of adults with the disorder (Bollée et al., 2010; Runolfsdottir et al., 2016). Although a significant number of patients are asymptomatic at diagnosis, APRT deficiency has first been diagnosed after kidney transplantation in a number of cases (Kaartinen et al., 2014; Nasr et al., 2010; Runolfsdottir et al., 2020). Treatment of APRT deficiency with a xanthine oxidoreductase (XOR; previously xanthine oxidase and xanthine dehydrogenase) inhibitor, allopurinol or febuxostat, effectively reduces urinary DHA excretion and alleviates or even prevents disease progression (Edvardsson et al., 2012).

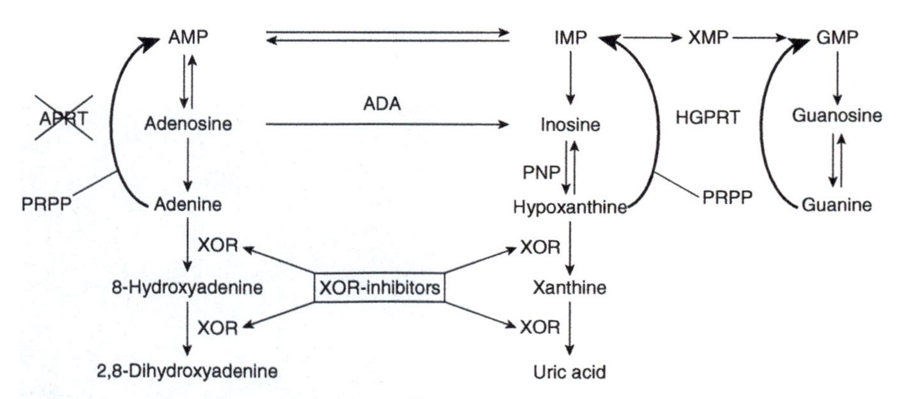

Figure 8.1 Overview of adenine metabolism. Adenine is oxidized by xanthine oxidoreductase (XOR) to 2,8-dihydroxyadenine via 8-hydroxyadenine when APRT is nonfunctional. ADA, adenosine deaminase; AMP, 5′-adenosine monophosphate; APRT, adenine phosphoribosyltransferase; GMP, guanosine monophosphate; HPRT, hypoxanthine phosphoribosyltransferase; IMP, inosine monophosphate; PNP, purine nucleoside phosphorylase; PRPP, 5-phosphoribosyl-1-pyrophosphate; XMP, xanthine monophosphate.

8.1.2 Diagnosis of Adenine Phosphoribosyltransferase Deficiency

The diagnosis of APRT deficiency is traditionally confirmed by demonstration of absent APRT enzyme activity in red cell lysates or biallelic pathogenic mutations in the *APRT* gene (Edvardsson et al., 2012). Detection of the characteristic DHA crystals by urine microscopy (Figure 8.2) is the most common finding that raises the suspicion of APRT deficiency. However, microscopic examination of the urine sediment has not proven to be reliable enough for diagnostic purposes as the DHA crystals are often misidentified or overlooked. The effectiveness of pharmacotherapy has traditionally been monitored by urine microscopy, with the absence of DHA crystals considered suggestive of adequate treatment. Unfortunately, it remains unclear if paucity or absence of crystals in the urine sediment truly reflects levels of DHA excretion that are low enough to prevent progression of chronic kidney disease. As briefly mentioned earlier, we have developed and optimized a UPLC-MS/MS assay for absolute quantification of urine DHA, using the chemometric approach design of experiment. Linearity was achieved over the concentration range of 100–5,000 ng/mL, with acceptable accuracy and precision on three different days. Processed urine samples were stable in the autosampler for up to 18hr at 22 °C. The UPLC-MS/MS urinary DHA assay allows for more precise monitoring of pharmacotherapy and additionally appears to be diagnostic of the disorder in patients with APRT deficiency (Runolfsdottir et al., 2019).

In the following section, we will describe the studies that form the basis for our UPLC-MS/MS urinary DHA assay.

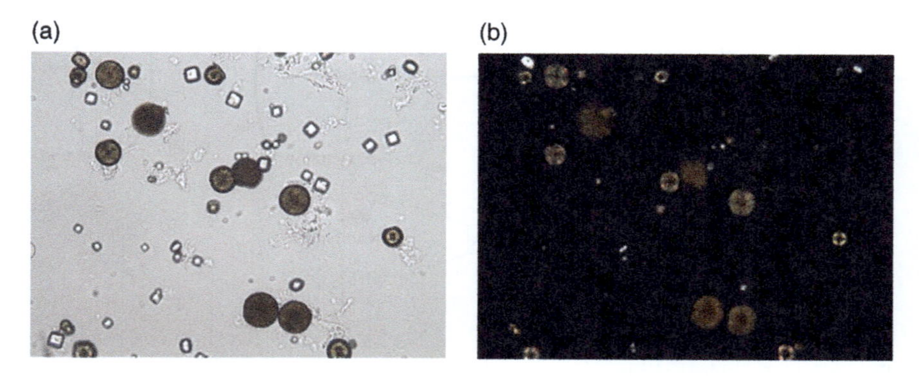

Figure 8.2 Urinary 2,8-dihydroxyadenine crystals. (a) The characteristic medium-sized crystals are brown, featuring a dark outline and central spicules. (Original magnification ×400.) (b) The same field viewed with polarized light microscopy, revealing the small- and medium-sized crystals that appear yellow in color and produce a central Maltese cross pattern. (Original magnification ×400).

8.2 Methods

8.2.1 Sample Collection

High-quality datasets and matching biological samples from patients enrolled in the APRT Deficiency Registry and Biobank of the Rare Kidney Stone Consortium (http://www.rarekidneystones.org/) were used for the purpose of a study of urinary DHA quantification. The registry and biobank were approved by the National Bioethics Committee of Iceland (NBC 09-072). In this case series, the objective was to study the correlation between the DHA-to-Cr ratio in first-morning void urine samples and 24-hr urinary DHA excretion in patients with APRT deficiency. The study consisted of three periods: (a) first, a seven-day washout period where the participants were taken off XOR inhibitor therapy; (b) this was followed by a two-week treatment period where all subjects received allopurinol as a single daily dose of 400 mg; and (c) finally another 2-week treatment period where all participants received febuxostat as a single daily dose of 80 mg. All urine samples were obtained on the last day of each of the three study periods.

The participants were provided with urine containers and both verbal and written instructions for the sample collection. They were urged to adhere to their normal diet and physical activities throughout the study period. The urine collection was initiated in the morning, and the participants were asked to empty their bladder into a separate container, marking the date and time as the beginning of the sample collection. The patients were instructed to collect all voids in the next 24 hr into the collection bottle, with the last sampling being the first urine voided the following morning, at the exact same time as the previous morning's first voided sample. The patients were advised to store the urine collection bottle at room temperature and were asked to bring the first-morning void sample and the 24-hr collection to the hospital laboratory at their earliest convenience. All urine samples were placed in the biobank for storage at −80 °C.

8.2.2 Preparation of Urine Samples for Analysis

The analytical standard DHA and the internal standard 2,8-dihydroxyadenine-2-^{13}C-1,3-$^{15}N_2$ (2,8-DHA-2-^{13}C-1,3-$^{15}N_2$) were synthesized by our collaborator in the Department of Chemistry at the University of Iceland. Stock solutions of DHA and 2,8-DHA-2-^{13}C-1,3-$^{15}N_2$ were prepared in 100 mM ammonium hydroxide (NH_4OH) at a concentration of 100 µg/mL and stored at −20 °C. Calibration curve and quality control (QC) samples were prepared fresh for each analysis at a concentration of 100, 150, 250, 400, 600, 1,000, 3,000, and 5,000 ng/mL for the calibration curve and 300, 800, and 2,000 ng/mL for the QC samples. Intermediate working solutions for the calibration curve, lower limit of quantification (LLOQ) samples, and QC samples were prepared in 10 mM NH_4OH at a concentration of 200, 300, 500, 600, 800, 1,200, 1,600, 2,000, 4,000, 6,000, and 10,000 ng/mL. A working solution of the internal standard was prepared in 10 mM NH_4OH at a concentration of 8,000 ng/mL. Urine samples donated by healthy volunteers, diluted 1:15 (v/v) with 10 mM NH_4OH, were used as a blank matrix. The urine samples that had been collected by the participants at home were prepared for measurement in the following manner:

1) 1 mL of each urine sample was aliquoted into a 5 mL Eppendorf tube in the hospital laboratory.
2) The 5 mL Eppendorf tubes were brought to the mass spectrometry laboratory and immediately placed at −80 °C for storage.
3) Prior to UPLC-MS/MS analysis, the frozen urine samples were thawed at room temperature and processed as follows:
   a) The samples were diluted 1:15 (v/v) with 10 mM NH_4OH and inverted three times. High dilution factor is necessary to increase the solubility of DHA.
   b) 50 μL of the diluted samples were pipetted into a 96-well plate.
   c) 100 μL of 10 mM NH_4OH were added to all samples in the 96-well plate.
   d) 50 μL of the internal standard working solution were added to all samples in the 96-well plate.
   e) Samples were subsequently mixed in the plate for 3 min and centrifuged at 1,719 × g for 10 min at 4 °C.
   f) 5 μL of each sample were injected into the UPLC-MS/MS system.
4) Blank, zero blank, LLOQ, calibration standards and QC urine samples were processed as follows:
   a) 50 μL of 1:15 (v/v) diluted blank sample were pipetted into the 96-well plate.
   b) 100 μL of appropriate working solutions for calibration standards and QC's were added to the diluted blank samples, except for the total blank to which 150 μL of 10 mM NH_4OH were added and 100 μL of 10 mM NH_4OH to the zero blank sample.
   c) 50 μL of the internal standard working solution were added to all samples in the 96-well plate, except for the blank sample to which 50 μL of 10 mM NH_4OH were added.
   d) Samples were subsequently mixed in the plate for 3 min and centrifuged at 1,719 × g for 10 min at 4 °C.
   e) 5 μL of each sample were injected into the UPLC-MS/MS system.

8.2.3 The UPLC-MS/MS Urinary 2,8-Dihydroxyadenine Assay

A Waters ACQUITY UPLC instrument coupled to a Quattro Premier™ XE tandem quadrupole mass spectrometer (Waters Corporation, Milford, MA, US) was utilized for the analysis of urine samples, using multiple reaction monitoring in

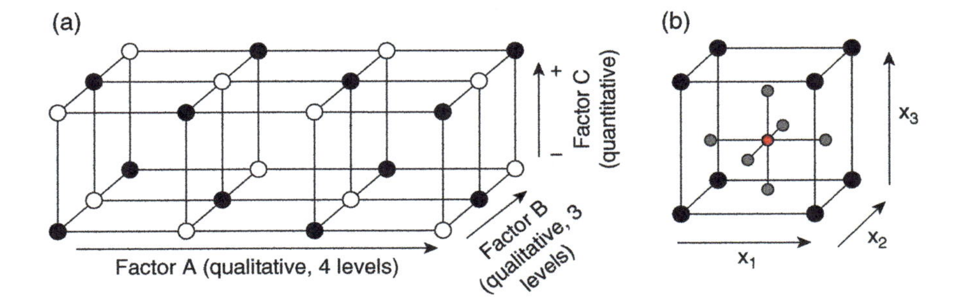

Figure 8.3 D-optimal design (a) and central composite face (CCF) design (b).

positive electrospray ionization mode. Design of experiment was used for development and optimization of the UPLC-MS/MS urinary DHA assay with the aim of increasing the sensitivity and chromatographic retention of DHA, and thus, peak area, peak height, and retention time were chosen as responses for the modeling. Experimental screening was performed with D-optimal design (Figure 8.3a), which allows for simultaneous optimization of both quantitative and qualitative factors (de Aguiar et al., 1995). Significant factors from the D-optimal design were optimized with central composite face (CCF) design (Figure 8.3b) and related to the peak area, peak height, and retention time of DHA using partial least square regression. Data is only shown for the peak height and retention time. The selected factors and their levels used in the D-optimal and the CCF design are shown in Table 8.1. The optimized experimental conditions for the UPLC-MS/MS assay for quantification of DHA in urine are provided in Table 8.2. Following optimization, the assay was prevalidated. The following parameters were investigated during the prevalidation: intra- and interday accuracy and precision, calibration function, selectivity, sensitivity, auto-sampler stability, carryover, dilution integrity, and matrix effect. For a detailed description on the assay's development, peak area data, and validation, the reader is referred to the original publication (Thorsteinsdóttir and Thorsteinsdóttir, 2021; Thorsteinsdottir et al., 2016).

8.3 Results

8.3.1 Assay Development and Optimization

The UPLC-MS/MS urinary DHA assay was successfully developed and optimized utilizing design of experiment. The principal goal was to improve the sensitivity and to increase chromatographic retention of DHA. The regression coefficient plots in Figure 8.4 revealed that capillary voltage, flow rate, type of gradient, pH of the mobile phase, as well as several factor interactions had significant effect on

Table 8.1 Experimental Factors and Settings for the Design of Experiments.

Design	Qualitative factor	Settings	
D-optimal design	pH of mobile phase	pH 2.8; pH 6.7	
	Organic solvent	Methanol; acetonitrile	
	Type of gradient slope[a]	4; 6; 8	
	Quantitative factor	**Lower limit**	**Upper limit**
	Capillary voltage (kV)	0.5	3.5
	Gradient steepness (min)	1.5	3.5
	Flow rate (mL/min)	0.4	0.6
CCF design[b]	Capillary voltage (kV)	0.4	2.0
	Cone voltage (V)	20	35
	Flow rate (mL/min)	0.35	0.45

[a] Slope 4 is concave downward; slope 6 is linear; slope 8 is concave upward.
[b] CCF design is a central composite face design.

Table 8.2 Experimental Conditions for the Optimized UPLC-MS/MS urinary 2,8-Dihydroxyadenine (DHA) Quantification Assay.

Instrument/conditions	Details		
UPLC-MS/MS	Waters ACQUITY-Quattro Premier™ XE		
Analytical column	Waters Acquity HSS T3 2.1 × 100 mm, 1.8 μm		
Column temperature	35 °C		
Autosampler temperature	22 °C		
Flow rate	0.35 mL/min		
Mobile phase	A: 2 mM ammonium acetate at pH 6.7, B: Acetonitrile		
Gradient slope	**Time (min)**	**%A**	**%B**
	0.0	99.5	0.50
	1.0	99.5	0.50
	2.5	80.0	20.0
	5.0	20.0	80.0
	5.5	20.0	80.0
	5.7	99.5	0.50
	6.5	99.5	0.50
Gradient slope	Concave downward		
Sample injection	5 μL		
Ionization mode	Positive electrospray ionization		
Capillary voltage	0.4 kV		
Cone voltage	35 V		

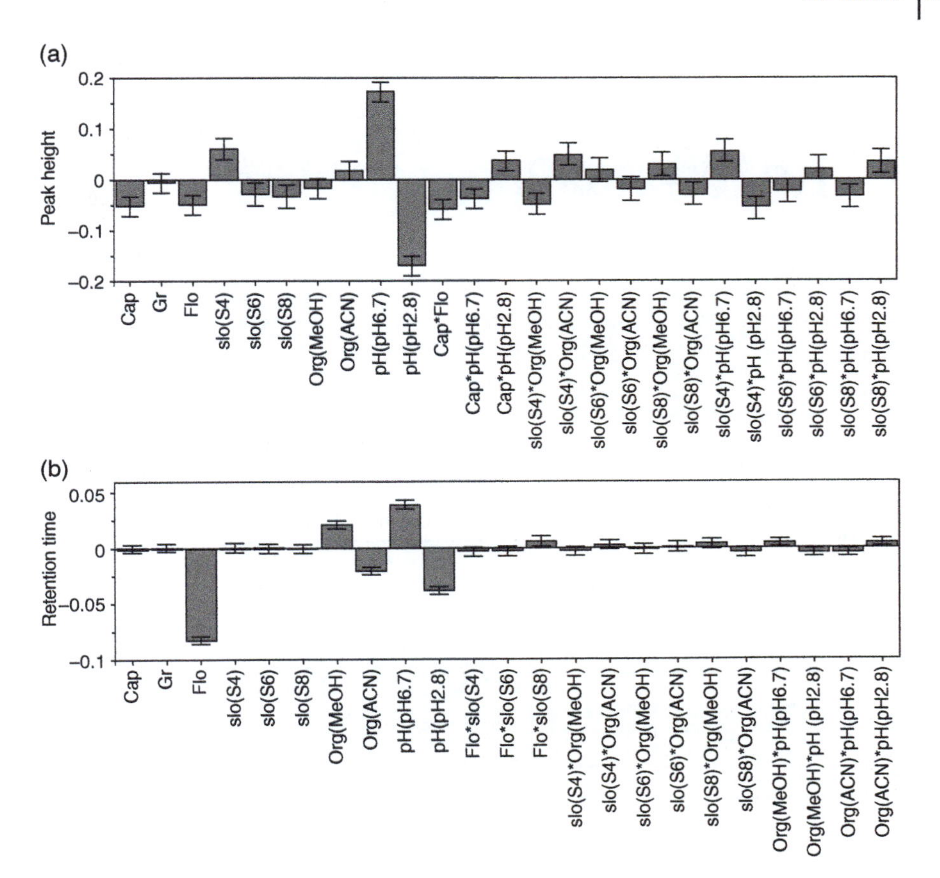

Figure 8.4 Regression coefficients plot scaled and centered for peak height (a) and retention time (b) of 2,8-dihydroxyadenine (DHA). The *x*-axis displays experimental factors and factor interactions that have significant effect on the response of DHA and the *y*-axis denotes the peak height (a) and retention time (b). Error bars represent the 95% confidence interval. ACN, acetonitrile; Cap, capillary voltage (kV); Flo, flow rate (mL/min); Gr, gradient steepness (min); MeOH, methanol; Org, organic mobile phase; Slo, slope. *Indicates interaction effect between two experimental factors, e.g. Cap*Flo indicates interaction effect between capillary voltage and flow rate. A positive value for the regression coefficient means that increasing the factor will increase the response, while a negative value indicates that the factor should be decreased to achieve a higher response.

the peak height of DHA (Figure 8.4a). Moreover, the flow rate, organic solvent, and pH of the mobile phase had a significant effect on the retention time of DHA (Figure 8.4b), whereas interaction effects were not significant. The counter plot shown in Figure 8.5a revealed that the sensitivity of the assay for DHA could be increased by reducing the flow rate and capillary voltage and by selecting

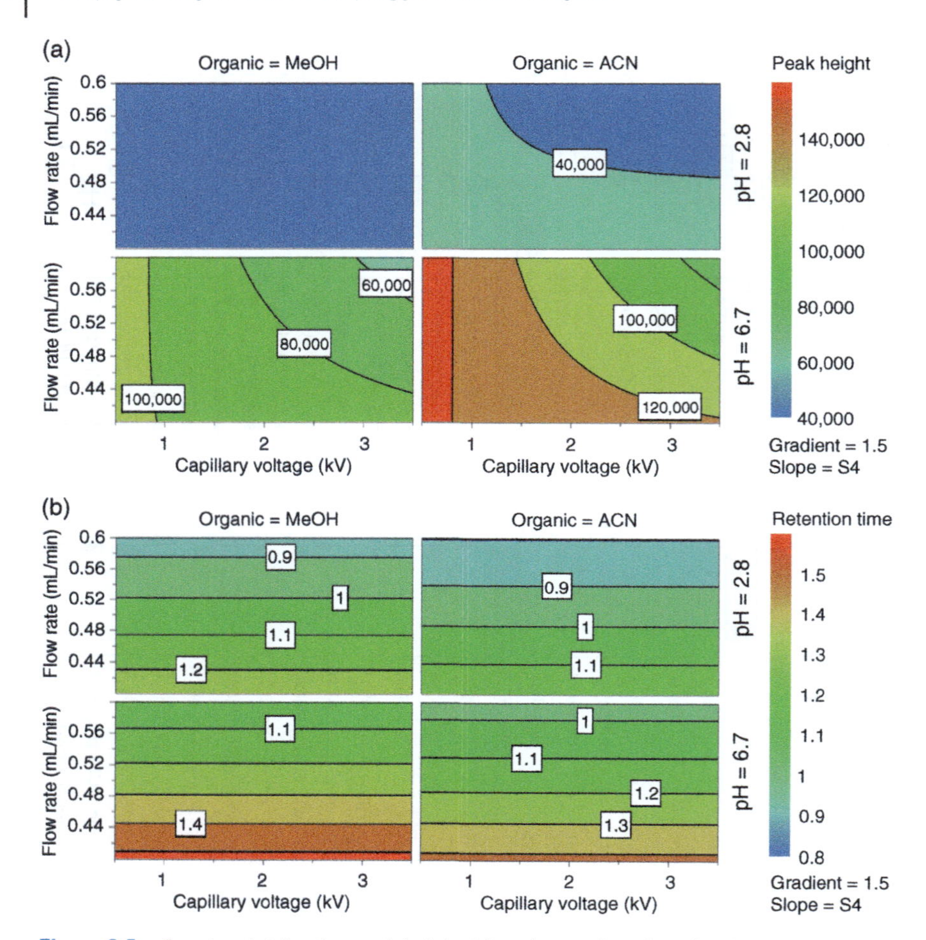

Figure 8.5 Counterplot for the peak height (a) and retention time (b) of 2,8-dihydroxyadenine (DHA), according to flow rate, capillary voltage, pH of the mobile phase and the organic modifiers, methanol (MeOH), and acetonitrile (ACN). The color scale to the right explains the value of each color on the plot. Red indicates the highest values and green the lowest values.

acetonitrile as organic modifier and mobile phase A at pH 6.7. Chromatographic retention indicated by the retention time was increased by reducing the flow rate, selecting methanol as organic modifier and mobile phase A at pH 6.7 (Figure 8.5b). According to the regression model, increasing the sensitivity and retention time of DHA required similar conditions, and the only compromise made was selecting acetonitrile instead of methanol as the organic modifier, since this resulted in much higher sensitivity as well as adequate chromatographic retention for

DHA. The final optimized conditions for the UPLC-MS/MS urinary DHA assay are shown in Table 8.2.

Urine samples from APRT deficiency patients and healthy controls were analyzed using the optimized UPLC-MS/MS urinary DHA assay. A peak corresponding to DHA was detected in urine samples from an untreated patient with APRT deficiency (Figure 8.6a) and a patient on allopurinol therapy (Figure 8.6b), while DHA was not detected in a patient treated with febuxostat (Figure 8.6c) and a healthy control subject (Figure 8.6d).

Results from the prevalidation study of the UPLC-MS/MS urinary DHA assay are shown in Table 8.3. Acceptable selectivity was demonstrated without any interfering peaks at the retention time for DHA in human urine samples from six

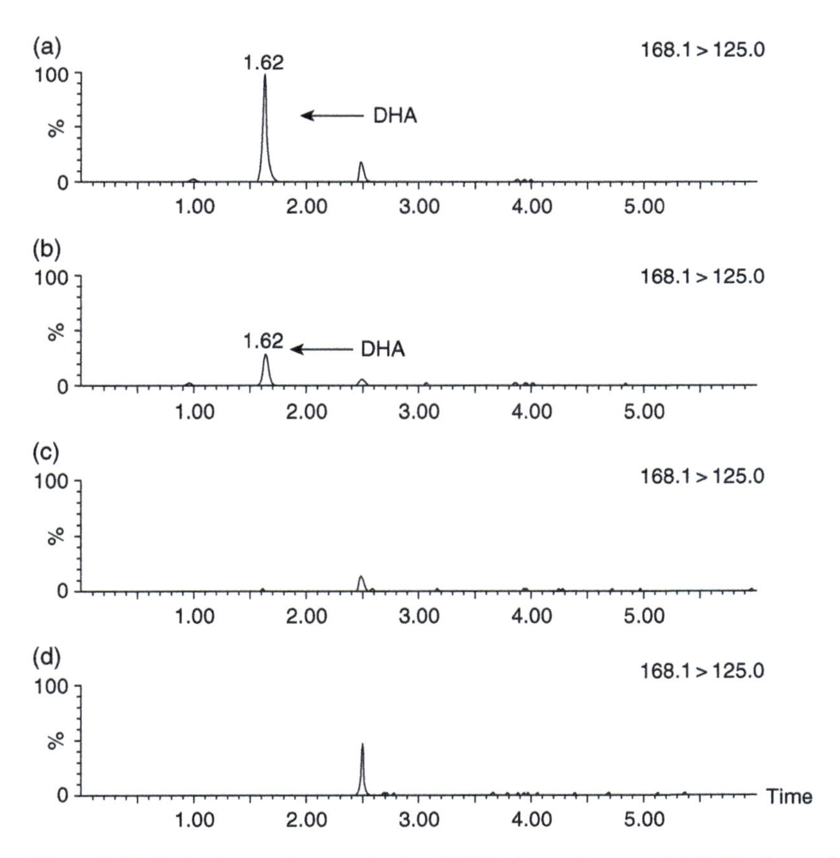

Figure 8.6 Multiple reaction monitoring (MRM) chromatogram of 2,8-dihydroxyadenine (DHA) in urine samples from (a) untreated patient with APRT deficiency, (b) patient with APRT deficiency treated with allopurinol (300 mg/day), (c) patient with APRT deficiency treated with febuxostat (80 mg/day), and (d) a healthy control. *Y*-axis are linked.

Table 8.3 Intra- and Interassay Precision and Accuracy of 2,8-Dihydroxyadenine (DHA) Measurements in Urine Samples.

QC level	DHA concentration (ng/mL)	Intraday (n = 6)		Interday (n = 3)	
		%CV[a]	%Bias[b]	%CV[a]	%Bias[b]
LLOQ	100	8.3	3.3	7.8	0.5
QC-low	300	5.7	−2.0	5.4	−4.6
QC-medium	800	4.7	6.5	7.4	3.1
QC-high	2,000	4.5	8.5	3.4	4.4

Calibration range: 100–5,000 ng/mL, $R^2 > 0.99$ with weighing factor $1/X$.
CV, coefficient of variation; QC, quality control; LLOQ, lower limit of quantification.
[a] % RSD = (standard deviation/mean) × 100.
[b] %Bias = (mean calculated concentration − nominal concentration)/(nominal concentration) × 100.

different sources. Furthermore, no significant endogenous levels of DHA were found, since DHA was not detected in any of the six sources of urine matrices analyzed. The back-calculated concentration of all calibration standards was within the ±15% acceptable limits of the nominal value in the concentration range of 100–5,000 ng/mL. Data are only shown for the LLOQ. Intraday precision (coefficient of variation, %CV) and accuracy (%bias) were 4.5–5.7% and −2.0–8.5%, respectively, for all QC samples. Intraday precision for LLOQ was 8.3% with accuracy of 3.3%. Interday precision and accuracy were 3.4–7.4% and −4.6–4.4%, respectively, for all QC samples and 7.8 and 0.5%, respectively, for LLOQ. Thus, intra- and interday precision and accuracy for the QC samples and LLOQ were within the ±15% acceptable limits. Processed urine samples ($n = 6$) were stable up to 18hr at 22 °C in the autosampler with CV and bias for all QC samples within the ±15% acceptable limits. No carryover was detected, sample integrity during the dilution integrity study was acceptable (within ±15% of nominal values), and no significant matrix effect was observed. For a detailed description of the assay's validation, the reader is referred to the original publication (Thorsteinsdottir et al., 2016).

8.3.2 Comparison of First-Morning Void Urine Specimens and 24-hr Urine Collections for Assessment of 2,8-Dihydroxyadenine Excretion

Forty-four paired first-morning void urine specimens and 24-hr urine samples were collected during the same 24-hr period at home by 11 patients (7 females). Their clinical characteristics are described in Table 8.4. The median (range) age at the time of collection was 38.0 (28.2–66.6) years. Six patients had a history of kidney stones, and two patients had chronic kidney disease.

Table 8.4 Clinical Characteristics of Patients with Adenine Phosphoribosyltransferase Deficiency Who Collected First-Morning Void and 24-hr Urine Samples.

Patient	Sex	Age at presentation (years)	Age at first collection (years)	eGFR (mL/min/1.73 m^2)	Major clinical feature	Number of urine sample pairs[a]
1	Female	3.5	35.4	58	Kidney stones, CKD	3
2	Male	23.4	60.6	90	Crystalluria	4
3	Male	24.2	51.9	99	Kidney stones	7
4	Male	0.8	28.2	103	Kidney stones	3
5	Female	52.2	66.6	37	Kidney stones, CKD	4
6	Female	33.1	60.5	83	Kidney stones	7
7	Female	3.3	38.0	87	LUTS	4
8	Male	32.9	55.8	80	Kidney stones	4
9	Female	21.6	33.1	92	Crystalluria	3
10	Female	0.5	32.5	103	Crystalluria	3
11	Female	36.9	36.9	81	Crystalluria	2

[a] Sample pair: a first-morning void urine specimen and a 24-hr urine sample, collected in the same 24-hr period.
CKD, chronic kidney disease; LUTS, lower urinary tract symptoms.

The median (range) urinary DHA excretion was 119.5 (68.3–289.0) mg/24 hr off XOR inhibitor therapy and 20.7 (8.3–112.0) mg/24 hr while receiving treatment with allopurinol or febuxostat. The DHA-to-Cr ratio was 14.1 (8.2–36.1) mg/mmol in untreated patients and 2.3 (0.8–8.4) mg/mmol in patients on allopurinol or febuxostat therapy. The results suggest a robust lowering effect of pharmacotherapy on urinary DHA excretion. The median (range) collection time was 1,360 (930–1,670) min. Six collections (13.6%) went beyond the 24-hr period, while 24 collections (54.5%) were more than an hour short of the desired collection time. When corrected for the collection time, the median (range) DHA excretion was 119.0 (68.2–268.1) mg/24 hr in patients off therapy and 21.1 (7.6–105) mg/24 hr in those treated with allopurinol or febuxostat.

Thirty-six of the 44 sample pairs were analyzed with the UPLC-MS/MS urinary DHA assay in a single run. The correlation between the DHA-to-Cr ratio in the first-morning void specimens and 24-hr urinary DHA excretion was highly significant ($r_s = 0.78$, $p < 0.001$; Figure 8.7). Nineteen of the 44 sample pairs were obtained while the patients were off pharmacotherapy, and 17 pairs were collected during treatment with an XOR inhibitor.

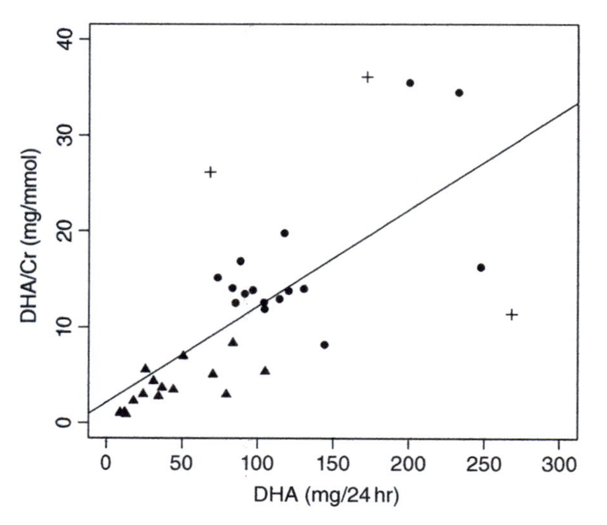

Figure 8.7 Scatter plot of 24-hr urinary 2,8-dihydroxyadenine (DHA) excretion versus the urinary DHA-to-creatinine ratio (DHA/Cr) in first-morning void urine samples from patients with APRT deficiency; untreated (•) and treated (▲) with an XOR inhibitor. *The correlation analysis was performed with the exclusion of outliers (+).

8.4 Discussion

Our novel UPLC-MS/MS urinary DHA assay has been successfully used for the detection and quantification of DHA in human urine samples. The assay can readily be used for the diagnosis of APRT deficiency (Figure 8.6a) as the urinary DHA excretion is high in untreated patients and is invariably below the limit of detection in healthy subjects (Figure 8.6d). A strong correlation between the DHA-to-Cr ratio in first-morning void urine samples and the 24-hr urinary DHA excretion has been demonstrated, suggesting that first-morning void specimens can replace at-home timed collections for diagnosis and monitoring of the effect of pharmacotherapy and treatment adherence (Figure 8.6b, c). Thus, the UPLC-MS/MS urinary DHA assay is a promising tool for both diagnosis and TDM in APRT deficiency patients.

Based on our findings, the determination of 24-hr urinary DHA excretion could be replaced by DHA-to-Cr ratio in first-morning void urine samples for monitoring the effectiveness of pharmacotherapy and treatment adherence, both in the clinic and in research studies. Indeed, considering the challenges of 24-hr urine collections, both in terms of accuracy and practicality, using first-morning void urine samples appears to be a preferable option.

Even though the urine sample handling procedure for the UPLC-MS/MS urinary DHA assay is relatively simple, the aliquoting of the urine samples prior to transfer to the mass spectrometry laboratory has been prone to error. In some instances, the wrong type of tubes has been used and/or an incorrect volume of the urine sample has been aliquoted into the tubes. In these cases, the urine samples must be aliquoted twice. Since a small variation in collection, preprocessing and storage of clinical samples can have a significant effect on the stability of the analytes, and thereby reduce the reliability of the analytical results, each step of the sample handling procedure must be standardized and strictly adhered to. Careful standardization will reduce experimental variations and ensure reliability of the results (Bi et al., 2020).

One of the drawbacks of collecting urine samples in a traditional manner is the substantial urine volume that is generally collected and stored. For the UPLC-MS/MS urinary DHA assay, 5 mL Eppendorf tubes are used, and the samples are placed at −80 °C for long-term storage. The tube racks take up significant freezer space which frequently is a limited resource. The use of microsampling strategies, including dried urine spots or volumetric absorptive microsampling, could solve this problem, since these microsamples require less storage space than traditional urine samples (Velghe et al., 2016; FDA (Food and Drug Administration), 2018). Other advantages of microsampling techniques include improved stability of many analytes in microsamples as the desiccation can increase analyte stability (Protti et al., 2021). Furthermore, transport of microsamples is more convenient as they can be stored and shipped at room temperature resulting in a reduced risk of damage to the samples during transport (Avataneo et al., 2019; Nys et al., 2017). However, the use of urine microsamples for DHA measurements must be studied to confirm the accuracy of this sampling method.

8.5 Conclusions and Future Directions

The urinary DHA assay has become instrumental for the clinical diagnosis and TDM of patients with APRT deficiency. To our knowledge, we are the only laboratory worldwide that offers absolute quantification of urine DHA, using the UPLC-MS/MS assay. Based on our findings, TDM in patients with APRT deficiency can be greatly simplified by replacing at-home 24-hr urine collections with first-morning void urine specimens, as a reliable and accurate method in the clinic and in research studies.

As highlighted in the present chapter, one of the shortcomings of traditional urine sampling for our assay is the freezer space required for urine sample storage. Moving toward microsampling could be an attractive solution to this problem, since storage of microsamples require significantly less space and can be stored at room temperature where stability of the analytes has been demonstrated. For TDM, plasma samples are preferred over urine samples, and microsampling offers a less invasive sampling procedure compared to traditional blood sampling. As our laboratory is currently the only laboratory worldwide that offers an analysis of urinary DHA excretion, we regularly receive samples from abroad. If successfully developed, microsampling would offer a more convenient way of sample shipping and storage. The introduction of this technique into clinical practice would be of significant benefit for both patients and hospital staff. Hence, we plan to include microsamples in future studies of urine and plasma DHA measurements.

References

Avataneo, V., D'Avolio, A., Cusato, J., Cantù, M., & De Nicolò, A. (2019). LC-MS application for therapeutic drug monitoring in alternative matrices. *Journal of Pharmaceutical and Biomedical Analysis*, *166*, 40–51. doi:https://doi.org/10.1016/j.jpba.2018.12.040.

Bi, H., Guo, Z., Jia, X., Liu, H., Ma, L., & Xue, L. (2020). The key points in the pre-analytical procedures of blood and urine samples in metabolomics studies. *Metabolomics*, *16*(6), 1–15. doi:https://doi.org/10.1007/s11306-020-01666-2.

Bollée, G., Dollinger, C., Boutaud, L., Guillemot, D., Bensman, A., Harambat, J., Deteix, P., Daudon, M., Knebelmann, B., & Ceballos-Picot, I. (2010). Phenotype and genotype characterization of adenine phosphoribosyltransferase deficiency. *Journal of the American Society of Nephrology*, *21*(4), 679–688. doi:https://doi.org/10.1681/ASN.2009080808.

Cartier, P., Hamet, M., & Hamburger, J. (1974). A new metabolic disease: The complete deficit of adenine phosphoribosyltransferase and lithiasis of 2,8-dihydroxyadenine. *Comptes rendus hebdomadaires des seances de l'Academie des sciences. Serie D: Sciences naturelles*, *279*(10), 883–886.

Corder, C. J., Rathi, B. M., Sharif, S., & Leslie, S. W. (2002). 24-hour urine collection. In *StatPearls [Internet]*. Treasure Island (FL): StatPearls Publishing. Retrieved 8 August, 2022, from https://www.ncbi.nlm.nih.gov/books/NBK482482/.

Côté, A.-M., Firoz, T., Mattman, A., Lam, E. M., von Dadelszen, P., & Magee, L. A. (2008). The 24-hour urine collection: Gold standard or historical practice? *American Journal of Obstetrics and Gynecology*, *199*(6), 625.e1–625.e6. doi:https://doi.org/10.1016/j.ajog.2008.06.009.

de Aguiar, P. F., Bourguignon, B., Khots, M. S., Massart, D. L., & Phan-Than-Luu, R. (1995). D-optimal designs. *Chemometrics and Intelligent Laboratory Systems*, *30*, 199–210. doi:https://doi.org/10.1016/0169-7439(94)00076-X.

Edvardsson, V. O., Sahota, A., & Palsson, R. (2012; updated 2019). Adenine phosphoribosyltransferase deficiency. In M. P. Adam, G. M. Mirzaa, R. A. Pagon, S. E. Wallace, L. J. H. Bean, K. W. Gripp, A. Amemiya (Eds.), *GeneReviews®[Internet]*. Seattle, WA: University of Washington, 1993–2022. https://www.ncbi.nlm.nih.gov/books/NBK100238/.

Fishel Bartal, M., Lindheimer, M. D., & Sibai, B. M. (2022). Proteinuria during pregnancy: Definition, pathophysiology, methodology, and clinical significance.

American Journal of Obstetrics and Gynecology, *226*(2S), S819–S834. doi: https://doi.org/10.1016/j.ajog.2020.08.108.

French, D. (2017). Advances in clinical mass spectrometry. In G. S. Makowski (Ed.), *Advances in clinical chemistry* (1st ed., pp. 153–198). Elsevier Inc. https://doi.org/10.1016/bs.acc.2016.09.003.

Grant, R. P., & Rappold, B. A. (2018). Development and validation of small molecule analytes by liquid chromatography-tandem mass spectrometry. In N. Rafai, A. R. Horvath, & C. T. Wittwer (Eds.), *Principles and applications of clinical mass spectrometry: Small molecules, peptides, and pathogens* (pp. 115–179). Elsevier Inc. https://doi.org/10.1016/B978-0-12-816063-3.00005-0.

Jensen, J. S., Clausen, P., Borch-Johnsen, K., Jensen, G., & Feldt-Rasmussen, B. (1997). Detecting microalbuminuria by urinary albumin/creatinine concentration ratio. *Nephrology Dialysis Transplantation*, *12 Suppl 2*, 6–9.

Justesen, T. I., Petersen, J. L. A., Ekbom, P., Damm, P., & Mathiesen, E. R. (2006). Albumin-to-creatinine ratio in random urine samples might replace 24-h urine collections in screening for micro- and macroalbuminuria in pregnant woman with type 1 diabetes. *Diabetes Care*, *29*(4), 924–925. doi:https://doi.org/10.2337/diacare.29.04.06.dc06-1555.

Kaartinen, K., Hemmilä, U., Salmela, K., Räisänen-Sokolowski, A., Kouri, T., & Mäkelä, S. (2014). Adenine phosphoribosyltransferase deficiency as a rare cause of renal allograft dysfunction. *Journal of the American Society of Nephrology*, *25*(4), 671–674. doi:https://doi.org/10.1681/ASN.2013090960.

Kelley, W. N., Levy, R. I., Rosenbloom, F. M., Henderson, J. F., & Seegmiller, J. E. (1968). Adenine phosphoribosyltransferase deficiency: A previously undescribed genetic defect in man. *The Journal of Clinical Investigation*, *47*(10), 2281–2289. doi:https://doi.org/10.2215/CJN.02320312.

Nasr, S. H., Sethi, S., Cornell, L. D., Milliner, D. S., Boelkins, M., Broviac, J., & Fidler, M. E. (2010). Crystalline nephropathy due to 2,8-dihydroxyadeninuria: An under-recognized cause of irreversible renal failure. *Nephrology Dialysis Transplantation*, *25*(6), 1909–1915. doi:https://doi.org/10.1093/ndt/gfp711.

Nys, G., Kok, M. G. M., Servais, A. C., & Fillet, M. (2017). Beyond dried blood spot: Current microsampling techniques in the context of biomedical applications. *Trends in Analytical Chemistry*, *97*, 326–332. doi:https://doi.org/10.1016/j.trac.2017.10.002.

Online Mendelian Inheritance in Man, OMIM®. (2016). Johns Hopkins University, Baltimore, MD. MIM Number:102600:07/09/2016. World Wide Web URL: https://omim.org/.

Price, C. P., Newall, R. G., & Boyd, J. C. (2005). Use of protein:creatinine ratio measurements on random urine samples for prediction of significant proteinuria: A systematic review. *Clinical Chemistry*, *51*(9), 1577–1586. doi:https://doi.org/10.1373/clinchem.2005.049742.

Protti, M., Sberna, P. M., Sardella, R., Vovk, T., Mercolini, L., & Mandrioli, R. (2021). VAMS and StAGE as innovative tools for the enantioselective determination of clenbuterol in urine by LC-MS/MS. *Journal of Pharmaceutical and Biomedical Analysis, 195*. doi:https://doi.org/10.1016/j.jpba.2020.113873.

Runolfsdottir, H. L., Palsson, R., Agustsdottir, I. M. S., Indridason, O. S., & Edvardsson, V. O. (2016). Kidney disease in adenine phosphoribosyltransferase deficiency. *American Journal of Kidney Diseases, 67*(3), 431–438. doi:https://doi.org/10.1053/j.ajkd.2015.10.023.

Runolfsdottir, H. L., Palsson, R., Agustsdottir, I. M. S., Indridason, O. S., Li, J., Dao, M., Knebelmann, B., Milliner, D. S., & Edvardsson, V. O. (2020) Kidney transplant outcomes in patients with adenine phosphoribosyltransferase deficiency. *Transplantation, 104*(10), 2120–2128. doi:https://doi.org/10.1097/TP.0000000000003088.

Runolfsdottir, H. L., Palsson, R., Thorsteinsdottir, U. A., Indridason, O. S., Agustsdottir, I. M. S., Steinunn Oddsdottir, G., Thorsteinsdottir, M., & Edvardsson, V. O. (2019). Urinary 2,8-dihydroxyadenine excretion in patients with adenine phosphoribosyltransferase deficiency, carriers and healthy control subjects. *Molecular Genetics and Metabolism, 128*(1-2), 144–150. doi:https://doi.org/10.1016/j.ymgme.2019.05.015.

Sahota, A. S., Tischfield, J. A., Kamatani, N., & Anne Simmonds, H. (2001). Adenine phosphoribosyltransferase deficiency and 2,8-dihydroxyadenine lithiasis. In C. R. Scriver, A. L. Beaudet, W. S. Sly, & D. Valle (Eds.), *The metabolic and molecular bases of inherited disease* (8th ed., pp. 2571–2584). New York, NY: McGraw-Hill Education.

Stone, J. (2017). Sample preparation techniques for mass spectrometry in the clinical laboratory. In H. Nair, & W. Clarke (Eds.), *Mass spectrometry for the clinical laboratory* (pp. 37–62). Elsevier Inc. https://doi.org/10.1016/B978-0-12-800871-3.00003-1.

Thorsteinsdottir, M., Thorsteinsdottir, U. A., Eiriksson, F. F., Runolfsdottir, H. L., Agustsdottir, I. M. S., Oddsdottir, S., Sigurdsson, B. B., Hardarson, H. K., Kamble, N. R, Sigurdsson, S. T., Edvardsson, V. O., & Palsson, R. (2016). Quantitative UPLC–MS/MS assay of urinary 2,8-dihydroxyadenine for diagnosis and management of adenine phosphoribosyltransferase deficiency. *Journal of Chromatography B, 1036–1037*, 170–177. doi:https://doi.org/10.1016/j.jchromb.2016.09.018.

Thorsteinsdóttir, U. A., & Thorsteinsdóttir, M. (2021). Design of experiments for development and optimization of a liquid chromatography coupled to tandem mass spectrometry bioanalytical assay. *Journal of Mass Spectrometry, 56*(9), 1–19. doi:https://doi.org/10.1002/jms.4727.

U.S. Department of Health and Human Services Food and Drug Administration Center for Drug Evaluation and Research (CDER) Center for Veterinary Medicine (CVM). (2018). *Bioanalytical Method Validation: Guidance for Industry*. https://www.fda.gov/media/70858/download.

van der Gugten, J. G. (2020). Tandem mass spectrometry in the clinical laboratory: A tutorial overview. *Clinical Mass Spectrometry, 15*, 36–43. doi:https://doi.org/10.1016/j.clinms.2019.09.002.

Velghe, S., Capiau, S., & Stove, C. P. (2016). Opening the toolbox of alternative sampling strategies in clinical routine: A key-role for (LC-)MS/MS. *Trends in Analytical Chemistry, 84*, 61–73. doi:https://doi.org/10.1016/j.trac.2016.01.030.

Wells, D. A. (2018). Sample preparation for mass spectrometry applications. In N. Rafai, A. R. Horvath, & C. T. Wittwer (Eds.), *Principles and applications of clinical mass spectrometry: Small molecules, peptides, and pathogens* (pp. 67–91). Elsevier Inc. https://doi.org/10.1016/B978-0-12-816063-3.00003-7.

9

Utilization of Patient Centric Sampling in Clinical Blood Sample Collection and Protein Biomarker Analysis

Jinming Xing[1], *Joseph Loureiro*[2], *Dmitri Mikhailov*[1], *and Arkady I. Gusev*[1]

[1] *Biomarker Development, Novartis Institutes for BioMedical Research, Inc., Cambridge, MA, USA*
[2] *Disease Area X, Novartis Institutes for Biomedical Research, Inc., Cambridge, MA, USA*

9.1 Introduction

9.1.1 Challenges with the Current Clinical Trial Model

Patient transit to traditional clinical trials is a barrier that negatively affects clinical trial enrollment and patient retention. Technical and operational advances enabling at home patient sample and data collection will reduce the number of patient visits to clinical sites and expand the trial candidate pool. A recent study showed a median distance of 25.8 miles traveled by 1,600 cancer trial participants from home to trial center (Borno et al., 2018), with each visit requiring scheduling and often assistance from family or professional caretakers. A study conducted by Sully and colleagues assessed clinical trials recruiting between 2002 and 2009. It was observed that 55% of trials achieved their originally specified target sample size, and 45% of trials faced significant delays due to slow recruitment and received time or financial extensions (Sully et al., 2013). Among 419 National Cancer Institute-sponsored oncology clinical trials initiated between 2000 and 2004, 37.9% failed to achieve minimum accrual goals (Cheng et al., 2010). In addition, the complexity of clinical trials has expanded considerably over the past 15 years, resulting in an increased number of procedures and clinical visits for trial participants (Getz & Campo, 2017). This has resulted in an increased need for invasive assessments such as blood collections, hence, more burden on trial participants. It is considered that reducing the travel burden required to participate in clinical trials will improve patient recruitment and retention.

Patient Centric Blood Sampling and Quantitative Bioanalysis, First Edition.
Edited by Neil Spooner, Emily Ehrenfeld, Joe Siple, and Mike S. Lee.

To overcome these issues for clinical trials, a number of solutions are being adopted. Through the application of decentralized trial design, including telemedicine, remote clinical visits, innovative digital devices, and secure electronic data transfer, a lot of progress is being made to reduce patient burden, while ensuring data quality. A pioneer decentralized trial under an Investigational New Drug (IND) application was introduced in 2011 by Pfizer (Orri et al., 2014). The phase IV study recruited, screened, and enrolled patients with overactive bladder from nine states in the United States through web-based questionnaires and consent forms. Investigational product was shipped directly to the patients, while lab assessments were conducted through local laboratories or visiting phlebotomist. Patients were asked to enter e-diary data into a mobile phone provided by the study staff. The study's successful recruitment demonstrated the feasibility of a decentralized trial model utilizing innovative technologies and logistics (Orri et al., 2014). A 2017 remote trial conducted in a nursing home in Ireland demonstrated reduced burden on trial participants. However, caregiver and nursing home staff experienced more burden associated with administrative tasks and data collection, indicating areas for improvement (Donnelly et al., 2018). A substudy within a phase III trial was conducted by Tarolli and colleagues, where patients with Parkinson's disease were asked to complete three video-based remote visits apart from their in-person visits. The study showed excellent correlation in patient-reported outcome between remote and in-person visits. Researchers found remote visits are less time-consuming and are generally more attractive to the study participants (Tarolli et al., 2020). Considering the significance of clinical sample testing, safe and effective self-administrated blood collecting devices will serve an important function in the distributed clinical trial model.

9.1.2 Clinical Trial Conduct Faced Unprecedented Challenges Brought by COVID-19

On January 30, 2020, the WHO Director General declared that the outbreak of COVID-19 constitutes a global health emergency. As of May 25, 2021, WHO reports a total of 166,352,007 confirmed cases of COVID-19, including 3,449,189 confirmed deaths.

The various preventive measures adopted by different countries included travel restrictions (domestic and/or international), social distancing, restrictions on mass gatherings, adherence to personal hygiene, use of facemasks, and quarantine. Although effective in controlling the spread of the virus, these preventive strategies have posed unprecedented challenges and complications for the conduct of clinical research. Since the start of the pandemic, many clinical research sites have experienced lower rates of follow-up and higher likelihood of patient dropout in the ongoing clinical trials, have faced difficulties in recruiting new trial participants, or have paused trial recruitment altogether. The potential challenges, such as social distancing and quarantines, may lead to difficulties in meeting protocol-specified procedures, including getting blood draws at the investigational site, at scheduled study visits and/or follow-up.

The disruption caused to clinical trial conduct calls for more innovative methods of sample collection, especially methods that allow remote sampling with minimum contact between trial participant and clinical study staff. The pandemic also drew attention to the urgent need of self-sampling under quarantine for COVID-19 treatment or diagnostic purpose (James et al., 2020).

9.2 Current Patient Centric Sampling Landscape

A number of patient centric sampling technologies have been introduced to reduce trial participant burden. Patient centric sampling (aka—microsampling) can minimize subject discomfort, eradicate risks of needle sticks, and enable sample collection at convenient locations, including at study participants' homes (Denniff & Spooner, 2014; Kothare et al., 2016; Roadcap et al., 2020).

Examples of patient centric sampling technologies include microtubes or absorbent materials that collect and retain finger-prick blood in either wet or dried format. Dried blood spots (DBSs) were first introduced in the 1960s to screen newborns for metabolic disorders, where a small amount of blood gathered from a heel prick is applied to an absorbent filter paper (Fingerhut et al., 2009; Guthrie & Susi, 1963). However, the impact of hematocrit (Hct) has brought analytical challenges in the field of quantitative DBS analysis (Denniff & Spooner, 2010). As described in manuscripts by Abu-Rabie and Velghe, Hct-based assay bias has multiple contributing factors, including Hct-based area bias, Hct-based recovery bias, and Hct-based matrix effect bias (Abu-Rabie et al., 2015; Velghe et al., 2019). As suggested by Fan and Lee, volumetric DBS collection would eliminate the variation from nonhomogeneity and spreading (Fan & Lee, 2012).

Volumetric absorptive microsampling (VAMS) by Neoteryx is another well-known patient centric blood sampling technology. The Mitra VAMS sample tip is made of absorbent materials and collects a fixed amount of dried blood (Denniff & Spooner, 2014). After drying on the absorbent media, the DBS and VAMS samples can be extracted and quantified with immunoassay or liquid chromatography with tandem mass spectrometry (LC-MS/MS). Technologies similar to that of VAMS have been integrated into patient centric blood sampling devices. For example, the Tasso OnDemand device consists of a lancet and microfluidic channels to collect a blood sample in liquid or proprietary tips similar to VAMS depending on clinician needs.

Touch-Activated Phlebotomy devices—also known as TAP (YourBio Health, formally Seventh Sense Biosystems, Medford, MA, USA)—were introduced as a more convenient and less invasive alternative blood collection method than venipuncture or finger stick by automating virtually painless blood draw (Blicharz et al., 2018). The TAP device received FDA approval to be self-administered at home by a layperson without medical training (US FDA, 2019). It is important to note that TAP devices can only collect liquid samples, contrasting with the dried samples obtained by DBS or VAMS collection methods. Similar to the TAP device, other patient centric sampling technologies such as Tasso, Loop, and Drawbridge also simplify blood sample collection with the push of a button (Drawbridge, 2022; Loop, 2022; Tasso, 2021). Several patient centric sampling devices have also received various regulatory approval status ranging from class I exempt to CE mark, including TASSO-M20 OnDemand (Tasso, Inc., Seattle, WA, USA), OneDraw (Drawbride Health, Inc., Menlo Park, CA, USA), hemaPEN (Trajan Scientific and Medical, Australia), Mitra (Neoteryx, Torrance, CA, USA), and HEMAXIS (HemaXis, Switzerland) (Drawbridge, 2022; HemaXis DB10, 2022; Tasso, 2021; Trajan, 2020).

Interestingly, several "in vitro" experiments have been conducted to evaluate the analytical performance of patient centric blood collection devices by ex vivo sample collection, followed by sample processing and extraction (Sen et al., 2020). Sample QCs were prepared by spiking acetaminophen (paracetamol) into fresh blood before being collected by hemaPEN device (Sen et al., 2020). The LC-MS/MS results from the hemaPEN samples showed acceptable precision and accuracy across QC concentrations while in agreement with paired DBS results (Sen et al., 2020). This in vitro approach offers a way to fine-tune analytical method and to evaluate analysis feasibility of a particular analyte outside of clinical studies. However, real-world experience using patient centric sampling technologies in clinical setting remain critical in understanding study setup, technology robustness, logistics considerations, and patient experience, among others.

9.3 Clinical Proteome Profiling Technologies for Testing Patient Centric Microsampling Devices

9.3.1 Biomarker and Profiling Can Be Used to Benchmark Patient Centric Sampling Technologies

Clinical analysis of blood by venipuncture is a core testing procedure used in clinical trials. Measuring those blood biomarkers integral to hospital patient care and at home is critical to the development of novel therapeutic products. Understanding the clinical utility and limitations of patient centric blood sampling technologies can be evaluated using reference biomarkers. It is especially important to introduce the human element and stress test such technologies in a relevant setting to better gauge device performance and real-world utility. Additionally, broad protein profiling technologies further probe capabilities and limitations of patient centric sampling technologies without significant investment in full clinical qualification and bioanalytical validation of individual biomarkers.

Comparing blood collection methods from the same subjects provides a controlled, holistic view of the sampling technology's clinical utility across a variety of peptides and proteins under investigation. However, the sample volume from the microsampling collection approaches used is oftentimes limited compared to phlebotomy. Small sample volume limits the utility of singleplex ligand binding assays (LBAs) for systematic comparison of peptides/proteins between different samples and collection approaches. Multiplexed immunoassays could allow accurate measurement of up to 100 proteins; however, antibody cross-reactivity often reduces sensitivity compared to singleplex immunoassays (Ellington et al., 2009; Schwenk et al., 2007). Therefore, broader proteome profiling technologies offer additional advantages for studying patient centric microsampling technologies, especially when investigators are constrained by very small sample volumes.

In a conference proceeding authored by Evans and colleagues, the levels of protein extracted from samples derived from different blood collection methods were examined using Olink panels—a high-throughput proteomic profiling platform using proximity extension assay technology (Evans et al. 2019; Lundberg et al., 2011). These investigators found that 96.74% of the 92 proteins in the cardiometabolic (CM) panel were detectable from the Tasso VAMS Tips in samples obtained from the upper arm ($n = 36$). This compares well to 97.83% of proteins in the CM panel that were detectable from the VAMS tips that were dipped in venous blood ($n = 24$). Lower percentages of proteins were detectable in the inflammation (IFN) panel, with 63.04 and 65.22% for Tasso VAMS tips ($n = 10$) and Venous VAMS tips ($n = 7$), respectively. Correlation coefficient between plasma and both methods of VAMS tips are around 0.5, which suggest care is needed to understand these measurements in relation to standard of care, venipuncture-based measurements (Evans et al., 2019).

SomaScan by SomaLogic is another clinical proteome profiling platform (Gold et al., 2012; Mehan et al., 2013). The modified oligonucleotide aptamer-based proteomic assay provides broad, relative abundance assessment of the proteome from a sample volume of less than 60 μL. Orthogonal technologies, including ELISA and LC-MS, have been used to assess sensitivity and specificity of SOMAmers in the context of endogenous targets in blood. Several studies have compared SomaScan to standard clinical assays and validated immunoassays by plotting relative fluorescence unit against concentration (Tin et al., 2019).

9.3.2 Orthogonal Analysis Cultivates Confidence for Biomarker Test with Patient Centric Sampling

Proteomics discovery platforms are often semi-quantitative, providing investigators with measurements suitable for comparative analyses. At home blood draw and shipping adds another layer of variability to the measurements, and orthogonal analyses can buttress the discovery proteomic platform measurements to quantify the endogenous protein(s) identified by discovery proteomics experiments. A study conducted by Raffield and colleagues explored correlations among proteomic assay platforms, including SomaScan, Olink, Luminex, MSD, and other immunoassays, by comparing assay results derived from the same plasma or serum sample. Results of a 13-panel Luminex assay were compared with SomaScan 1.3k data, where correlation coefficients for the 63 proteins assessed between platforms were calculated. The median correlation coefficient was 0.57 and 0.46 in the two cohorts tested. Level of correlation was defined based on Spearman's rho ($r_S < 0.3$ for low correlation; $0.3 < r_S < 0.7$ for moderate correlation; $r_S > 0.7$ for high correlation). Among the 63 proteins, 13 (20%) had low correlation including intercellular adhesion molecule (ICAM1), interleukin 2 receptor subunit alpha (IL2RA), and vascular endothelial growth factor A (VEGF-A); 33 (52%) had moderate correlation including fibrinogen (FGB), IL8, IL16, and interleukin 18 binding protein (IL18BP); and 17 (27%) had high correlation including adiponectin (ADIPOQ), C-reactive protein (CRP), ferritin (FTH1), and IL6RA (Raffield et al., 2020).

According to Raffield and colleagues, similar analysis was conducted between SomaScan 1.1k and Olink panel to test 425 proteins among 48 samples from 10 patients. The correlation coefficients range between −0.58 and 0.93 among the 425 proteins, with 56 (13%) of proteins highly correlated ($r_S < 0.3$) and 179 (42%) proteins poorly correlated ($r_S > 0.7$). Example proteins with high correlation between SomaScan 1.1k and Olink panel include angiopoietin-1 (ANGPT1), interleukin-6 (IL6), renin (REN), insulin-like growth factor-binding protein 1 (IGFBP1), and interleukin-1 receptor-like 1 (IL1RL1). Example proteins with poor correlation include cystatin-C (CST3). Example proteins with medium correlation include angiogenin (ANG) with an r_S of 0.38. Confirmation of aptamers by orthogonal LC-MS analysis was associated with high correlation of SomaScan with both Luminex and Olink panel. Of note, endoglin (ENG) had an r_S of −0.05, pulmonary surfactant-associated protein D (SFTPD) had an r_S of −0.04, VEGF-A had an r_S of −0.21, galectin-3 (LGALS3) had an r_S of 0.55, metalloproteinase inhibitor 1 (TIMP1) had an r_S of 0.55, leptin (LEP) had a $0.60 \leq r_S < 0.70$, and ICAM1 had an r_S of −0.31. Comparison between SomaScan and ProterixBio ELISA showed that adiponectin (ADIPOQ) had an r_S of 0.71. Comparison between SomaScan and quotient bioresearch immunoassay showed CRP with an r_S of 0.94 (Raffield et al., 2020).

This study by Raffield and colleagues also demonstrated the general feasibility of using proteomic platforms such as SomaScan and Olink panel to assess protein abundance. However, a subset of the proteins may not correlate well between the proteomic platforms and immunoassay. The discrepancies may be due to assay specificity and difference in affinity and off-target binding effect of the antibody and aptamer being used. Based on these results, we can conclude that SomaScan results are protein target-dependent and are suitable for semi-quantitative comparisons. Ideally, results of hybridization-based protein measuring technologies (e.g. SomaScan, OLink, and ELISA) should be confirmed by orthogonal quantitative methods (e.g. LC-MS or LBA assays). Reference standards for clinical decision-making can be developed using proteome measurements from these devices. Data reported in this chapter were obtained utilizing this approach.

9.4 Clinical Sample Collection with Tap Device: A Clinical Case Study

9.4.1 Clinical Study with TAP Device

A number of studies have been conducted to compare samples collected from patient centric sampling technology against the gold standard venipuncture (De Kesel et al., 2015; Kita et al., 2019; Kothare et al., 2016). However, many of these studies were done in animal models or in an in vitro fashion where venous blood was applied to the patient centric sampling collection device. (De Kesel et al., 2015; Kita et al., 2019). Clinical studies conducted with patients or healthy volunteers are warranted to better compare the collection methods and provide deeper insights into the clinical utility of such technologies.

A clinical study was conducted by Novartis Institutes for Biomedical Research in Cambridge, United States, where patient centric sampling technology was applied to collect blood samples from healthy volunteers under Institutional Review Board (IRB)-approved protocol and informed consent. The self-reported baseline characteristics of the subjects are listed in Table 9.1. Aspects of the study have been previously reported. This chapter is an extension of the study published in *Bioanalysis* (Xing et al. 2020).

The study design is shown in Figure 9.1. Sixteen healthy volunteers were enrolled in the study, collecting matched venous and peripheral blood samples in a clinical setting using conventional venous phlebotomy and the TAP device, respectively. All TAP devices used in the study were K2EDTA TAP version 1 device under an investigational use only (IUO) label. From each participant, two of the four TAP samples were pooled and processed to EDTA plasma within 90 min of collection (TAP_0). The remaining two TAP samples were stored at ambient lab

Table 9.1 Study Subject Baseline Characteristics. Subject Baseline Characteristics were Recorded at the Beginning of the Clinical Visit. The Screening and Enrollment were Conducted to Promote Even Gender Distribution. Of Note, 62.5% (10 of 16) of the Study Participants are Female.

Number of subjects	Age	Gender	Seasonal allergy	Asthma	Caffeine use	Smoker	Non-steroidal anti-inflammatory drugs (NASID) use	Pregnancy
16	Range 22–62, average 37, median 37	F:M = 5:3	31.25% (5)	5.25% (1)	87.5% (14)	6.25% (1)	50% (8)	None

Reproduced from Xing et al. (2020)/with permission of Future Science.

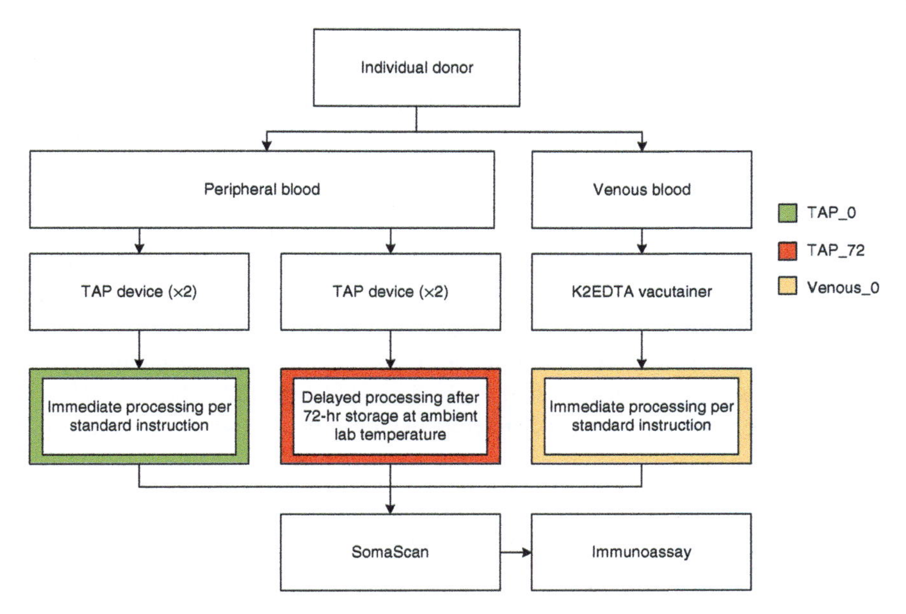

Figure 9.1 Study design and workflow. Healthy volunteer study design and sample collection. From each healthy volunteer donor, three sample types were collected: TAP_0 samples were processed within 90 min upon collection and correspond to clinic collection for point-of-care diagnostic testing. TAP_72 samples were processed after 72 hr ambient lab temperature stability and were designed to mimic remote collection and shipping condition. Venous_0 samples were processed following the existing standard plasma sample collection procedure. TAP: Touch activated phlebotomy. Reproduced from Xing et al. (2020)/with permission of Future Science.

temperature for 72 hr prior to being pooled and processed to EDTA plasma (TAP_72). The venous sample collected through BD K2EDTA vacutainer was processed to plasma within 90 min of collection (Venous_0). All plasma samples were processed by centrifuging blood at 4 °C and 2000 × g force for 15 min and immediately stored at −70 °C. In all, three sets of samples were collected from each subject labeled TAP_0, TAP_72 and Venous_0. The comparison between Venous_0 and TAP_0 evaluates blood from a TAP device to standard venipuncture. Comparison between TAP_0 and TAP_72 represents the variables introduced when TAP device is used at home and shipped to the analysis lab (72 hr of ambient temperature storage and shipment).

SomaScan is a human clinical proteome profiling technology that uses a pool of SOMAmers to probe the patient sample. The SomaScan version used makes 5,080 SOMAmer measurements mapping to 4,120 proteins. SomaScan reads out on a customized Agilent oligo microarray, where SOMAmer binding to the sample is measured by fluorescence intensity. SOMAmers are synthetic, ssDNA sequences with protein-mimetic modifications that tightly bind a specific protein target. Each of the SOMAmers in the SomaScan Assay was isolated based on in vitro sensitivity to a specific protein antigen using SomaLogic proprietary SELEX technology, chemistry, and methodology. Like antibodies, SOMAmers bind to epitopes in complex samples, requiring careful experimental design and/or deliberate follow-up to elucidate the basis for any biological signal(s) evident in the data. Median normalization, quantile normalization and SomaLogic custom normalization were applied to raw data, and all generated comparable univariate models. For example, modeling gender as a biological focus was comparable across all normalization approaches and was clearly evident in the raw data. For each SOMAmer, a Benjamini–Hochberg adjusted $p < 0.001$ and a fold change (FC) greater than 30% ($|\log 2\ FC| > 0.379$) is considered significantly changed. These thresholds are based on platform characteristics, evident on observation of technical replicates.

9.4.2 User Experience, TAP Device Performance, and Sample Hemolysis

For the TAP patient centric sampling study, the device manufacturer conducted in-person training for all clinical staff involved. The training included overview of the study design, device operation, and sample collection logistics. A total of 64 TAP samples were collected using 72 TAP devices. Eight of the TAP devices failed to properly collect the blood samples. Some of the devices' fill indicator failed to turn red upon 10 min after actuation, and others failed to deploy the microneedles. According to verbal feedback, the clinical staff suggested that the TAP device is easy to operate, while the participants suggested that the device collection is virtually painless. The overall feedback from the study staff has been positive with specific comments around painless collection and user friendliness.

Visual inspection was performed upon sample processing to assess sample hemolysis. All Venous_0 plasma were within the expected color of plasma with low to no hemolysis (Figure 9.2).

However, both TAP_0 and TAP_72 plasma were hemolyzed upon visual inspection (Figure 9.3).

No color difference can be seen between TAP_0 and TAP_72 plasma. Interestingly, separation of blood components was observed when transferring a blood sample that had been kept within TAP device for 72 hr (Figure 9.4).

The separation could be due to blood clotting or blood fractionation. No such separation was observed in blood sample taken from TAP device within 90 min of collection (i.e. TAP_0).

Hemolysis was quantified by measurement of hemoglobin concentration which was calculated based on a spectrophotometric method (Harboe method). By measuring light absorbance at different wavelengths, the Harboe method is

Figure 9.2 Typical Venous_0 plasma samples. The venous sample collected through K2EDTA vacutainer was processed to plasma within 90 min of collection. The venous samples have the typical yellow color of regular plasma.

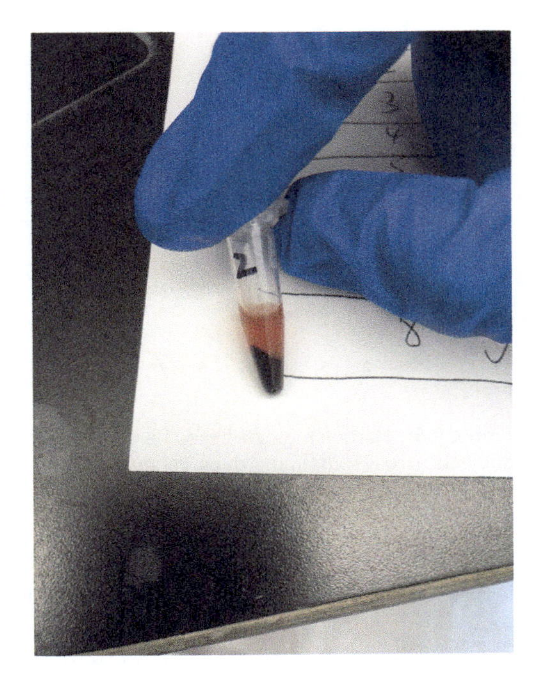

Figure 9.3 Typical TAP_72 plasma samples post-centrifugation. TAP-collected blood samples were stored at ambient lab temperature for 72 hr prior to being pooled and processed to EDTA plasma. Hemolysis was noted in the resulted plasma samples.

considered one of the most reproducible methods used to measure plasma-free hemoglobin (Cookson et al., 2004). Hemolysis between 13.71 and 38.64 (average 26.12 ± 6.56) mg/dl hemoglobin was observed among TAP_0 and TAP_72 samples. Hemolysis between 0.72 and 2.96 (average 1.42 ± 0.55) mg/dL hemoglobin

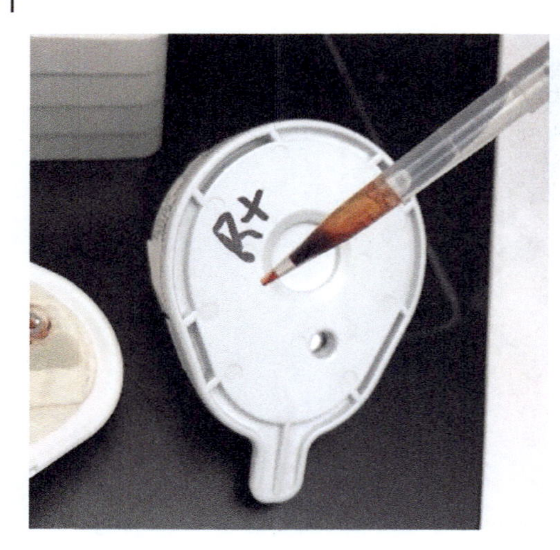

Figure 9.4 Typical TAP_72 blood samples upon transferring from TAP device.

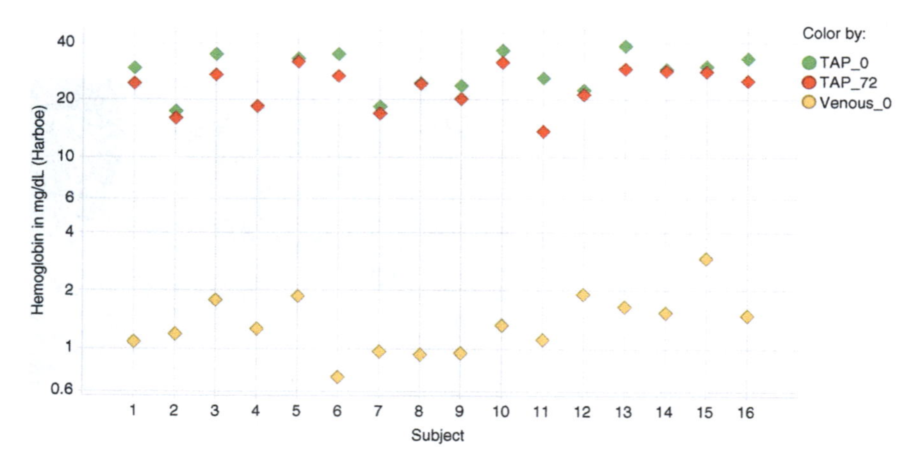

Figure 9.5 Harboe method-based hemoglobin concentration comparison between subjects across sampling groups, Venous_0, TAP_0, and TAP_72. Reproduced from Xing et al. (2020)/with permission of Future Science.

was observed among Venous_0 samples. The plasma generated from TAP has consistently higher amount of hemolysis than the venous plasma (Figure 9.5).

9.4.3 SomaScan Blood Proteome Profiling Landscape

An unsupervised method to compare SomaScan proteome measurements in aggregate is using principal component analysis (PCA). Each point shown in Figure 9.6 corresponds to a patient sample measured using SomaScan. The three

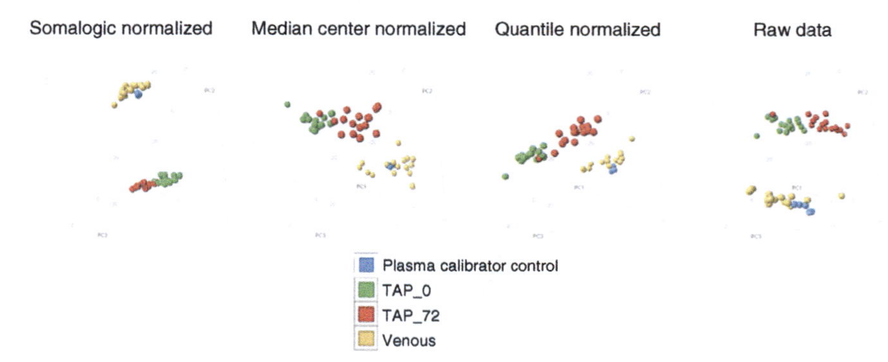

Figure 9.6 Flatted projections of first three principal components comparing TAP_0, TAP_72, and Venous_0. The distance between samples reflects the extent of similarity across the vector of 5,080 SOMAmers measurements per sample. Shorter distances reflect greater similarity between two samples. Samples derived from TAP collection showed distinctive global properties when compared with paired venous samples. This pattern is evident in raw data and persists even when the dataset is normalized by distinct standard microarray normalization methods. Expanded analysis of work published in Bioanalysis (2020).

sample types (Venous_0, TAP_0, TAP_72), along with plasma sample calibrator controls, are shown with different normalization schemes to evaluate how assay design and normalization method control for assay variability. Calibrator samples are technical replicates of matched sample matrix pooled from a collection of healthy volunteers used for data normalization and assessment of technical variability (Candia et al., 2017). Note that the Venous_0 plasma samples coincide with the calibrator plasma samples in a region distinct from the TAP_0 and TAP_72 samples. Within the TAP group, there is a more subtle separation of TAP samples separated by processing time. Overall, PCA shows a clear separation between TAP and venous plasma indicating that TAP plasma is different from venous plasma, but their global proteome profiles also do have comparable elements observed in the data, as determined by linear modeling described below.

Further investigation into the top principal components reveals interesting findings consistent with the known biology of the samples. Figure 9.7a demonstrates how a plasma proteome gender model can be developed from the data through factor loading analysis. The first two principal components comprise the largest variance in the data, which likely include elements of the overall experimental including the different blood drawing technologies and the deliberate sample handling aspects. Interestingly, a separation in male and female attributes can be observed in principal component 3 (PC3), indicating that patient clinical attributes are retained across both technologies. In Figure 9.7b, all features within PC3 are plotted based on their weight and ranking and the features with the

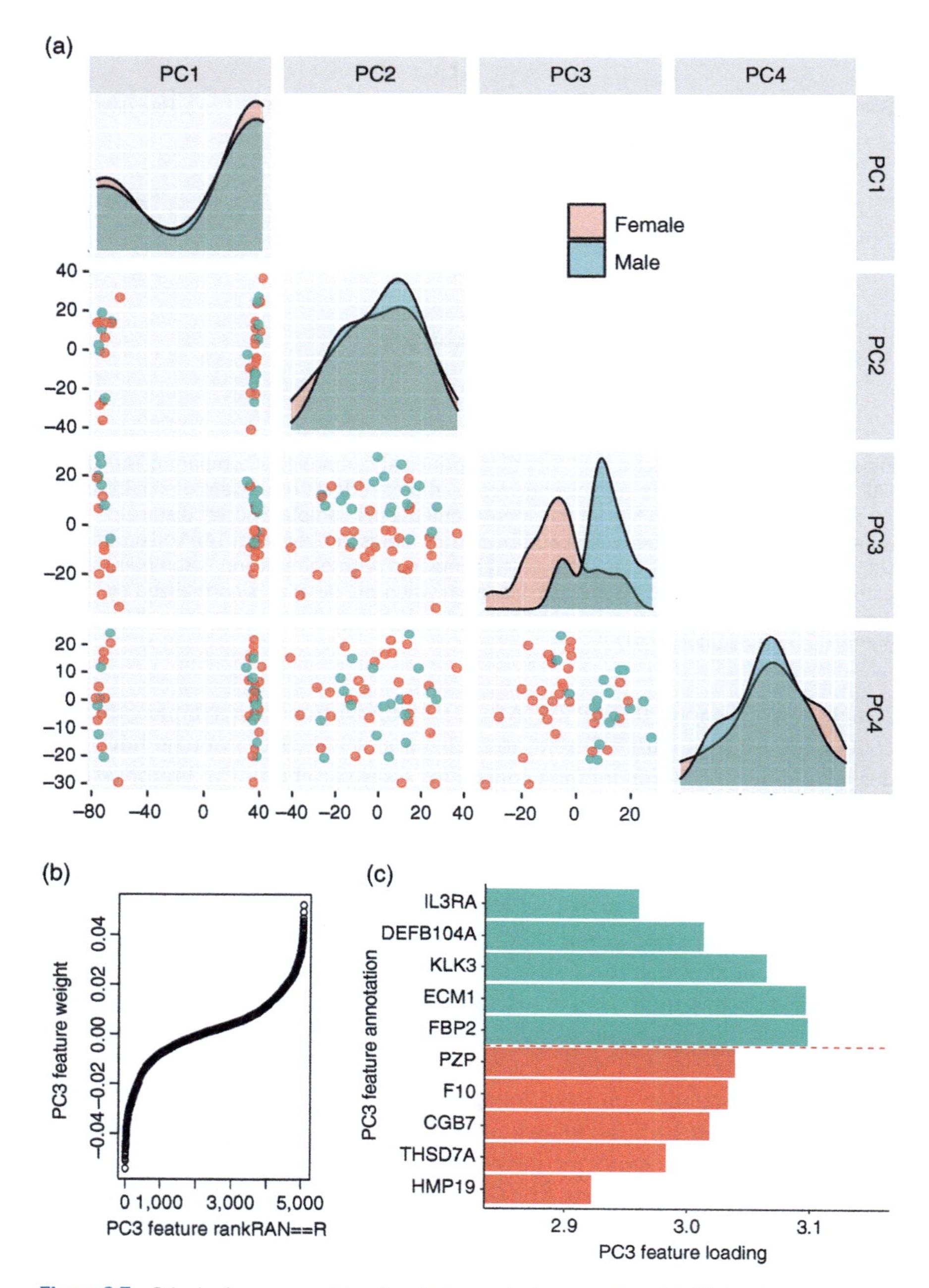

Figure 9.7 Principal component loading factor analysis comparison highlights known gender markers. (a) First four principal components labeled by sample gender. Gender separation can be observed in PC3. Male is represented by color teal, and female is represented by color salmon. (b) The variables with the highest loading are the most important contributors to the given principal component. (c) Protein annotations to the highest feature loading scores for each gender within PC3 are shown. Expanded analysis of work published in Bioanalysis (2020).

highest loading scores on either end are the top contributors in PC3. Many of these features are gender-specific biomarkers, including the male biomarker KLK3 (prostate-specific antigen), female biomarkers PZP (pregnancy zone protein), and CGB7 (choriogonadotropin subunit beta) among the top contributors to PC3. These examples affirm the human biology captured by SomaScan proteome profiling. They also confirm relevance of SomaScan results in assessing patient centric sampling collection and biomarker applicabilities.

The global differences in proteomic composition between TAP and venous-derived plasma can be further inspected using Pearson correlation plots shown in Figure 9.8. All 5,080 SOMAmers from the 16 subjects are plotted in logarithmic scale in each graph, and a linear best fit to the points is plotted with the coefficient of determination (r^2) shown on each graph, which is another indicator for the degree of similarity between the two axes. Strong correlation ($r^2 = 0.97$) can be observed in Figure 9.8a when comparing TAP_0 and TAP_72, confirming our findings in Figures 9.6 and 9.9a that the overall protein profiles of TAP_0 and TAP_72 are highly concordant. Although there is still a strong positive correlation, there are more modest values ($r^2 = 0.62$ and $r^2 = 0.60$, respectively) when comparing either set of TAP samples to venous samples (Figures 9.8b and 9.8c). This confirms our findings in Figures 9.8b and 9.8c that although a number of endogenous proteins differ between TAP and paired venous samples, many proteins do have concordance between sample types (Venous_0, TAP_0, and TAP_72). The concordance seen between the Pearson correlation curves in Figure 9.8 can be further explored using volcano plots (Figure 9.9).

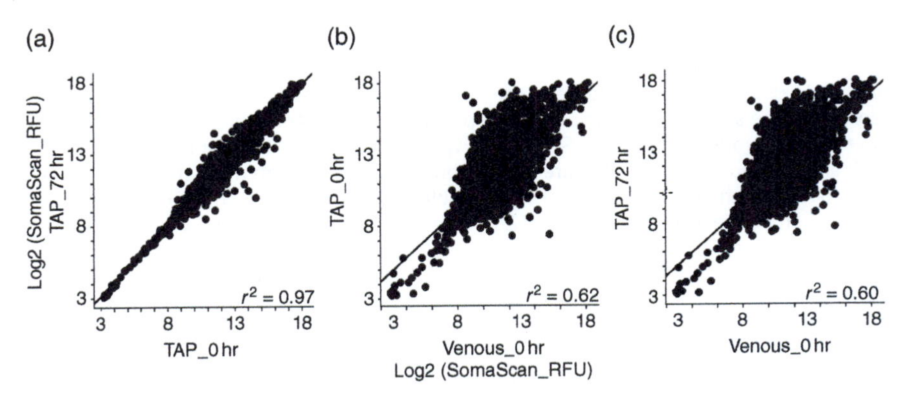

Figure 9.8 SomaScan profiling data comparison between sampling groups (Venous_0, TAP_0 and TAP_72). Each dot represents the SomaScan signal comparison of a single SOMAmer in logarithmic scale. A strong correlation ($r^2 = 0.97$) was observed in Figure 9.8a comparing TAP_72 and TAP_0. Weaker correlations ($r^2 = 0.62$ and 0.60, respectively) were observed in (a) and (c) comparing TAP_0 and TAP_72 to Venous_0. Reproduced from Xing et al. (2020)/with permission of Future Science.

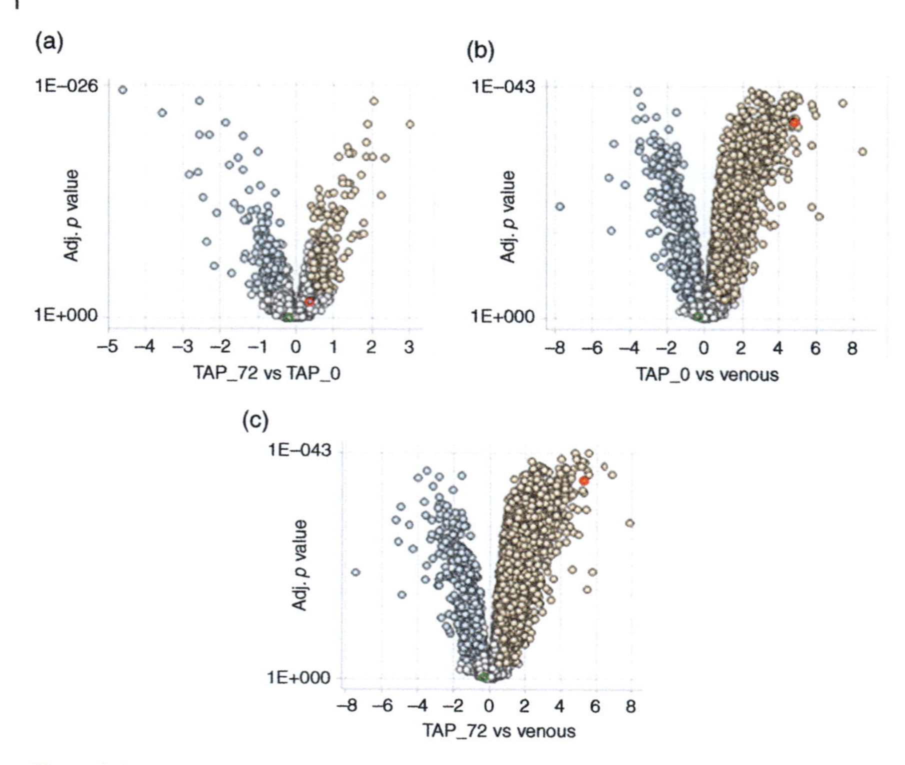

Figure 9.9 Volcano plots comparing the entire 5,080 SOMAmer profile between sample groups. The *X* axis represents beta-coefficient when comparing between SOMAmers (i.e. protein biomarkers) in different groups. A negative value on the *X* axis indicates a decrease in SOMAmer signal as compared between sample groups, while a positive value on the *X* axis indicates an increase in the relative abundance of signal. The *Y* axis represents the adjusted *p*-value of the comparison, and SOMAmers higher on the *Y* axis have a smaller adjusted *p*-value. SOMAmers with more than 30% signal decrease and with an adjusted *p*-value <0.001 are colored blue. SOMAmers with more than 30% signal increase and with an adj. *p*-value <0.001 are colored yellow. (a) The SomaScan comparison between TAP_0 and TAP_72. (b) The SomaScan comparison between TAP_0 and Venous_0. (c) The SomaScan comparison between TAP_72 and Venous_0. Expanded analysis of work published in Bioanalysis (2020). Adapted from Xing et al. (2020).

Figure 9.9 shows the overall SomaScan results comparison between TAP_0, TAP_72, and Venous_0 groups in volcano plots. All 5,080 SOMAmers are shown with each point representing the aggregate performance of a single SOMAmer across 16 subjects. Volcano plots can be used to visualize concordance between groups and identify proteins in TAP device use comparable to venous measurement. Figure 9.9a suggests that majority (4,667 of 5,080, colored gray) of SOMAmers had little or no significant change in their relative protein abundance when comparing TAP_0 and TAP_72, suggesting the corresponding proteins are stable at ambient temperature for 72 hr, and that a subset of SOMAmers (1,906 of 5,080, colored gray) had no significant change when comparing TAP_0 and Venous_0. This suggests that some protein measurements in peripheral blood are predictive of the same in venous blood. This subset of proteins could be good candidates for patient centric sampling followed by venous confirmation. Figure 9.9c suggests that a subset (1,823 of 5,080, colored gray) of SOMAmers had little or no significant change when comparing TAP_72 and Venous_0. This gray area in the volcano plots defined by the absence of change is referred in this paper as a "sweet spot" for comparison of TAP-derived measurements to venous standards. Based on SomaScan readout, a high concordance between TAP and venous collections are expected for proteins within the sweet spot.

To further illustrate the cutoff criteria, we selected two SOMAmers from different regions on the SomaScan volcano plots. SOMAmer α colored green in Figure 9.9 represents CRP. Its position in Figure 9.9a, 9.9b, and 9.9c (within the sweet spot) suggests the protein is stable under 72 hr at ambient temperature, and the endogenous levels in venous and paired TAP samples show high concordance. SOMAmer β colored red in Figure 9.9 represents SFTPD (surfactant protein D). Its position in Figure 9.9a suggests the protein is relatively stable under 72 hr at ambient temperature. Its position in Figure 9.9b (outside of the sweet spot) suggests the protein has higher abundance in TAP_0 than paired venous sample. Its position in the data in Figure 9.9c suggests that when compared with paired venous sample, the protein abundance is higher when blood is collected through TAP device and stored for 72 hr at ambient temperature.

9.4.4 Orthogonal Confirmation Provided by Quantitative Immunoassay

Orthogonal quantitative immunoassays were performed to further investigate and quantify SomaScan profiling findings. We selected a few proteins from all three groups: blue (decrease of signal in TAP_72 vs. Venous_0), yellow (increase of signal), and gray (sweet spot, nonsignificant changes in TAP_72 vs. Venous) and conducted quantitative ELISA confirmation of individual proteins.

As part of the ELISA analysis, we are examining the r^2 and the slope, representing how well data fit on the linear regression line and the change in protein concentration from Venous_0 to TAP_72. The Pearson correlation coefficient r would inform the strength of correlation between TAP_72 and Venous_0 samples. For example, according to the critical value table of Pearson correlation for a two-tailed hypothesis and 14 degrees of freedom ($N-2$), r^2 larger than 0.55 would suggest statistically significant relationship between TAP_72 and Venous_0 with 99.9% confidence.

CRP, ADIPOQ, ANG, and ICAM1 were first selected to be analyzed using ProteinSimple immunoassay on Ella platform (model CyPlexLS-12). The selection was based on a few criteria. First of all, they are all located within the "sweet spot" of SomaScan volcano plot suggesting minimum change between TAP_72 and Venous_0. In addition, the proteins must have relatively high concentration endogenously, allowing for a higher dilution factor, which is needed due to sample volume limitations. Minimum required dilution was established at 2000-fold through parallelism experiments. All 48 samples were serial diluted to 2000-fold, and 50 μL of each diluted sample was loaded onto Ella for analysis. Three technical replicates per sample were produced for each protein measurement. Final calculated concentrations reported by Simple Plex Explorer software were used for ELISA data analysis.

Figure 9.10 shows the correlation curves of ELISA concentration results (in pg/mL) between TAP_72 and Venous_0 for CRP, ADIPOQ, ANG, and ICAM1. The coefficients of determination r^2 are shown in Table 9.2. The comparison in ELISA results correlates well with the corresponding SomaScan comparison, suggesting minimum change in protein abundance between TAP_72 and Venous_0 sample groups. Significant correlation between TAP_72 and Venous_0 sample as well as slopes that are close to 1 in quantitative ELISA assay demonstrate that these proteins/biomarkers can be good candidates for TAP use in clinical studies. The results also demonstrate the feasibility of using SomaScan data as a roadmap to gauge the clinical utility of the TAP device for specific proteins.

9.4.5 Extending Protein Biomarkers Tested by Quantitative Immunoassay Beyond the Zone of Highest Concordance (Negative Controls)

In addition to the initial protein set chosen from the SomaScan "sweet spot," a few proteins outside that area have been selected for quantitative ELISA analysis. Chromogranin A (CHGA), Endoglin (CD105), Osteopontin (OPN), and Surfactant protein D (SP-D) were measured. CHGA, CD105, and OPN are located in the blue quadrant of SomaScan volcano plot in Figure 9.9, suggesting the protein abundance in TAP_72 is >30% lower than in Venous_0. Whereas SP-D is located in the upper right yellow quadrant of Figure 9.9, suggesting the protein abundance in TAP_72 is >30% higher than in Venous_0.

CHGA, CD105, OPN, and SP-D were analyzed using immunoassay on Ella platform with ProteinSimple multi-analyte cartridge at dilution factor of 10.

Figure 9.11 shows the correlation curves of ELISA concentration results (in pg/mL) between TAP_72 and Venous_0 for CHGA, CD105, OPN, and SP-D. The coefficients of determination r^2 are shown in Table 9.3. ELISA data suggest significant correlation between TAP_72 and Venous_0 samples for CHGA, OPN, and

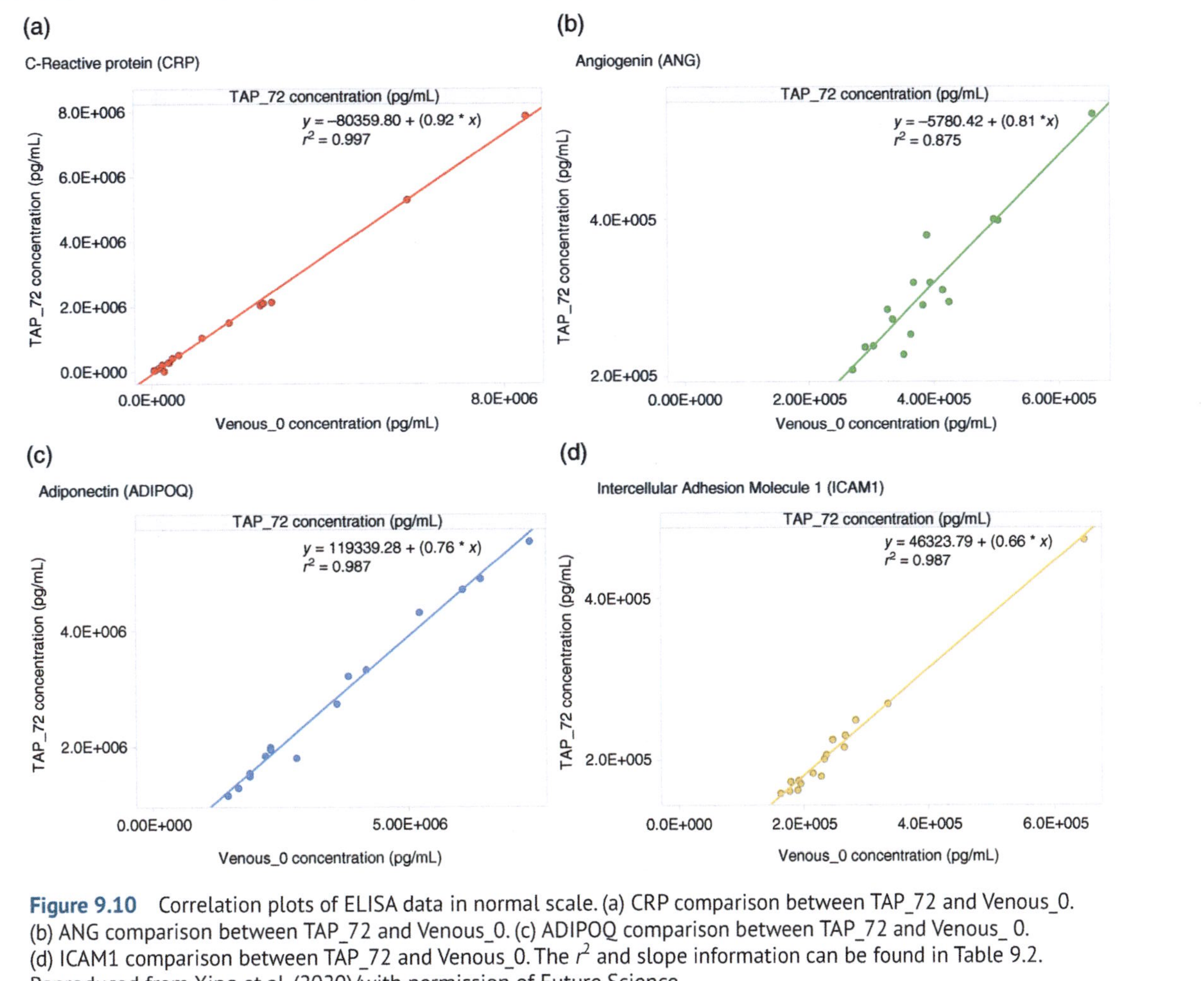

Figure 9.10 Correlation plots of ELISA data in normal scale. (a) CRP comparison between TAP_72 and Venous_0. (b) ANG comparison between TAP_72 and Venous_0. (c) ADIPOQ comparison between TAP_72 and Venous_ 0. (d) ICAM1 comparison between TAP_72 and Venous_0. The r^2 and slope information can be found in Table 9.2. Reproduced from Xing et al. (2020)/with permission of Future Science.

Table 9.2 Pearson Correlation Curve Characteristics of Different Proteins Seen in SomaScan and ELISA. Using a Two-Tailed t-Test with Degrees of Freedom of 14, $r^2 > 0.55$ Suggest Statistically Significant Relationship Between Sampling Groups (TAP_72 and Venous_0) with 99.9% Confidence.

Analyte	ELISA r^2 (TAP_72 vs. Venous_0)	ELISA slope (TAP_72 vs. Venous_0)
CRP	1.00	0.92
ADIPOQ	0.99	0.76
ANG	0.88	0.82
ICAM1	0.99	0.66

SP-D (r^2 = 0.898, 0.742, and 0.949) and insignificant correlation for CD105 (r^2 = 0.114). Although the findings for SP-D, OPN, and CHGA are concordant between SomaScan and ELISA, the CD105 measurement is less so and care must be taken to fully characterize individual SOMAmer or ELISA measurements for clinical decisions.

Hepcidin (HAMP), interleukin 6 signal transducer (IL6ST), galectin-3 (LGALS3), and tissue inhibitor matrix metalloproteinase 1 (TIMP1) were also selected for subsequent immunoassay analysis. HAMP is located in the SomaScan sweet spot, therefore acting as positive control. IL6ST, LAGLS3, and TIMP1 are outside of the SomaScan sweet spot, therefore serving as negative controls. They were analyzed on Ella platform with ProteinSimple multi-analyte cartridge at dilution factor of 100. HAMP is located in the gray area of the SomaScan volcano plot in Figure 9.9, suggesting minimum protein abundance change between TAP_72 and Venous_0. IL6ST is located in the blue quadrant of SomaScan volcano plot in Figure 9.9, suggesting the protein abundance in TAP_72 is >30% lower than in Venous_0. Whereas LGALS3 and TIMP1 are located in the upper right yellow quadrant of Figure 9.9, suggesting the protein abundance in TAP_72 is >30% higher than in Venous_0.

Figure 9.12 shows the correlation curves of ELISA concentration results (in pg/mL) between TAP_72 and Venous_0 for HAMP, IL6ST, LGALS3, and TIMP1. The coefficients of determination r^2 and slope can be found in Table 9.4. The HAMP data suggest great correlation between TAP_72 and Venous_0 ($r^2 = 0.986$). Therefore, HAMP can be a good candidate for clinical sample collection with TAP device. While the correlation seen for IL6ST, LGALS3, and TIMP1 are weaker but are still considered significant ($r^2 > 0.55$), indicating that these proteins may potentially be collected with the TAP device in clinical studies.

Based on quantitative results presented in this chapter, we can conclude that all proteins tested (5 out of 5) from the SomaScan "sweet spot" (1,623) can be good candidates for clinical studies with TAP device sample collection. Interestingly,

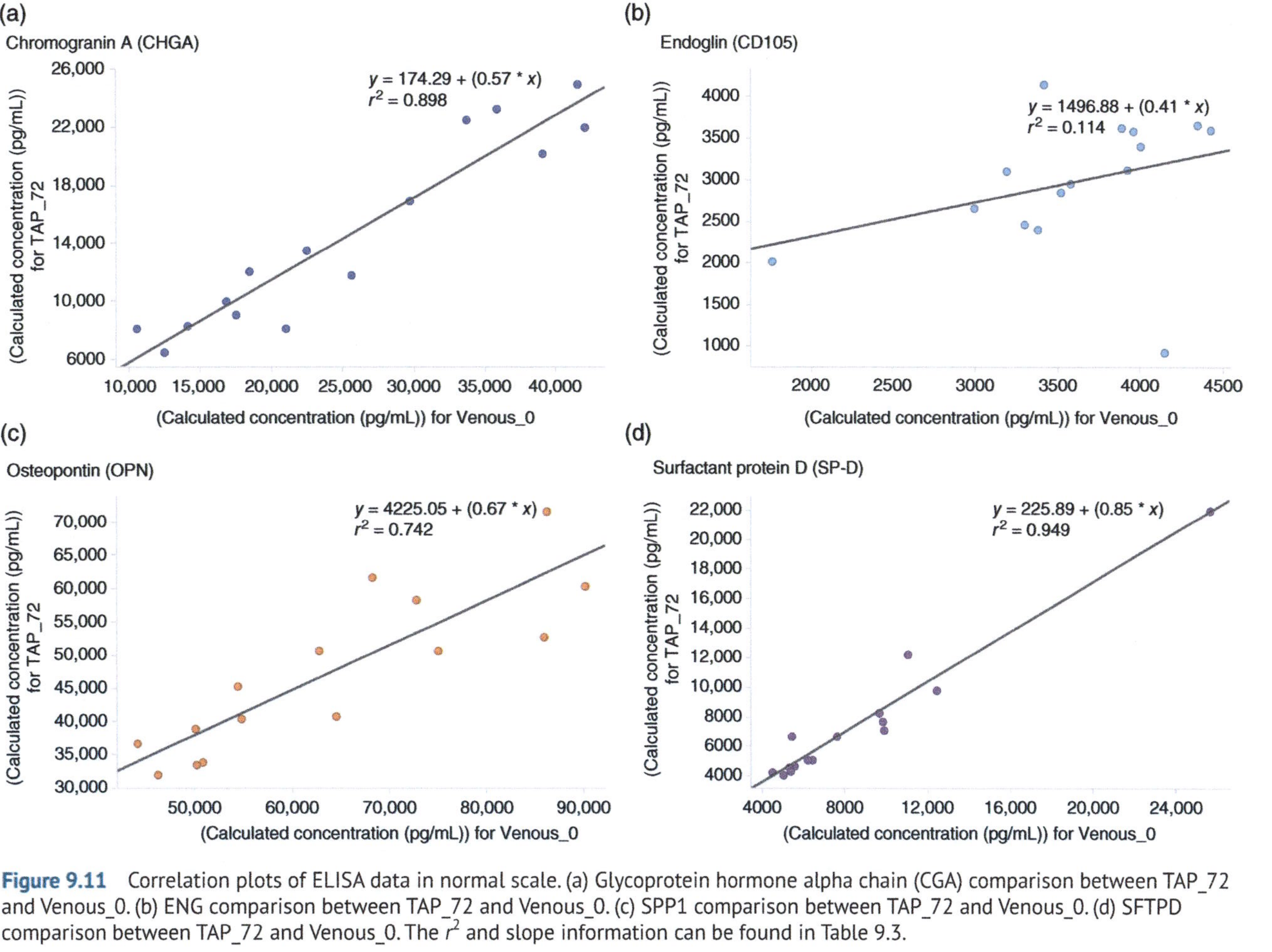

Figure 9.11 Correlation plots of ELISA data in normal scale. (a) Glycoprotein hormone alpha chain (CGA) comparison between TAP_72 and Venous_0. (b) ENG comparison between TAP_72 and Venous_0. (c) SPP1 comparison between TAP_72 and Venous_0. (d) SFTPD comparison between TAP_72 and Venous_0. The r^2 and slope information can be found in Table 9.3.

Table 9.3 Pearson Correlation Curve Characteristics of Different Proteins Seen in SomaScan and ELISA. Using a Two-Tailed *t*-Test with Degrees of Freedom of 14, $r_2 > 0.55$ Suggest Statistically Significant Relationship Between Sampling Groups (TAP_72 and Venous_0) with 99.9% Confidence. Expanded Analysis of Work Published in Bioanalysis (2020).

Analyte	ELISA r^2 (TAP_72 vs. Venous_0)	ELISA slope (TAP_72 vs. Venous_0)
CHGA	0.90	0.57
CD105	0.11	0.41
OPN	0.74	0.67
SP-D	0.95	0.85

Adapted from Xing et al. (2020).

Table 9.4 Pearson Correlation Curve Characteristics of Different Proteins Seen in SomaScan and ELISA. Using a Two-Tailed *t*-Test with Degrees of Freedom of 14, $r_2 > 0.55$ Suggest Statistically Significant Relationship Between Sampling Groups (TAP_72 and Venous_0) with 99.9% Confidence.

Analyte	ELISA r^2 (TAP_72 vs. Venous_0)	ELISA slope (TAP_72 vs. Venous_0)
HAMP	0.99	1.70
IL6ST	0.66	0.55
LGALS3	0.61	3.65
TIMP1	0.64	1.19

proteins coming outside of SomaScan sweet spots (6 out of 7) may still work with TAP devices. SomaScan data could be used to guide the understanding of TAP clinical utility for select protein biomarkers.

9.5 Discussion

9.5.1 Utility of Patient Centric Sampling for Clinical Proteome Sample Collection

The clinical study described above collected paired blood samples using standard venous phlebotomy and a novel capillary blood approach using TAP devices. The hypothesis generation and testing approaches were demonstrated by SomaScan profiling followed by individual biomarker ELISA quantification. Semi-quantitative results generated from proteomic profiling provided global overview of TAP capabilities and utilities. The dataset enables the determination of the confidence level

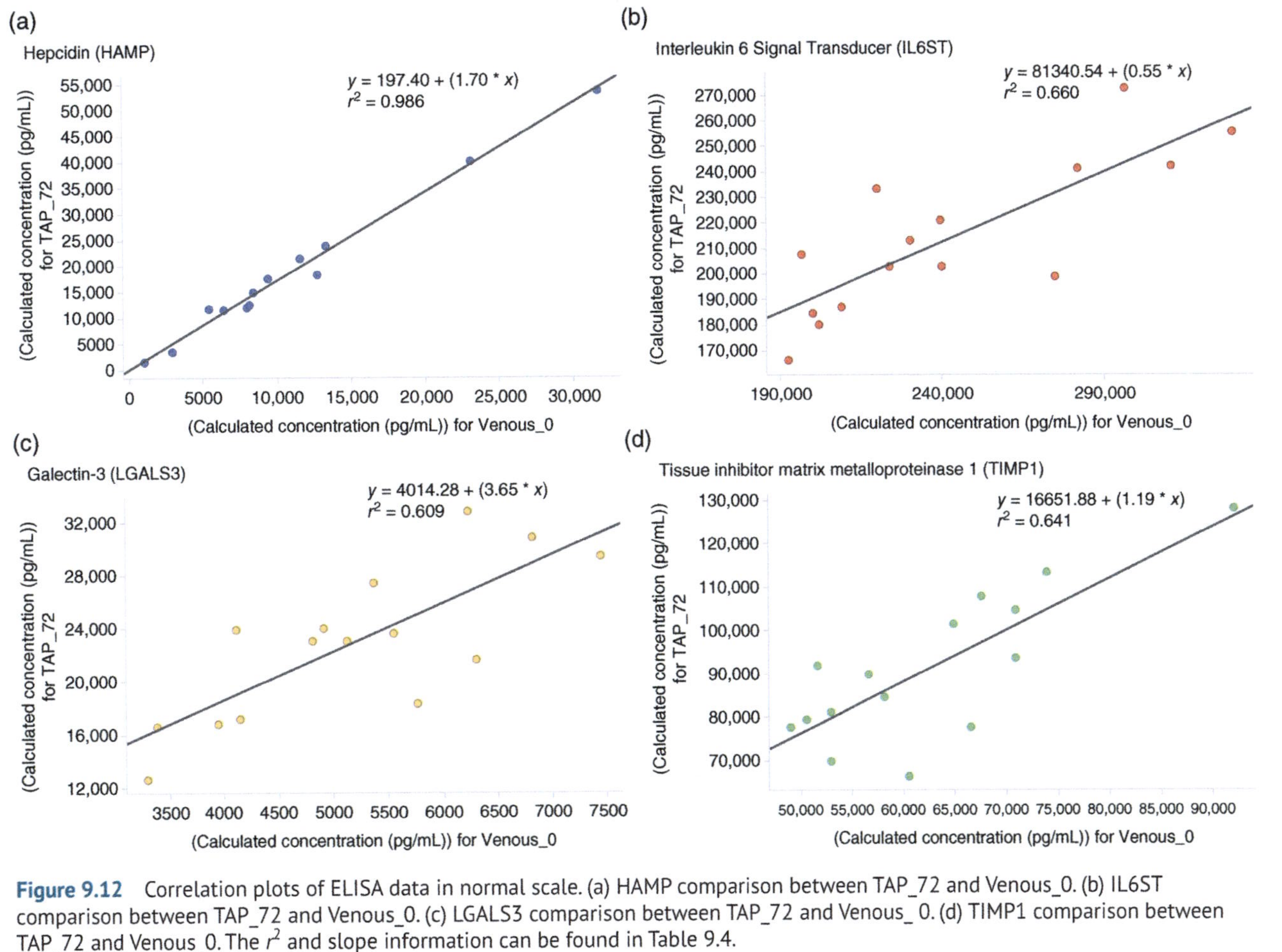

Figure 9.12 Correlation plots of ELISA data in normal scale. (a) HAMP comparison between TAP_72 and Venous_0. (b) IL6ST comparison between TAP_72 and Venous_0. (c) LGALS3 comparison between TAP_72 and Venous_ 0. (d) TIMP1 comparison between TAP_72 and Venous_0. The r^2 and slope information can be found in Table 9.4.

of biomarker agreement between phlebotomy and the TAP device. Individual ELISA confirmation can then be performed to quantitatively validate profiling findings. This approach helps to efficiently identify potential biomarker candidates for patient centric sampling in clinical studies.

Through our findings from SomaScan and ELISA experiments, the utility of patient centric sampling for clinical biomarker sample collection can be better understood. SomaScan data showed that a subset (1,906 of 5,080) of SOMAmers had little to no significant change when comparing TAP_0 and Venous_0 (different collection but with immediate sample processing), suggesting the corresponding protein concentrations are in agreement between different collection approaches. This subset of proteins could be good candidates for patient centric sampling collection, followed by point-of-care testing in clinic or patient's home. Improved comfort and reduced blood volume are among the advantages of such a patient centric sampling collection method.

When comparing TAP_72 and Venous_0 (different collection and different processing time), a subset (1,823 of 5,080) of SOMAmers had little to no significant change—suggesting this group of proteins could be good candidates for remote clinical visit or patient self-collection where patient centric sampling collections are followed by shipment to testing laboratories.

Based on the immunoassays conducted on the 12 selected biomarkers, it is evident that false-negative readings could occur in SomaScan as seen for SP-D and OPN. In other words, SP-D and OPN were located outside of the SomaScan sweet spot but showed significant correlation between venous and TAP collection methods when ELISA quantification was used. However, no false-positive readings from SomaScan have been encountered thus far, i.e. all proteins located in the SomaScan sweet spot demonstrated significant correlation between venous and TAP collection using quantitative ELISA. Hemolysis in TAP samples may have a stronger impact on SomaScan results compared to ELISA, and a larger subset of proteins (even outside of SomaScan sweet spot) may have concordance between venous and TAP samples. Therefore, the collection of biomarker samples through patient centric sampling is feasible for a large set of protein biomarkers in the clinical trial setting.

9.5.2 Considerations for Patient Centric Sampling Implementation in Clinical Trials

Depending on the biomarker's context of use, patient centric sampling could enable remote patient prescreening, rapid safety monitoring, event-driven measurements, and long-term follow-up in clinical trials. As the decentralized trial model is shifting the clinical trial paradigm, there is an urgency to bridge the gap of blood sample collection. Home health services are widely used in such settings to collect samples and perform other clinical procedures in the patient's home. However, the cost of such service provision may hinder its wide adoption, especially in multi-country late phase studies and post-marketing surveillance and research. Patient centric sampling solutions such as the TAP device offer an elegant way of blood sample collection to enable a wide range of biomarker analysis in various settings and reveal new opportunities in clinical research.

Prior to its deployment in a clinical trial, the patient centric sampling device performance including success rate, collection time and sample volume, level of discomfort, and level of hemolysis should be evaluated. This can be achieved through in vitro and healthy volunteer studies. Such studies will also be pivotal in the development and validation of analytical methods to be later deployed in conjunction with the corresponding patient centric sampling device as well as building skills, knowledge base, and confidence across the organization.

Through engaging with several clinical trial teams, the challenges of implementing patient centric sampling have become apparent. Interestingly, some feedback from clinical teams was that patient centric sampling may result in additional burden to trial participants and introduce additional complexity to the study conduct until the necessary validation work is completed to support its replacement of the standard venous collection. Comparison of paired venous and patient centric sampling collection approaches will be needed to evaluate concordance, to establish reference interval, and to build up bridging datasets. Therefore, it is important to reach alignment with the study team during the early phase of the clinical program. The goal is to establish enough bridging data to support patient centric sampling collection in replacement of venous collection before the program progresses to the later stage.

Another challenge is the ever-changing nature of the devices available and the regulatory landscape surrounding their use. Most early patient centric sampling devices, including TAP, will be replaced by newer models with improved features such as higher sample volume and reduced hemolysis. It would be beneficial to use the same patient centric sampling method throughout multiple clinical trial programs to avoid the need for additional bridging studies. The complex regulatory environment also poses a challenge to patient centric sampling implementation. For example, the Medical Device Directive (MDD) in the EU was transitioned to Medical Device Regulation (MDR) as of May 25, 2021 (European Commission, 2021). Under the new regulation, it may be required for patient centric sampling devices to have CE marking to be deployed in clinical trials in EU countries. Different regulatory status of the various patient centric sampling devices may also influence device and country selection in clinical trial setting.

Integrating patient centric sampling devices into a clinical trial workflow may also add complexity and require sophisticated trial operations. Ideally, the devices will be part of the central laboratory visit kits distributed to clinical sites. The devices should be labeled properly to streamline clinical site workflow and to facilitate sample tracking and reconciliation at the central laboratory. For remote patient centric sampling collections, shipment logistics must be considered based on analyte stability and country feasibility. If immediate sample processing is required, home health service or a portable centrifuge such as the TorqZDrive MR (Sandstone Diagnostics Inc, Pleasanton, CA, USA) may be used in patient's home. Home shipment pickup and cold chain logistics may also need to be considered to ensure sample integrity is well maintained for central laboratory testing.

Sample volume limitation could also hinder the wide adoption of patient centric sampling. Clinical studies routinely collect 3–10 mL of blood compared to 100–200 μL generally currently available with patient centric sampling devices. Furthermore, standard clinical laboratory tests, such as clinical chemistry and routine safety measurements, mostly rely on automated platforms such as Roche Cobas, which may have a high dead volume.

9.5.3 Future Outlook for Patient Centric Sampling

Patient centric sampling will no doubt provide valuable data otherwise unattainable by the clinical trial team. As the clinical trial paradigm is shifting more toward a decentralized model, the need for remote sample collection will only intensify. The emergence of digital biomarkers and the wide adoption of integrating wearable devices in clinical trials will broaden applicability of patient centric sampling, e.g. collect samples outside of scheduled clinical visits. There is much work to be done to get to the point where patient centric sampling is widely adopted and accepted by clinical trial sponsors, sites, central laboratories, regulatory bodies, and the patients. Patient centric sampling devices need to be optimized to better fit clinical trial needs. Clinical trial sponsors need to take more risk in adopting more novel patient centric sampling technologies in their studies. Central laboratories need to be more flexible in integrating remote visits and sourcing appropriate shipment logistics without compromising trial participant's confidentiality and privacy, such as name and address. Regulatory bodies need to be more proactive in providing relevant guidance and set forth clear requirements for patient centric sampling-driven data to support primary and secondary objectives. Clinical trial sponsors, device companies, bioanalytical instrument manufacturers, central laboratories, regulatory bodies, and other stakeholders would need to continue collaborating building knowledge, confidence, and success stories. Dedicated conference sessions and consortiums will help to navigate in the uncharted waters.

Acknowledgements

The authors thank M. Cusano, R. Hillenbrand, S. Richards, S. Lewitzky, J. Jacob, and A. Madar for contributing to the TAP device study design. The authors thank T. Blicharz and T. Richards for providing insightful discussions and device training. The authors thank J. Katz for facilitating IRB approval and subject enrollment. The authors thank A. Kinhikar and L. Yu for providing insightful discussions around ELISA selection. The authors thank M. T. Patel for assistance in generating hemolysis data.

References

Abu-Rabie, P., Denniff, P., Spooner, N., Chowdhry, B. Z., & Pullen, F. S. (2015). Investigation of different approaches to incorporating internal standard in DBS quantitative bioanalytical workflows and their effect on nullifying hematocrit-based assay bias. *Analytical Chemistry*, *87*(9), 4996–5003.

Blicharz, T. M., Gong, P., Bunner, B. M., Chu, L. L., Leonard, K. M., Wakefield, J. A., Williams, R. E., Dadgar, M., Tagliabue, C. A., El Khaja, R., Marlin, S. L., Haghgooie, R., Davis, S. P., Chickering, D. E., & Bernstein, H. (2018). Microneedle-based device for the one-step painless collection of capillary blood samples. *Nature Biomedical Engineering*, *2*(3), 151–157.

Borno, H. T., Zhang, L., Siegel, A., Chang, E., & Ryan, C. J. (2018). At what cost to clinical trial enrollment? A retrospective study of patient travel burden in cancer clinical trials. *The Oncologist*, *23*(10), 1242–1249.

Candia, J., Cheung, F., Kotliarov, Y., Fantoni, G., Sellers, B., Griesman, T., Huang, J., Stuccio, S., Zingone, A., Ryan, B., Tsang, J., & Biancotto, A. (2017). Assessment of variability in the SOMAscan assay. *Scientific Reports*, *7*(1), 14248.

Cheng, S. K., Dietrich, M. S., & Dilts, D. M. (2010). A sense of urgency: Evaluating the link between clinical trial development time and the accrual performance of cancer therapy evaluation program (NCI-CTEP) sponsored studies. *Clinical Cancer Research*, *16*(22), 5557–5563.

Cookson, P., Sutherland, J., & Cardigan, R. (2004). A simple spectrophotometric method for the quantification of residual haemoglobin in platelet concentrates. *Vox Sanguinis*, *87*(4), 264–271. https://ec.europa.eu/health/sites/default/files/md_newregulations/docs/timeline_mdr_en.pdf.

De Kesel P. M., Lambert W. E., & Stove C. P. (2015). Does volumetric absorptive microsampling eliminate the hematocrit bias for caffeine and paraxanthine in dried blood samples? A comparative study. *Analytica Chimica Acta*, *881*, 65–73.

Denniff, P., & Spooner, N. (2010). The effect of hematocrit on assay bias when using DBS samples for the quantitative bioanalysis of drugs. *Bioanalysis*, *2*(8), 1385–1395.

Denniff, P., & Spooner, N. (2014). Volumetric absorptive microsampling: A dried sample collection technique for quantitative bioanalysis. *Analytical Chemistry*, *86*(16), 8489–8495.

Donnelly, S., Reginatto, B., Kearns, O., Mc Carthy, M., Byrom, B., Muehlhausen, W., & Caulfield, B. (2018). The burden of a remote trial in a nursing home setting: Qualitative study. *Journal of Medical Internet Research*, *20*(6), e220.

Drawbridge. (2022). OneDraw blood collection device. Retrieved January 31, 2022, from https://www.drawbridgehealth.com/onedraw/.

Ellington, A. A., Kullo, I. J., Bailey, K. R., & Klee, G. G. (2009). Antibody-based protein multiplex platforms: Technical and operational challenges. *Clinical Chemistry*, *56*(2), 186–193.

European Commission. (2021). Transition timelines from the directives to the medical devices regulation. Retrieved January 31, 2021, from https://ec.europa.eu/health/sites/default/files/md_newregulations/docs/timeline_mdr_en.pdf.

Evans B. R., Anderson M., Wang N., Holder D., Robinson M., Spellman D. S., & Bateman K. P. (2019). Applying proximity extension assays for proteomic profiling of dried blood microsamples to enable patient-centric sampling for endogenous proteins levels in clinical trials. *Clinical & Pharmaceutical Solutions through Analysis USA 2019* (28–31 October 2019), Langhorne, PA.

Fan, L., & Lee, J. A. (2012). Managing the effect of hematocrit on DBS analysis in a regulated environment. *Bioanalysis*, *4*(4), 345–347.

Fingerhut, R., Ensenauer, R., Röschinger, W., Arnecke, R., Olgemöller, B., & Roscher, A. A. (2009). Stability of acylcarnitines and free carnitine in dried blood samples: Implications for retrospective diagnosis of inborn errors of metabolism and neonatal screening for carnitine transporter deficiency. *Analytical Chemistry*, *81*(9), 3571–3575.

Getz, K. A., & Campo, R. A. (2017). Trends in clinical trial design complexity. *Nature Reviews Drug Discovery*, *16*(5), 307–307.

Gold, L., Walker, J. J., Wilcox, S. K., & Williams, S. (2012). Advances in human proteomics at high scale with the SOMAscan proteomics platform. *New Biotechnology*, *29*(5), 543–549.

Guthrie, R., & Susi, A. (1963). A simple phenylalanine method for detecting phenylketonuria in large populations of newborn infants. *Pediatrics*, 32, 338–343.

HemaXis DB10. (2022). HEMAXIS DB10 whole blood collection device (25 February 2019). Retrieved January 31, 2022, from https://hemaxis.com/products/hemaxis-db10/.

James, C. A., Barfield, M. D., Maass, K. F., Patel, S. R., & Anderson, M. D. (2020). Will patient-centric sampling become the norm for clinical trials after COVID-19? *Nature Medicine*, *26*(12), 1810–1810.

Kita, K., Noritake, K., & Mano, Y. (2019). Application of a volumetric absorptive microsampling device to a pharmacokinetic study of tacrolimus in rats:

Comparison with wet blood and plasma. *European Journal of Drug Metabolism and Pharmacokinetics*, *44*(1), 91–102.

Kothare, P. A., Bateman, K. P., Dockendorf, M., Stone, J., Xu, Y., Woolf, E., & Shipley, L. A. (2016). An integrated strategy for implementation of dried blood spots in clinical development programs. *The AAPS Journal*, *18*(2), 519–527. https://pubmed.ncbi.nlm.nih.gov/26857396/

Loop. (2022). Loop medical blood collection simplified (31 January 2022). Retrieved January 31, 2022 from https://www.loop-medical.com/.

Lundberg, M., Thorsen, S. B., Assarsson, E., Villablanca, A., Tran, B., Gee, N., Knowles, M., Nielsen, B. S., González Couto, E., Martin, R., Nilsson, O., Fermer, C., Schlingemann, J., Christensen, I. J., Nielsen, H.-J., Ekström, B., Andersson, C., Gustafsson, M., Brunner, N., & Stenvang, J. (2011). Multiplexed homogeneous proximity ligation assays for high-throughput protein biomarker research in serological material. *Molecular & Cellular Proteomics: MCP*, *10*(4), M110.004978.

Mehan, M. R., Ostroff, R., Wilcox, S. K., Steele, F., Schneider, D., Jarvis, T. C., Baird, G. S., Gold, L., & Janjic, N. (2013). Highly multiplexed proteomic platform for biomarker discovery, diagnostics, and therapeutics. *Advances in Experimental Medicine and Biology*, *735*, 283–300.

Orri, M., Lipset, C. H., Jacobs, B. P., Costello, A. J., & Cummings, S. R. (2014). Web-based trial to evaluate the efficacy and safety of tolterodine ER 4mg in participants with overactive bladder: REMOTE trial. *Contemporary Clinical Trials*, *38*(2), 190–197.

Raffield, L. M., Dang, H., Pratte, K. A., Jacobson, S., Gillenwater, L. A., Ampleford, E., Barjaktarevic, I., Basta, P., Clish, C. B., Comellas, A. P., Cornell, E., Curtis, J. L., Doerschuk, C., Durda, P., Emson, C., Freeman, C. M., Guo, X., Hastie, A. T., Hawkins, G. A., & Herrera, J. (2020). Comparison of proteomic assessment methods in multiple cohort studies. *Proteomics*, *20*(12), e1900278.

Roadcap, B., Hussain, A., Dreyer, D., Carter, K., Dube, N., Xu, Y., Anderson, M., Berthier, E., Vazvaei, F., Bateman, K., & Woolf, E. (2020). Clinical application of volumetric absorptive microsampling to the gefapixant development program. *Bioanalysis*, *12*(13), 893–904.

Schwenk, J. M., Lindberg, J., Sundberg, M., Uhlén, M., & Nilsson, P. (2007). Determination of binding specificities in highly multiplexed bead-based assays for antibody proteomics. *Molecular & Cellular Proteomics*, *6*(1), 125–132.

Sen, A., Gillett, M., Weaver, L., Barfield, M., Singh, P., Lapierre, F., & Spooner, N. (2020). In vitro testing of the hemaPEN microsampling device for the quantification of acetaminophen in human blood. *Bioanalysis*, *12*(24), 1725–1737.

Sully, B. G. O., Julious, S. A., & Nicholl, J. (2013). A reinvestigation of recruitment to randomised, controlled, multicenter trials: A review of trials funded by two UK funding agencies. *Trials*, *14*(1), 166.

Tarolli, C. G., Andrzejewski, K., Zimmerman, G. A., Bull, M., Goldenthal, S., Auinger, P., O'Brien, M., Dorsey, E. R., Biglan, K., & Simuni, T. (2020). Feasibility, reliability, and value of remote video-based trial visits in Parkinson's disease. *Journal of Parkinson's Disease*, *10*(4), 1779–1786.

Tasso. (2021). *Tasso-M20 device earns CE mark certification for at-home, self-sampling blood collection* (19 May 2021). Retrieved January 28, 2021, from https://www.tassoinc.com/press-releases/2021/05/19/2021-5-19-tasso-m20-device-earns-ce-mark-certification-for-at-home-self-sampling-blood-collection.

Tin, A., Yu, B., Ma, J., Masushita, K., Daya, N., Hoogeveen, R. C., Ballantyne, C. M., Couper, D., Rebholz, C. M., Grams, M. E., Alonso, A., Mosley, T., Heiss, G., Ganz, P., Selvin, E., Boerwinkle, E., & Coresh, J. (2019). Reproducibility and variability of protein analytes measured using a multiplexed modified aptamer assay. *The Journal of Applied Laboratory Medicine*, *4*(1), 30–39.

Trajan. (2020). CE mark for Trajan's hemaPEN blood microsampling device now available for diagnostic use across EU and UK (7 May 2020). Retrieved January 28, 2021 from https://www.trajanscimed.com/blogs/news/ce-mark-for-trajan-s-hemapen-blood-microsampling-device-now-available-for-diagnostic-use-across-eu-and-uk.

US FDA. (2019). *TAP blood collection device 510(k) premarket notification*. Retrieved January 27, 2021, from https://www.accessdata.fda.gov/scripts/cdrh/cfdocs/cfpmn/pmn.cfm?ID=K190225.

Velghe, S., Delahaye, L., & Stove, C. P. (2019). Is the hematocrit still an issue in quantitative dried blood spot analysis? *Journal of Pharmaceutical and Biomedical Analysis*, *163*, 188–196.

Xing, J., Loureiro, J., Patel, M. T., Mikhailov, D., & Gusev, A. I. (2020). Evaluation of a novel blood microsampling device for clinical trial sample collection and protein biomarker analysis. *Bioanalysis*, *12*(13), 919–935.

10

Enabling Patient Centric Sampling Through Partnership*: A Case Study

Christopher Bailey[1], *Cecilia Arfvidsson*[2], *Stephanie Cape*[3], *Paul Severin*[3], *Silvia Alonso Rodriguez*[4], *Robert Nelson*[5], *and Catherine E. Albrecht*[5]

[1] *Integrated Bioanalysis, Clinical Pharmacology and Safety Sciences, R&D, AstraZeneca, Cambridge, UK*
[2] *Integrated Bioanalysis, Clinical Pharmacology and Safety Sciences, R&D, AstraZeneca, Gothenburg, Sweden*
[3] *Labcorp Drug Development, Madison, WI, USA*
[4] *Translational Science and Experimental Medicine Early R&I, Biopharmaceuticals R&D, AstraZeneca, Cambridge, UK*
[5] *Labcorp Drug Development, Geneva, Switzerland*

10.1 Introduction

In this chapter, we describe how the AstraZeneca–Labcorp Drug Development clinical laboratory partnership has been and continues to be integral to the development and implementation of patient centric sampling (PCS) strategies within both businesses. The challenges and successes experienced along our journey will be presented as well as how our experiences are driving new ways of working within AstraZeneca and Labcorp Drug Development.

10.1.1 The Partnership

The AstraZeneca–Labcorp Drug Development clinical laboratory partnership is a novel delivery model that supports the regulated clinical bioanalysis of pharmacokinetic (PK) samples from AstraZeneca's small-molecule portfolio and central laboratory services (clinical trial kit supply and central laboratory testing) and

---

* Disclaimers used in this chapter, the terms *partnership* and *alliance* do not denote a legal business entity that has been formed between AstraZeneca and Labcorp Drug Development.

Patient Centric Blood Sampling and Quantitative Bioanalysis, First Edition.
Edited by Neil Spooner, Emily Ehrenfeld, Joe Siple, and Mike S. Lee.

specialty clinical laboratory management for the whole portfolio (Arfvidsson et al., 2017; Bailey and Goodwin, 2017). This alliance has realized clear benefits to both companies and harmonized, efficient and standard ways of working have been established.

The partnership now supports all AstraZeneca-sponsored clinical trials, with approximately 200+ ongoing at any one time with most, but not all, including a PK component that requires bioanalysis of a small-molecule drug. These studies support thousands of patients across the globe in different disease populations. The collaboration also creates access to a much wider pool of scientific and operational expertise for the two companies. This relationship offers the opportunity for broader engagement across disease areas and researchers, to glean diverse perspectives when developing and implementing clinical development strategies. The partnership also offers an ideal platform to consolidate the cost of consumables, specialized reagents, and equipment as well as to develop and implement innovative approaches more efficiently. The partnership and unifying business and laboratory strategy has brought real benefits to patients much sooner, with the adoption of PCS sampling approaches now becoming the norm rather than the exception in AstraZeneca's clinical trials.

10.1.2 AstraZeneca's Evolving PCS Approach

AstraZeneca's PCS approach seeks to reduce a patient's overall sampling burden, thereby improving their on-study experience (Bailey et al., 2020). This includes looking to provide flexibility in our sampling and analytical protocols that might allow different sampling approaches to be available for patients to enable them to choose how and where their samples are collected throughout a study's life cycle. As such, the aim is not necessarily to look at replacing existing venous-based sampling procedures, but to complement these. Thus, allowing for occasions when alternative approaches may not be appropriate or indeed times when a patient prefers to choose a venous collection.

10.1.3 Why Change?

Clinical trials require a lot of effort and financial commitment to deliver, so it is no surprise that researchers try to glean as much information and data as possible to assist in critical decision-making or to simply provide more understanding of a new candidate drug. This often leads to science teams taking the maximum blood volume allowed from a patient, not only to support determining the critical decision-making end points of a study but also to have samples available for exploratory end points, including those that are yet to determined (i.e. samples taken to be bio-banked for future use).

To adopt a patient centric mindset, it is necessary to challenge current ways of working. For example, due to technological advances many assays now require only microliters (often <100 μL) of sample matrix, yet the historical approach of collecting milliliters of blood from the patient remains common. It is imperative that we challenge the status quo to progress solutions that removing burden from the patient, while also preserving data quality.

In addition, clinical trial designs are becoming increasingly complex, both scientifically and operationally, but have sampling protocols aimed primarily to be delivered by trained healthcare professionals (e.g. phlebotomists) in hospital and/or clinic locations. These site visits can be disruptive to a patient's day-to-day life, with the burden of travel and being away from home, school, or work. The inconvenience of this may potentially impact their decisions to either enroll or subsequently to stay on a clinical study for its full duration. Thus, delaying the completion of a study or program of studies can ultimately delay the availability of new life-changing treatments. It therefore makes sense for research groups requesting samples (such as PK scientists and bioanalysts), to also explore alternative sampling schemes and procedures that may give patients the opportunities and flexibility to either be sampled or to self-sample, in both local and remote clinical and nonclinical settings (including the home), to reduce the disruption in their daily life and the burden of participating in a trial that may otherwise cause them to consider leaving a trial before its completion.

10.1.4 Patient Choice

For a sampling approach to be truly patient centric, engagement with the patient with respect to sampling is critical. The patient being part of the decision-making processes before and during a trial should not just be restricted to their treatment options but should also be extended to their choosing of how and where they provide their samples at any stage during a study. To facilitate this, bioanalytical and PK scientists need to consider this when developing assays and sampling workflows in clinical study protocols and instructions. For ultimate flexibility of choice, the viability of providing a selection of interchangeable sampling options, rather than sticking with the current standard "venous draw at the clinic" model, should be considered on a study-by-study, analyte-by-analyte and, if possible, patient-by-patient basis, balancing the breadth of choice options with the practicalities of the logistics, quality of the resultant data, regulatory compliance, and costs.

10.1.5 The Challenges—Why Isn't PCS Already the Norm?

The challenges to making PCS approaches the norm can probably be broken down into economic, scientific, operational delivery, and regulatory quality considerations.

From an economic perspective, any new ways of working can be expected to have added costs associated with them, at least initially. This is often the main barrier to researchers adopting new approaches, since funding is finite and those holding the purse strings are often reluctant to commit without evidence of business benefit. Bringing on board a PCS approach it is difficult, at the moment, to provide such evidence and convince payers that all the perceived advantages of PCS will eventually manifest themselves in cost savings. This is because those areas where it might be expected for PCS to positively impact costs the most, such as improved subject recruitment and/or retainment, on-study clinical site costs and investigator/patient compensation (e.g. for overnight stays) are difficult to capture quickly and are dependent on the design of each individual study. One PCS device manufacturer (Mitra Resources I User Guides, Applications and How to Guides, 2023) has developed an on-line cost comparator tool (Neoteryx, 2023) that attempts to estimate cost savings using their products. Using this tool gives some insight into how adopting a PCS approach might impact study costs with the primary driver for pricing differences appearing to be the frequency and destination of shipments during a study, which in turn is driven by the phase of drug development the study is in and the number of samples being generated. There are little to no savings evident in small early-phase studies, but potentially significant savings, relatively speaking, observed in later phase III trials. These costs might expand further when patients are given more choices with a significant amount of kit wastage (more unused kits) expected, but this is where determining what is the right balance between being patient centric and what is economically viable will always drive the desire for, and speed of, change.

To gain broader acceptance of new ways of working, it is essential to demonstrate that a new approach provides benefits over any "standard" practice. A good way to demonstrate this is to, where possible, operate new workflows in parallel with existing ones, enabling more direct and meaningful comparisons of different approaches until sufficient data are generated to justify any changes.

The following sections describe our assessment of one emerging quantitative microsampling technique, volumetric absorptive microsampling (VAMS®) utilizing the Mitra™ device (Mitra Resources I User Guides, Applications and How to Guides, 2023), as part of a PCS approach. The case study highlights the scientific and operational challenges of setting up and delivering a "trial@home" and how working in partnership with different groups of subject matter experts in the scientific and operational space, within and across different businesses has facilitated implementation of a more patient centric approach to the sampling and analytical procedures for supporting the PK and pharmacodynamics (PD) end points in a clinical trial.

10.2 Pre-Study Considerations

As mentioned previously, PCS can be differentiated from traditional blood sampling (typically by venepuncture) in that it puts the patient and their needs at the center of the process. The aim is to reduce the patient's sampling burden, thereby improving their on-study experience. PCS does not, however, necessarily mean that traditional venous sampling needs to be replaced. Keeping the patient in focus is really about giving the patient the choice and to understand that one solution does not fit all. It is about providing the flexibility of having different sampling approaches available that ultimately allow the patient to choose where and how their samples are collected throughout the study's life cycle, while ensuring operational feasibility.

In order to be able to offer that level of flexibility to each patient in how their PK and/or PD samples are collected during a study's life cycle, it should be expected that additional time, effort, and money, outside of that required for existing approaches, will have to be invested to ensure:

- maintenance of high-quality sampling procedures and storage processes at all the different locations and under all environmental conditions that the samples might be exposed to;
- availability of suitably validated assays for the quantitative determination of each analyte (for PK and/or PD end points) for each matrix collected (plasma, serum, whole blood, dry whole blood, etc.);
- data concordance/equivalence between the different matrices can be demonstrated either directly or via application of a conversion factor.

10.2.1 Early Engagement

The main rationale for introducing PCS sampling into a clinical development program or a particular study is likely to vary, depending on the program's target patient population and/or its current development phase. However, whatever the scenario, early engagement with the drug project team to facilitate full buy-in for the investment required to evaluate the new sampling approach is critical. In our experience, the project team's engagement and commitment should not be underestimated. Not only is this critical to ensure that any of the required additional budget is approved in a timely manner, but it also facilitates the identification and onboarding of subject matter experts from the different business functions (internal and external) to be part of the early feasibility assessments.

For the AstraZeneca–Labcorp Drug Development partnership, this involves not only bringing together scientific and operational subject matter experts from across both businesses but also with other partner organizations involved in the delivery of the clinical trial, such as the clinical contract research organization (CRO). One of the great benefits from working in our particular partnership model is the easy access to an extended expert panel that can complement and bridge any internal knowledge gaps. In addition, the partnership offers well-established lines of communications and procedures to facilitate the setup of new ways of working and novel approaches to delivery. Roles and responsibilities are already clearly defined within a well-established network of teams, which allows each group to be focused on the actions to be delivered as part of early engagement and any feasibility work.

Our experiences show it is well worth the time and effort to secure the investment and resources to have a cross-functional team setup as soon as possible to participate in the early engagement and feasibility evaluation phases. Figure 10.1 illustrates what can be a complex stakeholder interaction map when looking to implement a PCS approach in a clinical trial, calling out different project, study, and partnership roles (Labcorp Drug Development roles = Pink, AstraZeneca roles = other colors) and who they may be required to consult with during the pre-study early engagement and feasibility study setup or study conduct phases of a project.

10.2.2 Feasibility Assessment

The next critical aspect for a successful implementation of a PCS approach for PK and PD evaluations is a proper feasibility evaluation. This should involve both scientific and operational considerations, some of which are captured in Table 10.1.

By having a cross-functional team of scientific and operational experts that closely collaborates throughout the feasibility evaluation, a proper balance between the scientific requirements and what is feasible from an operational perspective can be achieved.

10.3 The Case Study

10.3.1 Background

In the case study, the drug project team was planning a phase I study in an adolescent patient population, where in addition to the standard end points of the study, they wanted to investigate the challenges and opportunities of running a trial with both clinical site visits and home nursing visits. The "at-home" premise was aimed at minimizing impact of trial participation on school attendance and more general family life, while maintaining the safety of the patients and quality of the

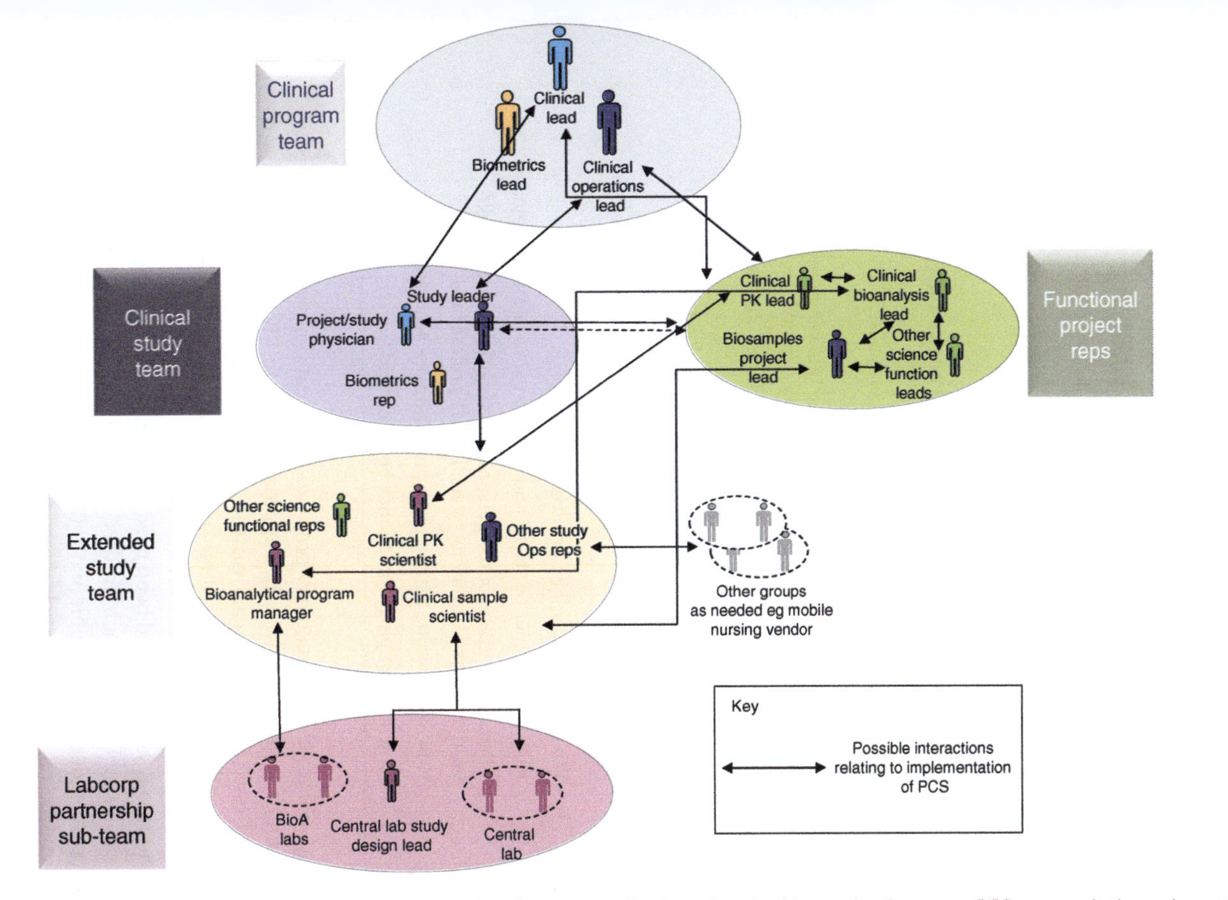

Figure 10.1 The complexities of interactions in a large organisation when looking to implement a PCS approach through partnership. This figure calls out different project, study, and partnership roles (Labcorp Drug Development roles = Pink, AstraZeneca roles = other colors) and who they may be required to consult with during the pre-study early engagement and feasibility study setup or study conduct phases of a project. In our experience, the AZ Clinical Bioanalysis Lead in close collaboration with the AZ clinical PK lead usually initiates the PCS activity and continues to drive it throughout the life cycle of the study and program.

Table 10.1 Considerations When Assessing the Feasibility of a PCS Approach.

Scientific	Which PK or PD assessments would benefit from PCS approach?
	What exposure/concentration levels are expected?
	Any impact of sample physiology, hematocrit, blood:plasma partition, etc.?
	Will changing the sample matrix require a data bridge?
	What dataset size would be required for the bridging activities?
	What sample volumes are available?
	Can the sensitivity requirements be met?
	How to optimize sample stability?
	Could more than one analyte and/or end point be assessed from the same sample—multi-analyte assay an option?
	Can standard laboratory workflows be maintained/is new instrumentation required?
Operational	Can the sampling device be incorporated into standard central lab kit designs, is it commercially available in all regions?
	What limits on sampling need to be applied for the intended patient population?
	What sample collection volume is feasible/tolerated?
	What frequency of sampling can be tolerated?
	How can the known sample stability be used to facilitate sample logistics?
	What countries will be involved in the trial?
	How will samples be stored and transported in different regions?
	What is the skillset of the person collecting the samples—self-sampling versus sampling by healthcare professional?
	How will training for sampling procedures be provided?
	How will sampling compliance be monitored?

clinical data obtained. This scenario presented an ideal opportunity for the bioanalytical scientist to show how a PCS approach for both PK and PD sample collection could have positive impact on the overall concept, and the "at-home" experience of patients, their caregivers, and the healthcare professionals that the approach should be expected to affect.

Since traditional sampling had been used as the standard approach to this point, to introduce a new PCS approach, in this case utilizing microsampling of whole dried blood would not be feasible without first bridging between the new and previous sample collection procedures and matrices collected. It was therefore agreed that the evaluation of microsampling should be added into the clinical study protocol as an exploratory objective, in parallel with the main study objectives that used traditional sampling. The exploratory objectives added into the protocol were phrased in terms such as:

- To establish if microsampling can be used in clinical studies without compromising the quality of the samples taken, maintaining the integrity of any subsequent PK and PD data generated.
- To determine if PK and PD data can be adequately bridged across plasma and whole blood matrices.

Introducing microsampling as an exploratory objective into this study added additional complexity and put further constraints on the total blood volume being collected. However, the information and value that could be gained via this exploratory objective was considered to offset the added complexity and provided a convincing argument that this was an appropriate study to evaluate PCS. The study offered an opportunity not only to bridge matrix data but also to "cross-validate" sampling-at-home procedures with those carried out in the clinic. Furthermore, it presented an additional opportunity to evaluate, generically, a less invasive sampling technique in a potentially more relevant patient population (adolescents with asthma rather than healthy volunteers). A further benefit was that a successful implementation of PCS at this stage of the program's development could result in this approach being available as an option in the upcoming pediatric program as a whole.

In more detail, the case study was an open label, phase I study to assess PK, PD, and safety upon 2-week treatment of an inhaled drug in adolescents with asthma. The first home visit was planned pre-first dose to collect baseline safety and PD (6 time points to create a 12-hr profile) assessments. After 14 days of dosing, the second home visit was planned to be carried out. During this visit, the primary and secondary end points (8 PK and 6 PD time points to create 12 hr profiles) were assessed. In addition, the exploratory samples for the assessment of PK and PD in venous whole blood (at the expected maximum concentration (C_{max})), venous dry whole blood (6 time points to create a 12-hr profile), and capillary dry whole blood (fingertip sampling at expected C_{max}) plus an additional optional sample 4–6 hr after dose) were collected. Both visits were carried out by nurses who collected the PK and PD samples and performed additional safety assessments in the home setting.

10.3.2 Scientific Considerations

Since both PK and PD assessments were planned to be included in the schedule of events for the home visits, both assessments were also considered for microsampling to gain optimum usage of the PCS strategy in the future clinical development.

Dried blood spots (DBSs) have been extensively used for many years for the collection of samples for quantitative analysis, and their application in clinical development has been well documented and reviewed (Abbott et al., 2010; Amsterdam and Waldrop, 2010; Barfield et al., 2008; Edelbroek, et al., 2009; Enderle et al., 2016; Keevil, 2011; Kovač et al., 2018; Li and Tse, 2010; Li et al., 2018; Patel et al., 2019; Rowland and Emmons, 2010; Smith et al., 2011; Spooner, 2010; Spooner and Barfield, 2009; Stove and Spooner, 2015; Wickremsinhe et al., 2013; Wilhelm, et al., 2014; Xu et al., 2012, 2013). The well-known and reported issues with DBS have however stimulated alternative sampling devices to be developed. As part of the feasibility work and in collaboration with the operational and clinical teams, some of these alternative devices were evaluated by the bioanalytical laboratory. In this process, the joint experience from previous internal and/or partner microsampling work was instrumental and guided the evaluation team to selecting the Mitra™ device with VAMS® technology. This device is approved by the FDA for clinical use and had previously proved to be a simple and efficient fingerstick-sample-collection approach for a much simpler, healthy volunteer study for another project. Using the Mitra™device, with its adsorptive tip design, fixed, accurate volumes can be collected. The entire sample (the whole adsorptive tip) is subsequently assayed, thereby avoiding some of the previous issues observed with DBS assays, such as variability in hematocrit levels and spot homogeneity impacting quantitation (Denniff and Spooner, 2010; De Vries et al., 2013; Timmerman et al., 2013).

Once the device had been identified, additional time and effort were invested by the bioanalytical laboratory, in close collaboration with the clinical pharmacokineticist/pharmacometrician, to evaluate the assay sensitivity in whole blood with the aim to keep it as close as possible to the existing plasma assay lower limits of quantitation (LLOQs) (assuming 1:1 blood:plasma partitioning). Due to the low systemic exposure levels expected after an inhaled drug administration, the PK assay sensitivity requirements were high, with an LLOQ in the low pM range. As a result, all available Mitra™ tip volumes (10, 20, and 30 μL) were evaluated in an attempt to balance the sensitivity requirements with what was feasible from an operational perspective (commercial availability of the devices).

Despite several attempts to optimize the assay sensitivity in whole blood using only 10 μL or 20 μL sample volume, the feasibility data pointed to 30 μL tips as providing the levels of sensitivity that would ensure a successful validation and thereby a reproducible LLOQ throughout the lifetime of the new assay. One of the limitations with Mitra™ is the single-use sample context (due to the entire tip being used), and this should be considered when devising *a priori* procedures for repeat assay design and interpretation, such as for incurred sample reproducibility (ISR) experiments during in-study analysis.

Another consideration in the single-use sample context is that since Mitra™ is based on a fixed-volume collection, then this puts added emphasis on the individual collecting the sample to accurately follow collection instructions to ensure the tips are neither underloaded nor overloaded. With this in mind, since an at-home setting was planned as part of the microsampling evaluation in the study, all sampling procedures were kept as simple as possible, using training material and instruction videos available from Neoteryx (Mitra Resources | User Guides, Applications and How to Guides, 2023) to avoid or reduce the potential of introducing sampling errors that might impact quantification. In addition, due to the inhaled administration route, extra precautions were added to minimize the risk of contamination. These precautions are discussed in more detail in Section 1.3.5. Finally additional validation work for the existing PK and PD plasma assays was also undertaken to account for the new sampling setting including evaluations to:

- maximize whole blood stability (up to 48 hr)—to allow for limited sample processing in the home setting and to maximize the home sampling flexibility and logistics;
- minimize the plasma volume required for the PK plasma assay—to limit the constraints on the total blood volume and allow for the additional exploratory samples to be collected within the acceptable total blood volume.

A close collaboration and continuous communication through emails and weekly study team meetings between the operational study team and external partners, who represent those actually collecting the samples, and the bioanalytical scientists was of great importance to identify any potential sampling procedure challenges and to put in place mitigating measures to maximize our prospects of obtaining samples of the required quality. The additional plasma assay work built further flexibility into the sampling instructions, expanded storage conditions beyond those that would normally be investigated, and also allowed the collection of the exploratory samples without exceeding the set total blood volume limits for the patient population.

Last but not least, a few ethical/patient centric considerations were made as part of the feasibility evaluation, to make sure that the introduction of microsampling could be made without jeopardizing the study's operational feasibility, while keeping the patient in focus. These patient centric discussions between the scientific and operational teams raised questions around:

- the time and effort required to accurately fill multiple (at least two to assess both PK and PD and then, if possible, a backup tip for any re-analysis) 30 μL Mitra™ tips from a single fingerstick;
- how the adolescent patient population would perceive the fingerstick sampling compared to venepuncture if more than one fingerstick is required per time point. Would this still be considered a less-invasive and preferred sampling approach?

Through the cross-functional collaboration an elegant solution, beyond established procedures and organizational boundaries, was identified. In this case, the team was able to develop and validate methodology that allowed both drug and PD biomarker concentration levels to be assessed from the same Mitra™ tip in a simultaneous two-analyte assay.

The cross-functional collaboration within and between partnering organizations was critical during the feasibility and method development stages to align the assay requirements and the study design with what was feasible from an operational perspective and continued to play an instrumental role throughout the study conduct and implementation of the PCS approach. As will be described further in the operational section, the continuous cross-functional communication and collaboration was critical to adjusting, in a timely manner, the sampling and storage instructions and updating training needs as necessary to ensure sample quality.

10.3.3 Regulatory Agency Expectations Bridging

With the ultimate aim to be able to offer alternative more PCS approaches in the clinical development program, allowing for each patient to choose their sampling setting (location as well as sampling approach), before rolling out any new approach, it is necessary to first demonstrate that any data or outcomes generated from the new approach can be meaningfully compared with, or bridged to, those generated previously. When designing and interpreting the data from these "bridging studies," there is a need not only to consider any inherent analytical bias that an assay might introduce but also to consider the matrix itself. In the case of whole blood, this means evaluating differences due to variations in hematocrit levels, the blood:plasma (b/p) partitioning ratio and potentially, depending on the type of plasma assay data being used (free or total), plasma protein binding too. In our case study, this bridging involved comparing data from samples not only collected venously and via fingerstick but also collected in different settings, i.e. in the clinic and at-home. The primary, secondary as well as exploratory PK and PD samples collected are summarized in Table 10.2 and Figure 10.2.

10.3.4 Study Operations Considerations

As previously mentioned, this case study involved a hybrid delivery model that combined site and home visit assessments. This design presented operational delivery challenges, and the following sections describe the challenges encountered, solutions developed, and recommendations for future studies with similar settings. The necessity to engage early with the right group of experts has been

Table 10.2 PK and PD Samples Collected, Assays Required, and Use of the Data as Part of Bridging Activities.

PK and PD samples collected	Validated assay required	Purpose of the sample
Primary/secondary assessments		
PK plasma	PK plasma	To support the primary PK objective in the study
PD plasma	PD plasma	To support the secondary PD objective in the study
Exploratory assessment		
PK venous whole blood (wet)	PK whole blood	To take all physiological differences in the b/p partitioning ratio into account and bridge between plasma and whole blood PK concentration levels
PD venous whole blood (wet)	PD whole blood	To take all physiological differences in the b/p partitioning ratio into account and bridge between plasma and whole blood PD concentration levels
PK/PD, venous whole blood (dry)	PK/PD whole blood (dry)	To take any analytical differences due to difference in matrix state into account and bridge between wet and dry matrix. This is also investigated *in vitro* as part of the assay validation activities
PK/PD, fingertip whole blood (dry)	PK/PD whole blood (dry)	To capture any potential differences introduced by fingertip sampling, such as contamination and increased variability but also to consider any physiological differences between venous and peripheral blood

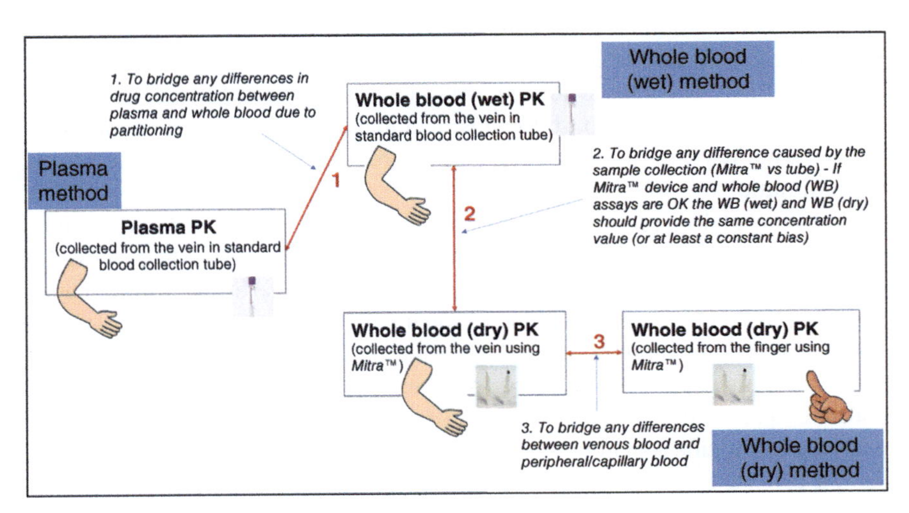

Figure 10.2 Bridging PK (and PD) involved comparing data from both plasma, wet whole blood and dry whole blood matrices, collected venously and via a fingerstick. The plasma—wet whole blood bridging—was made to consider any differences in drug concentration between plasma and whole blood due to partitioning. Bridging wet whole blood with dry whole blood was performed to capture any bias due to the sample collection procedure (Mitra™ vs. tube), and finally any differences between venous and capillary blood were bridged by comparing data from a sample collected from the vein with data from a sample collected from the fingertip. As a result, PK (and PD) assays had to be validated in all three matrices to complete the bridging activities.

discussed in the context of the scientific delivery, but this is even more important in the operations space. Operational staff may not be familiar with new sampling devices and may not fully understand the impact on analytical outcomes if they introduce a small procedural change as part of the rollout of a new device. The partnership between the two companies enabled quick access to a pool of experts from different backgrounds (scientific, operational, logistics, etc.), and the open and honest nature of the relationship allowed for timely, frank, and productive discussions before and during the study, facilitating speedy resolution to the challenges encountered during the setup and delivery of the study.

Based on our experiences during this case study, we have identified and elaborated on the challenges encountered and resolved, as well as how these learnings can be incorporated into processes for future studies where a PCS setup is being considered.

10.3.5 Route of Drug Administration and Potential for Sample Contamination

One of the first operational challenges encountered was when the sample bioanalysis showed a potential contamination issue of the Mitra™ fingertip samples. This was observed as a dose-dependent high variability in the determined drug concentrations from the fingertip samples, using the same Mitra™ venous concentration as a reference analyzed within the same run. This confirmed that the source of contamination was not from the analytical procedure and most likely occurred when the samples were being collected, and so the team began to investigate both the dosing and sampling instructions provided to the patients/caregivers and to work through to the root cause of the contamination.

Most inhaled drug products (similar to the one being investigated in this study) are designed to "fly" once airborne and "stick" once landed, so the team looked at the instructions on how to use the inhaler devices and also at the Mitra™ device sampling instructions. It soon became apparent that the probable cause of contamination was small amounts of drug being transferred through the air to the patient's hands/fingers during dosing and then being picked up during the Mitra™ collection procedure. The assay is highly sensitive so a relatively small amount of drug contamination would have a big impact on the analytical result, and so the sample collection and handling instructions were promptly amended to minimize the risk of continued contamination. These updates included, where possible, that drug administration should be carried out in a separate room from where the samples were going to be collected, gloves should be worn during administration then immediately discarded, and a thorough handwashing and fingertip cleaning procedure should be followed prior to all fingertip sampling. Subsequently, it was also found that the type of formulation used in the inhalers contributed to the potential for contamination. With a dry powder inhaler almost eliminating the potential, whereas when a nebulizer is used for inhalation, contamination was more likely especially at high-dose levels.

Prompt communication between the bioanalytical laboratory and operations teams enabled swift development and implementation of the modified instructions, thus minimizing the impact on the study. Furthermore, we would like to emphasize the importance of analyzing these exploratory samples on an ongoing basis so that any issues can be detected and mitigation actions taken as the study progresses.

10.3.6 Sample Handling

Pre-study feedback from the nurses who would perform the home visits indicated potential difficulties in switching from one collection device/sample type to another within the sample collection time window at some time points in the collection schedule, i.e. to have both the primary PK and PD samples and the exploratory bridging samples (wet and dry blood) collected within a 15-min time window (Table 10.2 summarizes the different sample requirements). This was part of the scientific study design and required to enable meaningful comparison of the data from the different sample types at the same time point in order to establish the bridging data. However, the feedback from the nurses highlighted that the scientific requirements were not compatible with operational practices. To secure the study design's operational feasibility, the bioanalytical team developed and validated a Mitra™ dry whole blood assay to measure both drug and PD biomarker concentrations simultaneously from the same Mitra™ sample (manuscript in preparation). Not only did this solution simplify the sample collection procedure, but it also reduced the number of Mitra™ samples collected from the patient by half. As a result of the combined PK/PD Mitra™ assay, the blood volume to be collected from each fingerstick was reduced from 120 to 60 μL, the number of analytical runs, the amounts of laboratory consumables being used (tubes, devices, etc.), and the analytical costs could also be reduced by around half.

Further pre-study feedback from the home visit nurses highlighted additional challenges in relation to the time they had to collect, process, and ship samples. This instigated further bioanalytical validation experiments (mentioned earlier in the chapter) to increase the whole blood room temperature stability (typically 2–4 hr) to 24 or even 48 hr, if stability allowed. This increased the nurses' flexibility and allowed more time between the sample collection and pickup, thereby reducing the shipment frequency and in some cases negating the need for "cold-chain" transport.

10.3.7 Training and Patient Recruitment Challenges

One challenge of the home sampling requirements was that the nurses needed to receive specific training on sample preparation for shipments on dry ice. During periods of the study when recruitment was slow over an extended time period, there was a need to provide consistent training on a recurring basis due to changes in mobile nursing staff personnel. The impact of long running studies on frozen sample handling training is therefore something to consider. We are developing other guidelines and recommendations for this aspect of logistics training for at-home trials, and these are outlined in the following paragraphs.

Having a clear understanding of the specific regulations for sample transportation is critical for safety protection and responsibility. Documented training of the person preparing the shipment is needed to ensure that the biological specimens are packaged to UN3373 Biological Substance, Category B shipping requirements, and when specimens are to be shipped frozen on dry ice, training on how to handle dry ice correctly is paramount. The patients were never expected to take on the accountability of shipper, and dry ice shipments were never left with the patient during nurse handovers. Couriers are available that can safely handle and package specimens on dry ice, though this would be an additional billable service.

Therefore, if self-collection is evaluated, samples should not be frozen and dry ice should not be handled by patients as it is considered a dangerous good. Samples that can be shipped and stored at either ambient or refrigerated temperature are good candidates to consider for PCS study designs, as this avoids the complications of dry ice handling.

10.3.8 Other Logistical, Data Protection, and Compliance Considerations

In preparation for a study with sampling in an at-home setting, it is important to include the entire sample life cycle into the logistical considerations. It is not only about considering how the samples will be shipped to the labs for processing and analysis but also about how the necessary material such as sample kits for the study should be made available for patients in their homes. A further important factor to consider is the data protection laws and how to best set up the logistical arrangements to protect the patient's identity as well as other personal data, such as the patient's home address. Under current data protection laws, central laboratories are not allowed to have patients' addresses stored in their databases to facilitate the shipment of kits and/or to collect samples from the patient's home. One solution suited to the mobile nursing model is for central laboratories to ship kits to the mobile nursing services hubs. The nurses can then collect the kits from the hubs and use them at the patients' homes. If the study is designed for noninvasive patient self-collection like urine or fecal sample collection and does not require home nursing services, special kits and materials can be shipped to the investigator sites. The investigators would then supply the kits to the patients as well as provide the necessary training and materials.

For sample collection and handling, the design of the home sampling protocol and kits needs to evaluate whether the home sampling will be performed by a nurse, carer, or by the patient him/herself. When home nursing services are planned, it is important to discuss the feasibility of sample collection with the vendor to ensure that the design is suitable for "at-home" collection and does not require specific handling steps which are not feasible or require special laboratory materials not suitable for a home setting. The materials provided to the patients for home sampling should not contain any chemical substances or items that could be dangerous for the patient or their caregiver. When temperature controlled storage is necessary, this requires special consideration as some patients might not have space or "equipment" (e.g. freezers or refrigerators) to adequately store biological samples at specific temperatures that can be maintained or monitored, in their homes.

If sampling time is critical to interpretation of the data generated from the sample, then how this information is recorded should also be assessed to determine if a nurse or other qualified person is required to be present at the time of collection, to avoid putting the compliance burden of recording the sample collection onto the patients. Self-collection for time critical end points still poses the greatest challenge in the compliance space, as the emphasis to capture accurate sampling dates and/or times falls upon the patients, who may either be unwilling or unable to perform this important task.

Training requirements also need to be considered carefully. Noninvasive sample collection (e.g. saliva, urine, feces, etc.), along with minimally invasive techniques that the newest microsampling devices allow, do not require extensive specialist training to ensure collection of a high-quality samples. However, providing collection instructions that are as simple as possible and presented in a number of formats, for example text, pictures, audio, video, or any combinations thereof, should be considered to maximise the probability of success.

Finally, specific country regulations or cultural differences should also be considered. For instance, stool collection at home is not culturally tolerated in some countries, where other countries see sample courier collections from personal addresses (patients' homes) as infringing privacy laws.

10.3.9 Study Participant Engagement

As discussed at the beginning of the chapter, engaging with patients and/or their caregivers is a crucial element when considering any PCS approach. A study questionnaire devised by AstraZeneca and our study delivery partners was provided to the case study participants with a broad range of questions relating to all aspects of the trial. These next sections summarize the feedback obtained for those questions with particular relevance to the PCS approaches taken during the study. The feedback provides valuable insight into patients' and caregivers' needs and preferences, to subsequently assist in the design of future trials for this patient population and potentially others, to remain as "Patient-Centred" as possible with sampling strategies.

The patients, as well as their caregivers, all responded to questions related to how and where samples were collected. A majority of the patients (66%) and their caregivers (77%) were comfortable or even very comfortable with having a nurse perform the blood sampling at home. Only 3% of the patients and 8% of the caregivers were very uncomfortable with home nursing. A majority of the patients (52%) also confirmed their preference to have a fingerstick sample collected rather than having a traditional venous blood draw from the arm. Twenty-one percent of the patients had no sampling preference, and 27% preferred to have traditional sampling.

The main conclusion of the feedback was that giving patients and/or their caregiver choices and more flexibility on how they participate in clinical trials, with respect to sampling, was a preferred approach.

It was a little surprising that only just over half of respondents (54% of the patients/52% of the caregivers) replied that they had opted to join the study because of reasons relating to furthering their own treatment or expanding research in the disease area (asthma) more generally. Of the remaining respondents, >40% gave reasons relating to the convenience and flexibility of the study conduct for deciding to enroll on the trial.

Further insight was obtained from patients regarding preferences relating to the microsampling procedure employed, with this mode of sampling, in the context of this study, appearing to be appealing for patients as an alternative to venous sampling. It was not possible to differentiate if the preference was a result of the different number of sampling occasions (<5 fingersticks vs. >10 venous samples), the less invasive nature of blood collection via fingerstick or a combination of the two, but it was clear patients felt as comfortable, if not more comfortable, in being sampled at home than at the clinical site, demonstrating that for most patients, home nurse visits would be considered a preferable alternative to site visits.

10.4 Summary

We have been able to employ a PCS approach that has allowed for at-home sampling which provided samples and data of the appropriate quality for us to realistically consider adopting similar approaches in other studies and programs across the portfolio, offering patients a choice in the sampling paradigm without having a detrimental impact on the generation and, importantly, the interpretation of the PK data generated. The case study presented numerous challenges and learning opportunities which have enhanced our and our partners' collective ability to overcome some of the perceived scientific and operational barriers of introducing PCS to clinical studies. In the future, this will allow us to better appraise the feasibility and value of introducing more PCS-aligned approaches and to better enable implementation and operation on a study-by-study, asset-by-asset, patient-by-patient basis.

Some of the important takeaway messages we believe are critical to keep in mind for future work include:

- adopting a collaborative, cross-functional mindset to overcome any barriers that may exist between different departments, especially in large organizations—do not assume you are the only experts in your locality;
- accepting any benefit realization may not always be related to the costs of delivery—the benefit may be balanced more toward patient satisfaction than to a monetary value;
- bioanalytical and PK science still have essential parts to play in developing solutions, e.g. in the areas of seeking alternative matrices, better assay sensitivity, simultaneous multi-analyte assays, PK modeling, and statistical approaches;
- clinical operational delivery subject matter experts are integral partners in developing and implementing any science-led solutions;
- understanding that one size does not fit all—if the aim is to give patients choices, then one device, one process or one study delivery model will not be the only solution.

The feedback received from patients, caregivers, and healthcare professionals was also sufficiently encouraging for us to continue to look to introduce PCS approaches, where feasible and economically viable, in more clinical development programs.

The application to increasing numbers of studies in more diverse therapeutic areas should help us build a dataset and experience base that can then be shared with the broader global clinical research community that are treading this path too. A much larger experimental database should subsequently make it easier to determine if indeed a PCS paradigm is achievable economically, scientifically, and operationally, thus leading to the approaches becoming more universally accepted in more general clinical settings and/or specifically within a clinical trial environment.

If proved beneficial, how quickly adoption of the new approaches can occur is difficult to determine. However, one outcome of the global Covid-19 pandemic is that patients, clinicians, and regulators are now, as a matter of necessity, a lot more familiar with the devices, processes, and procedures that are more patient centric with respect to sampling. The need for countries to better control access to hospitals, clinics, and doctors' offices, to keep them focused on treating those infected with Covid-19 or those needing emergency care for non-Covid-19-related illnesses, has meant many patients across the globe have been unable to attend the usual places where samples would typically be collected. Sampling has been brought closer to the patient, with mobile phlebotomists visiting patients' homes, local drop-in centeres, and mobile units (where a healthcare professional comes to a location that is more convenient to patients) becoming more commonplace, along with "self-sampling" (sampling carried out by patients, carers, or other non-healthcare professionals) coming more to the fore as well.

In the post-pandemic clinical trial space, patients and clinical investigators may now have an expectation that a patient will have the option of providing samples at those places mentioned above with mobile phlebotomists coming to patients, rather than all the sampling being clinical site-based (our case study has demonstrated that this can work in a clinical trial setting). The exposure during the pandemic of many millions of individuals to some aspects of "self-sampling," when no healthcare professional has been available, may smooth the way to the broader adoption of this approach to sampling in a clinical trial. To allow "self-sampling," researchers such as ourselves will need to look at alternative matrices for determining certain end points too. These matrices would need to be derived from less invasive, simple sampling procedures that are suitable for a caregiver or the patient themselves to use. The various blood microsampling procedures and devices currently available aim to fit in this category. We used such a system in the case study. The Mitra™ process requires a lancet to prick a finger to produce a few drops of blood collected, quantitatively, onto an absorbent tip that is subsequently assayed. Alternative "virtually painless" peripheral/capillary blood sampling systems are on the market, such as the Tasso-M20™ (TASSO-M20 — Tasso, Inc., 2023) liquid/dry blood collection platform and the TAP™ device (Virtually Painless Blood Collection Devices for Clinical Trials and Wellness Testing - YourBio Health, 2023) and offer alternative blood microsampling approaches. Research teams may also need to consider utilizing other surrogate test matrices, collected via noninvasive means, such as saliva, hair, sweat, tears, buccal and nasal swabs, and excreta (urine and feces). Whatever the sample type (blood microsample, saliva, swab, etc.) they would all need to allow for local/home storage and/or transportation (preferably at ambient temperatures), so there is no, or a reduced, need to use specialist equipment and/or couriers.

In the post-Covid-19 pandemic world, it is quite conceivable that sponsors that do not offer some elements of PCS, may it be flexibility in where sampling takes place or the type and frequency of sample taken, run the risk of subjects enrolling on other studies that offer more "patient-friendly" sampling options. This could lead to delays in completion of some studies, ultimately leading to delays in bringing those new medicines to patients. Those sponsors that do offer some PCS approaches may experience patient recruitment and retention and thereby benefit commercially as a result in getting their medicines approved sooner.

In conclusion, our initial efforts to develop and validate novel assays and implement new operational procedures to facilitate a PCS approach, fully utilizing the experience and expertise of our external partners, have given us a sound foundation to move forward with a broader adoption of PCS-based approaches across many more clinical trials. The work we performed to bridge across sample and matrix types provides solid scientific evidence for both internal/external stakeholders and regulatory agencies that PCS approaches can be taken without compromising the quality of data generated. The potential for a more extensive uptake across the wider healthcare research sector has been thrust forward by the events of the Covid-19 pandemic, with alternative sampling models getting widespread attention of patients, investigators, and regulators. With this increased awareness, it is hoped the rate of further exploration and adoption of current and future patient centric solutions will increase and facilitate the building of a large experimental database that can then be used to establish PCS approaches as the "new norm" across all clinical trials.

References

Abbott, R., Smeraglia, J., White, S., Luedtke, S., Brunet, L., Thomas, E., Globig, S., & Timmerman P. (2010). Conference report—European Bioanalysis Forum Workshop: Implementing dried blood spot sampling for clinical pharmacokinetic determinations. *Bioanalysis*, *2*(11), 1809–1816.

Amsterdam, P., & Waldrop, C. (2010). The application of dried blood spot sampling in global clinical trials. *Bioanalysis*, *2*(11), 1783–1786.

Arfvidsson, C., Severin, P., Holmes, V., Mitchell, R., Bailey, C., Cape, S., Li, Y., & Harter, T. (2017). AstraZeneca and Covance Laboratories Clinical Bioanalysis Alliance: An evolutionary outsourcing model. *Bioanalysis*, *9*(15), 1181–1194.

Bailey, C., Arfvidsson, C., Woodford, L., & de Kock, M. (2020). Giving patients choices: AstraZeneca's evolving approach to patient-centric sampling. *Bioanalysis*, *12*(13), 957–970.

Bailey, C., & Goodwin, L. (2017). The evolution of the Pharma–CRO working relationship: AstraZeneca and Covance. *Bioanalysis*, *9*(15), 1171–1174

Barfield, M., Spooner, N., Lad, R., Parry, S., & Fowles, S. (2008). Application of dried blood spots combined with HPLC-MS/MS for the quantification of acetaminophen in toxicokinetic studies. *Journal of Chromatography B: Analytical Technologies in the Biomedical and Life Sciences*, *870*(1), 32–37.

De Vries, R., Barfield, M., van der Merbel, N., Schmid, B., Siethoff, C., Ortiz, J., Verheij, E., van Baar, B., Cobb, Z., White, S., & Timmerman, P. (2013). The effect of hematocrit on bioanalysis of DBS: Results from the EBF DBS-microsampling consortium. *Bioanalysis*, *5*(17), 2147–2160.

Denniff, P., & Spooner, N. (2010). The effect of haematocrit on assay bias when using DBS for the quantitative bioanalysis of drugs. *Bioanalysis*, *2*(8), 1385–1395.

Edelbroek, P.M., van der Heijden, J., & Stolk, L.M. (2009). Dried blood spot methods in therapeutic drug monitoring: Methods, assays and pitfalls. *Therapeutic Drug Monitoring*, *31*(3), 327–326.

Enderle, Y., Foerster, K., & Burhenne, J. (2016). Clinical feasibility of dried blood spots: Analytics, validation, and applications. *Journal of Pharmaceutical and Biomedical Analysis*, *130*, 231–243.

Keevil, B. G. (2011). The analysis of dried blood spot samples using liquid chromatography tandem mass spectrometry. *Clinical Biochemistry*, *44*(1), 110–118.

Kovač, J., Panic, G., Neodo, A., Meister, I., Coulibaly, J. T., Schulz, J.D., & Keiser, J. (2018). Evaluation of a novel microsampling device, Mitra™, in comparison to dried blood spots, for analysis of praziquantel in Schistosoma haematobium-infected children in rural Côte d'Ivoire. *Journal of Pharmaceutical and Biomedical Analysis*, *151*, 339–349.

Li, C. C., Dockendorf, M., Kowalski, K., Yang, B., Xu, Y., Xie, I., Kleijn, H. J., Bosch, R., Jones, C., Thornton, B., Marcantonio, E. E., Voss, T., Bateman, K. P., & Kothare, P. A. (2018). Population PK analyses of ubrogepant (MK-1602), a CGRP receptor antagonist: Enriching in-clinic plasma PK sampling with outpatient dried blood spot sampling. *The Journal of Clinical Pharmacology*, *58*(3), 294–303.

Li, W., & Tse, F. L. (2010). Dried blood spot sampling in combination with LC-MS/MS for quantitative analysis of small molecules. *Biomedical Chromatograghy*, *24*(1), 49–65.

Neoteryx. (2023). "Cost Calculator for Clinical Trials (Beta Version)". https://calculator.neoteryx.com/.

Mitra Resources | User Guides, Applications and How to Guides (2023). https://www.neoteryx.com/mitra-vams-resources.

Patel, S.R., Bryan, P., Spooner, N., Timmerman, P., & Wickremsinhe, E. (2019). Microsampling for quantitative bioanalysis, an industry update: Output from an AAPS/EBF survey. *Bioanalysis*, *11*(7), 619–628.

Rowland, M., & Emmons, G. T. (2010). Use of dried blood spots in drug development: Pharmacokinetic considerations. *The AAPS Journal*, *12*(3), 290–293.

Smith, C., Sykes, A., Robinson, S., & Thomas, E. (2011). Evaluation of blood microsampling techniques and sampling sites for the analysis of drugs by HPLC-MS. *Bioanalysis*, *3*(2), 145–156.

Spooner, N. (2010). Dried blood spot sampling for quantitative bioanalysis. Time for a revolution?. *Bioanalysis*, *2*(11), 1781.

Spooner, N., & Barfield, M. (2009). Dried blood spots as a sample collection technique for the determination of pharmacokinetics in clinical studies: Considerations for the validation of a quantitative bioanalytical method. *Analytical Chemistry*, *81*(4), 1557–1563.

Stove, C., & Spooner, N. (2015). DBS and beyond. *Bioanalysis*, *7*(16), 1961–1962.

TASSO-M20 — Tasso, Inc. (2023). https://www.tassoinc.com/tasso-m20.

Timmerman, P., White, S., & Cobb, Z. (2013). Update on the EBF recommendation for the use of DBS in regulated bioanalysis, integrating the conclusions from the EBF DBS-microsampling consortium. *Bioanalysis*, *5*(17), 2129–2136.

Wickremsinhe, E. R., Huang, N. H, Abdul, B. G., Knotts, K., & Ruterbories, K. J. (2013). Preclinical bridging studies: Understanding dried blood spot and plasma exposure profiles. *Bioanalysis*, *5*(2), 159–170.

Wilhelm, A. J., den Burger, J. C. G., & Swart, E. L. (2014). Therapeutic drug monitoring by dried blood spot: Progress to date and future directions. *Clinical Pharmacokinetics*, *53*, 961–973.

Xu, G., Chen, J. S., Phadnis, R., Huang, T., Uyeda, C., Soto, M., Stouch, B., Wells, M. C., James, C. A., & Carlson, T. J. (2012). Application of DBS sampling in combination with LC-MS/MS for pharmacokinetic evaluation of a compound with specific species-species blood-to-plasma partitioning. *Bioanalysis*, *4*(16), 2037–2047.

Xu, Y., Woolf, E. J., Agrawal, N. G. B., Kothare, P., Pucci, V., & Bateman, K. P. (2013). Merck's perspective on the implementation of dried blood spot technology in clinical drug development: Why, when and how. *Bioanalysis*, *5*(3), 341–350.

Virtually Painless Blood Collection Devices for Clinical Trials and Wellness Testing - YourBio Health (2023). https://yourbiohealth.com/en-us/virtually-painless-blood-collection-devices-for-clinical-trials-and-wellness-testing.

11

Perspectives on Adopting Patient Centric Sampling for Pediatric Trials

Enaksha Wickremsinhe

Lilly Research Laboratories Eli Lilly and Company, Indianapolis, IN, USA

11.1 Overview and Why

This chapter provides an overview of the challenges and considerations when conducting clinical trials in pediatric patients. Pediatric patients are not "small adults," and therefore, processes and trial designs used in adult studies are not desirable. One of the processes that requires attention for pediatric clinical trials is the collection of blood samples for pharmacokinetic (PK) analysis as well as blood samples collected for routine clinical tests. The article discusses these challenges and the potential for adoption of blood collection techniques that are patient centric, designed to collect a small volume of blood without needing a phlebotomist, and therefore can even be performed at home.

11.1.1 Regulations and Legislation

Historically, in the absence of dose recommendations in pediatric populations, clinicians have generally used empirically derived doses to achieve the same dose exposure or PK profiles in children as in adults (Kern, 2009; Turner et al., 1999). Such "off-label" use (prescribing drugs to children when it is only approved for use in adults) can result in under or overexposures and potentially result in lack of efficacy or an adverse event that may require immediate medical attention or a visit to the emergency room (Roberts et al., 2003; Turner et al., 1999). Off-label use of drugs in children has been identified and highlighted as a critical need that

Patient Centric Blood Sampling and Quantitative Bioanalysis, First Edition.
Edited by Neil Spooner, Emily Ehrenfeld, Joe Siple, and Mike S. Lee.

needs to be addressed in order to protect public health and provide safe and effective medication for children. As a result, several guidances and directives have been implemented by regulators across the globe to require the conduct of pediatric trials and the approval of drugs for pediatric patients.

The first directive that defined pediatric clinical studies was issued in 2001 by the International Council for Harmonization of Technical Requirements for Pharmaceuticals for Human Use (ICH) and titled Clinical Investigation of Medicinal Products in Pediatric Population (also known as the E11 guidance) (ICH, 2001). Subsequently, legislatures and regulators in the USA issued the Best Pharmaceuticals for Childrens Act (BPCA) in 2002 (Congress, 2002) which grants "pediatric exclusivity," followed by the Pediatric Research Equity Act (PREA) in 2003 (Congress, 2003), which is also referred to as "the pediatric rule." Following their subsequent renewals (2007) and the Food and Drug Administration Safety and Innovation Act (FDASIA) in 2012, the BPCA and PREA were made permanent, with periodic updates provided to the US Congress (FDA, 2016). The European Pediatric Regulation (EC) 1901 was adopted in 2006 (EU, 2006), mandating the conduct of pediatric clinical trials to obtain pediatric labeling for the product (Christensen, 2012; EU, 2006).

11.1.2 Who are Pediatric Patients?

Pediatric patients range from preterm newborns to individuals under 18 years of age, per the ICH E11 guidance. They are further categorized into the following age groups: preterm newborn infants, term newborn infants (0–27 days), infants and toddlers (28 days–23 months), children (2–11 years), and adolescents (12 to 16–18 years, dependent on geographic region).

Growth and development occur rapidly during the first two years of life, where body weight doubles during the first 6 months and triples by the first year of life (Lu and Rosenbaum, 2014). Additionally, major organs and functions continue to mature through infancy and childhood. The stages of physiological development and the corresponding impact on enzymes, transporters, pathways, and mechanisms involved the absorption, distribution, metabolism, and excretion of drugs/xenobiotics can vary significantly between these age groups. For example, age and physiological development-related differences, such as gastric pH and gastric emptying times, can affect absorption and higher body water composition can affect distribution. Furthermore differences in drug-metabolizing enzymes, and drug-transporting enzymes can result in higher exposures and prolonged half-lives (Lu and Rosenbaum, 2014). Therefore, understanding these differences through the careful design and conduct of appropriate trials is paramount to ensure safety and efficacy across all pediatric patients and to determine age-specific dosing regimens from neonates to adolescents.

11.2 Challenges and Current Status

11.2.1 Conducting Pediatric Studies

Retrospective data shows that conducting pediatric trials has been challenging, approximately 40% of the trials conducted between 2008 and 2011 were never finished, or were finished but never published (Schmidt, 2017). The challenges associated with conducting and enrolling pediatric trials have been broadly categorized as falling under four classifications; ethical, physiological, pharmacometric, and economic (Kern, 2009).

Ethical challenges revolve around assent, consent, and balancing risk and benefit, keeping in mind that these interpretations can vary from country to country. Physiological challenges relate to the fact that children are not "small adults." Children go through a continuum of physiological development changes from birth till they reach adulthood, and this needs to be accommodated when designing a trial.

Pharmacometric challenges include understanding the adsorption, distribution, metabolism, and excretion in children, relative to adults. Since all pediatric studies are conducted in patients, study design and procedures must be carefully planned because each sample or data point collected is of extreme value, and these studies typically enroll a small number of patients (compared to early-phase studies in adults which are conducted in healthy volunteers).

Economic challenges can be traced to the cost of conducting pediatric studies – ranging from developing new pediatric-friendly formulations and multiple age-appropriate formulations (i.e. suspensions, rapidly disintegrating, and taste masking) to the actual execution of a successful trial. The BPCA and the PREA have afforded a "carrot and stick" approach, which provides financial incentives to conduct pediatric studies while also mandating them to do so. Although it is not the intent of this article to discuss these challenges in detail, key aspects related to patient centric sampling (PCS) are further discussed.

11.2.2 Ethics, Consent, and Assent

Unlike adults, children cannot consent (legally not permitted to) on their own behalf, and the consent of a parent or legal guardian is required. Therefore, decisions on acceptable risk–benefit profiles for proposed research fall to researchers, parents, and regulators. Study protocols and informed consent forms (ICF) are scrutinized by institutional review boards (IRBs) for approval. Enrollment in a trial requires the parent or legal guardian's consent by signing the ICF, which may be a challenging decision. The parent or legal guardian must (a) be convinced that the experimental drug is "better," (b) understand that the child may get enrolled in a placebo group, (c) understand and overcome fears of side effects, especially any fears that it may impact the child's growth and development, and (d) agree to trial-related procedures including invasive tests and blood draws.

Additionally, research in older children requires the assent of the child, which itself poses challenging and widely debated questions: What is assent? And what age does the assent requirement apply? What is the meaningful difference between assent and consent? What if the child and parent does not agree? (Laventhal et al., 2012). Discussing risk versus benefit can also be very challenging. Is the risk presented based on healthy children or sick children, where the "normal" life of a sick child involves invasive procedures and discomfort that the healthy child would not ordinarily experience (Laventhal et al., 2012).

11.2.3 Patient/Parent Burden

One of the key challenges with pediatric trials is to communicate the risks and benefits of the trial to the participant and parents/guardians and that the new "drug" is better and also in the case of a placebo controlled or blinded trial that the child may get enrolled in the placebo group. Additionally, the parents may have concerns on how the drug would impact the child—especially impact on the child's continued growth and development.

The design of the trial itself—the number of clinic visits, duration of each visit, including the total time away from home, school, and work, as well as the trial procedures themselves, especially invasive procedures such as numerous blood draws—can be a deterrent to patients and parents, resulting in noncompliance as well as dropouts after enrollment. A routine clinic visit may be a few hours, but visits to collecting blood samples for PK may require spending up to eight hours at the clinic; such long-time commitments can be challenging for parents, especially single parents or parents with multiple children who need to take time off from work. Time taken to attend clinic visits may contribute to loss of revenue for parents working an hourly paid job, paying for childcare for other children, transportation costs, and other expenses related to participation in the trial. Inconvenience and time away from home/work have been broadly reported as the key reasons for poor enrollment as well as trial dropouts.

11.2.4 Blood Sampling

Clinical trials require the routine collection of blood samples for standard clinical tests (also called safety labs) such as complete blood counts (CBC), comprehensive metabolic profile, and other disease-specific tests (i.e. liver function and C-reactive protein). These tests may be needed to be performed on a daily basis, depending on the trial, the indication, and the severity of the disease (e.g. oncology patients with advanced disease). Additionally, a series of blood samples need to be collected to understand the PK of the drug in the pediatric patient population.

Given the diversity in the ages and the corresponding stages of physiological development, such studies need to generate data from pediatric patients representing all age groups—from neonates to adolescents, as appropriate. The total blood volumes collected for PK analyses and the standard safety labs can add up to a significant volume of total blood, making the design of these studies more challenging than those performed in adult populations.

Though the volume of blood typically collected from a clinical trial is quite large, only a small volume of blood can be safely collected from most pediatric patients, especially from the very young age groups. The number of PK samples possible could be as few as one or two per patient. Sometimes PK sampling may not be feasible at all, and investigators may have to resort to using "salvage" samples which would-be left-over blood from the routine safety blood draws (pending acceptable stability of the analytes).

The smallest commercially available vacutainers are designed to collect one milliliter of blood. Other nonevacuated blood collection practices used to minimize blood volumes during pediatric trials include directly drawing approximately half a milliliter of blood using a syringe via an indwelling line, or similar access. Drawing blood directly via a syringe and squirting it into a collection tube could cause hemolysis (rupture of the red blood cells), which could affect the analysis and test results. From a practical standpoint, administration of drugs as well as blood sampling in neonates and younger pediatric patients may be performed via a single access line (catheter) during hospital stays. In such situations, if PK samples are collected via the same line, care must be taken to ensure the line is completely flushed with saline, and the samples are not contaminated with the drug and also ensure that the blood sample is not diluted with saline.

11.2.5 Blood Volume Limits

The volume of blood that can be safely drawn varies depending on geography and institution, typically implemented by local guidelines, categorized by age and total blood volume (TBV). In most situations, the volumes are limited to a maximum of 1% for a single blood draw and a maximum of 3% over a period of four weeks. The TBV and corresponding information ranging from pre-term neonates to 12-year-old children are captured in Table 11.1. For example, the restrictions on blood draw in a full-term neonate (weighing 4 kg and 320 mL TBV) could be a total of 3% of TBV over a period of 4 weeks, which translates to approximately 10 mL of blood and in a 3-year-old child (weighing 15 kg and 1.2 L TBV) a maximum of 36 mL of blood (3% of TBV) over a period of 4 weeks (Mulberg and Hausman, 2013).

Table 11.1 Estimated Pediatric Blood Volumes.

Cohort	Age	Weight (kg)	Estimated total blood volume	1–5% of total blood volume (mL)
24-week gestation, preterm neonate	NA	0.5	40 mL	0.4–2.0
Full-term neonate	NA	4	320 mL	3.2–16
Child	3 years old	15	1.2 L	12–60
Child	12 years old	40	3.2 L	32–160

Adapted from Clinical Laboratory Testing in Clinical Trials for Pediatric Subjects (Mulberg and Hausman, 2013) NA=not applicable.

In comparison, the TBV in an 80 kg adult (approximately 180 lbs) is approximately 5.7 L and is typically not a factor with regard to collecting blood for testing.

Exceeding these limits can cause light headedness, fatigue, tachycardia, etc., which could be indicative of anemia and could lead to the need for volume replacement, iron supplementation, and maybe even a blood transfusion. Infants (body weight less than 3 kg) can be at a higher risk. Performing a significant number of needle pricks or accessions to central lines/catheters can implicate additional risks, including the potential for infections. Additional caution is needed in children with conditions associated with inhibition of erythropoiesis, either as a result of the illness or children undergoing chemotherapy.

11.3 Solutions: How Do We Get This Done

11.3.1 Microsampling

Microsampling (typically referring to the collection of blood samples of less than 100 µL) to collect blood samples that are then analyzed to determine drug (and metabolite) concentrations has been demonstrated and successfully used by the pharmaceutical industry for select applications (Evans et al., 2015; Kothare et al., 2016). Significant advancements have been made over the past decade with analytical technologies that have enabled the quantification of very low drug concentrations using very small volumes of blood/plasma/serum. In general, PK concentrations can be determined from volumes as small as 10 µL of blood, negating the need for venipuncture and drawing milliliters of blood collected in vacutainers (Spooner et al., 2019).

The ability to collect small volumes of blood has also evolved with many novel techniques and devices developed during the past decade, enabling the adoption of microsampling for collecting PK samples as well blood samples for other types of analyses including general biomarkers, disease specific markers, and potentially for conducting standard clinical analyses such as the comprehensive metabolic panel (CMP) and related panels (which are typically performed at Clinical Laboratory Improvement Amendments (CLIA)-approved clinical laboratories). Several microsampling and PCS techniques and devices are available and depending on the device/technique they can collect volumes as small as 10 μL (a fraction of a drop of blood—for reference a drop of blood is approximately 50 μL) to several hundred microliters. These techniques are suitable for collecting "only what is needed for analysis," thus significantly reducing the total volume of blood collected each time (Spooner et al., 2019).

Although the adoption of microsampling to support drug development (especially its use during clinical trials) has been slower than anticipated, more recent data show that several of these approaches have been successfully incorporated into the workflow supporting both preclinical and clinical drug development (Patel et al., 2019; Spooner et al., 2019). The benefits of dried blood spot sampling for pediatric trials have been well documented, for over a decade (Pandya et al., 2011; Patel et al., 2010), but its adoption has also been very limited, possibly due to concerns related to the impact of variable/diverse hematocrit levels and its impact on the analyses.

11.3.2 Patient Centric Sampling

One of the major advantages of microsampling is the ability to collect blood samples without the need for a visit to the clinic and without the involvement of a nurse or phlebotomist, i.e. the sample can be collected at home. This would be a significant deviation from the current clinical trial operational paradigm. The transformation from a "clinic-centric" to a more "patient centric" operation has been viewed as a significant leap that would make clinical trials more attractive and "patient-friendly," which could lead to faster enrollment and potentially better patient compliance. A recent survey across a group of pediatric patients and parents showed overwhelming support and enthusiasm for adopting at-home blood sampling compared to having to visit the clinic for a blood draw (Wickremsinhe et al., 2020). The surveyed patients listed "convenience," "not having to travel to and spend time at the doctor's office," "not having to miss school," "not having to miss other activities" as what they liked most about the ability to be able to collect their blood samples at home. However; some noted that they were worried that they might "mess-up" if they had to collect the sample at home and also that they felt collecting it at the doctor's office was more "official" (Wickremsinhe et al., 2020).

11.3.3 Pediatric PCS Devices/Techniques

The research paradigm's obligations to children are clearly different from those of adults; similarly, the techniques and devices must be designed and customized to suit them. Any type of blood sampling will need a puncture of the skin, and this can appeal differently to children. Although finger sticks seem to be the most widely used blood "microsampling" technique, alternative sites need to be considered. The type of disease and severity, for example atopic dermatitis and other autoimmune conditions related to dermatology, neuropathy, etc., could affect sampling, as well as the patient's pain threshold.

Is using a lancet to puncture the skin, followed by collection in to a "device" acceptable or would a device/technique that can combine the puncture and collection processes preferable? The latter would alleviate any fears of "messing up" and improve the success of sampling. Several manufactures have recently marketed microsampling/PCS devices that incorporate the lancet into the collection device and is recommended for sampling from the upper arm (e.g. Tasso OnDemand [Tasso], 2021, YourBio TAP [YourBio], 2023 and Drawbridge OneDraw [Drawbridge], 2021). However, these devices are designed for use in adults and may not be appropriate for everyone, especially younger pediatric patients.

Continued innovation is needed such that the device/technique is easy to use, "fool-proof," designed to inflict minimal pain, and does not require any additional "processing" by the user. Once the sample is collected, it should also be tamper-proof. The site or location from which the blood is sampled may also vary depending on the type of sampling device/technique, coupled with the total volume of blood needed. Performing finger sticks (and heel sticks in newborns and neonates) using a lancet has been the time-tested approach to collect a small volume of blood, but this can be quite painful for some and therefore alternative or less-painful blood sampling options are desired, especially if a larger volume of blood is needed for testing, i.e. hundreds of microliters. Some device manufacturers have suggested designing and packaging the devices such that they appeal to kids, i.e. depicting cartoon characters on the device, including stickers in the sampling kit.

Ideally all microsampling or patient centric sampling devices should also be able to capture the time and date (preferably electronically) and maintain a "controlled" environment around the sample such that it is not exposed to extreme temperatures. From a sample integrity and chain-of-custody perspective, it will also be extremely beneficial to have the device be able to "track/monitor" the temperature and humidity it is exposed to, from the time the blood sample is collected to the time the sample is received at the testing laboratory.

11.3.4 COVID-19 Era

The COVID-19 pandemic has turned the world upside down; at the same time, accommodations made in response to the pandemic have "delivered" healthcare to our doorstep. Along with virtual doctor visits and "drive-through" COVID testing and vaccinations, many of us have got used to using home testing kits for COVID-19, though non-invasive. However, several COVID-19 clinical trials that required serology testing for antibodies have adopted home microsampling to safely collect blood samples for testing (Klumpp-Thomas et al., 2021; NIAID, 2020a, 2020b). Increasing familiarity with such techniques by the general population may encourage—if not demand—the broader adoption of "patient centric" approaches, especially for blood collection.

11.3.5 Training

As with any new technology or technique, training is critical for successful adoption and execution. As indicated previously, because of the COVID-19 pandemic, at-home sampling processes should be more familiar to the general population and patients as well as parents and caregivers will be more likely to take advantages of such capabilities (along with the convenience and the savings in time and effort).

Making the devices and techniques easy to use, painless, nonobtrusive, and foolproof are key to ensuring successful adoption as well as to ensure proper collection of the blood sample. Most device manufacturers have well-illustrated easy-to-follow instructions for the adult user. However, when these are being used within the pediatric population, the packaging and instruction should cater to the needs of the children. A set of printed instructions may not be the best medium for today's children, as they prefer alternative media, such as a simple interactive/animated video accessible on their personal mobile device. However, easy-to-follow, well-illustrated, step-by-step instructions will be needed in resource-constrained populations. We are beginning to see device manufacturers, as well as health/hospital networks, addressing these needs, especially given the need to collect samples safely in the midst of the COVID pandemic, and providing additional resources and training videos catered to pediatric patients (EMEESY 2021; Neoteryx 2021).

11.4 Summary

The 2019 COVID-19 pandemic has changed the world, forever, as we face a "new normal" going into the future. The pandemic has made the world cater to healthcare needs without the need to always travel to a hospital or a doctor's office, as we have experienced with drive-through testing and vaccinations, virtual doctor visits, tele-health, mobile clinics, and also becoming comfortable with receiving a test-kit in the mail, perform a simple finger-stick to collect a drop of blood, and mail it back to the laboratory for testing.

We should expect to see the new normal reflect a true transformation in healthcare—from a clinic- or hospital centric to be more patient- or home centric. This transformation should also provide access to a broader range of novel blood sampling and blood testing capabilities, for use in children as well as in adults, and the ability to collect a blood sample without the need for a visit to a hospital or clinic, but performed within the privacy of one's own home, and deliver test results that are of the same quality and standard expected by the medical community.

References

Christensen, M. L. (2012). Best pharmaceuticals for children act and pediatric research equity act: Time for permanent status. *Journal of Pediatric Pharmacology and Therapeutics*, *17*, 140–141.

Congress. (2002). Best Pharmaceuticals for Childrens Act. *107th US Congress*. Retrieved March 7, 2021, from https://www.congress.gov/107/plaws/publ109/PLAW-107publ109.pdf.

Congress. (2003). Pediatric Research Equity Act. *108th US Congress*. Retrieved March 7, 2021, from https://www.congress.gov/108/plaws/publ155/PLAW-108publ155.pdf.

Drawbridge. (2021). *OneDraw*. Retrieved March 15, 2021, from https://www.drawbridgehealth.com/onedraw/.

EMEESY. (2021). *CountOnMe—Information for patients and families*. Retrieved March 15, 2021, from https://www.emeesykidney.nhs.uk/resources-for-families/countonme.

EU. (2006). *Regulation (EC) No. 1901/2006 of the European Parliament and of the Council* (12 December 2006). Retrieved March 7, 2021, from https://ec.europa.eu/health//sites/health/files/files/eudralex/vol-1/reg_2006_1901/reg_2006_1901_en.pdf.

Evans, C., Arnold, M., Bryan, P., Duggan, J., James, C. A., Li, W., Lowes, S., Matassa, L., Olah, T., Timmerman, P., Wang, X., Wickremsinhe, E., Williams, J., Woolf, E., & Zane, P. (2015). Implementing dried blood spot sampling for clinical pharmacokinetic determinations: Considerations from the IQ Consortium Microsampling Working Group. *The AAPS Journal*, *17*, 292–300.

FDA. (2016). Best Pharmaceuticals for Children Act and Pediatric Research Equity Act—July 2016 status report to congress. *US FDA*.

ICH. (2001). Clinical investigation of medicinal products in pediatric population. *ICH E11*. Retrieved March 7, 2021, from https://www.ema.europa.eu/en/documents/

scientific-guideline/international-conference-harmonisation-technical-requirements-registration-pharmaceuticals-human-use_en-30.pdf.
Kern, S. E. (2009). Challenges in conducting clinical trials in children: Approaches for improving performance. *Expert Review of Clinical Pharmacology, 2*, 609–617.
Klumpp-Thomas, C., Kalish, H., Drew, M., Hunsberger, S., Snead, K., Fay, M. P., Mehalko, J., Shunmugavel, A., Wall, V., Frank, P., Denson, J. P., Hong, M., Gulten, G., Messing, S., Hicks, J., Michael, S., Gillette, W., Hall, M. D., Memoli, M. J., Esposito, D., & Sadtler, K. (2021). Standardization of ELISA protocols for serosurveys of the SARS-CoV-2 pandemic using clinical and at-home blood sampling. *Nature Communications, 12*, 113.
Kothare, P. A., Bateman, K. P., Dockendorf, M., Stone, J., Xu, Y., Woolf, E., & Shipley, L. A. (2016). An integrated strategy for implementation of dried blood spots in clinical development programs. *The AAPS Journal, 18*, 519–527.
Laventhal, N., Tarini, B. A., & Lantos, J. (2012). Ethical issues in neonatal and pediatric clinical trials. *Pediatric Clinics of North America, 59*, 1205–1220.
Lu, H., & Rosenbaum, S. (2014). Developmental pharmacokinetics in pediatric populations. *Journal of Pediatric Pharmacology and Therapeutics, 19*, 262–276.
Mulberg, A. E., & Hausman, E. D. (2013). Clinical laboratory testing in clinical trials for pediatric subjects. In G. Mulberg, D. Murphy, J. Dunne & L. L. Mathis (Eds.), *Pediatric drug development*. Wiley-Blackwell.
Neoteryx (2021). *At-home blood collection kits can reduce trauma for young patients*. Retrieved May 15, 2021, from https://www.neoteryx.com/microsampling-blog/pediatrics-at-home-blood-collection-kits-reduce-trauma-for-young-patients.
NIAID. (2020a). *COVID-19: Human epidemiology and response to SARS-CoV-2 (HEROS)*. Retrieved May 15, 2021, from https://clinicaltrials.gov/ct2/show/NCT04375761.
NIAID. (2020b). *SARS-COV2 Pandemic Serosurvey and Blood Sampling*. Retrieved May 15, 2021, from https://clinicaltrials.gov/ct2/show/NCT04334954.
Pandya, H. C., Spooner, N., & Mulla, H. (2011). Dried blood spots, pharmacokinetic studies and better medicines for children. *Bioanalysis, 3*, 779–786.
Patel, P., Mulla, H., Tanna, S., & Pandya, H. (2010). Facilitating pharmacokinetic studies in children: A new use of dried blood spots. *Archives of Disease in Childhood, 95*, 484–487.
Patel, S. R., Bryan, P., Spooner, N., Timmerman, P., & Wickremsinhe, E. (2019). Microsampling for quantitative bioanalysis, an industry update: Output from an AAPS/EBF survey. *Bioanalysis, 11*, 619–628.
Roberts, R., Rodriguez, W., Murphy, D., & Crescenzi, T. (2003). Pediatric drug labeling: Improving the safety and efficacy of pediatric therapies. *JAMA, 290*, 905–911.
Schmidt, C. (2017). *Many pediatric studies are a waste of time*. Retrieved May 15, 2021, from https://www.scientificamerican.com/article/many-pediatric-studies-are-a-waste-of-time/.

YourBio. (2023). *TAP*. Retrieved June 05, 2023, from https://yourbiohealth.com/.

Spooner, N., Anderson, K. D., Siple, J., Wickremsinhe, E. R., Xu, Y., & Lee, M. (2019). Microsampling: Considerations for its use in pharmaceutical drug discovery and development. *Bioanalysis*, *11*, 1015–1038.

Tasso. (2021). *Tasso-SST*. Retrieved May 15, 2021, from https://www.tassoinc.com/tasso-sst.

Turner, S., Nunn, A. J., Fielding, K., & Choonara, I. (1999). Adverse drug reactions to unlicensed and off-label drugs on paediatric wards: A prospective study. *Acta Paediatrica*, *88*, 965–968.

Wickremsinhe, E. S., Short, M., Talkington, B., & West, L. (2020). DIY blood sampling for pediatric clinical trials—The patients perspective. *Applied Clinical Trials*, *29*, 20–24.

Index

Note: Page numbers followed by "*f*" and "*t*" refers to figures and tables, respectively.

Patient Centric Blood Sampling and Quantitative Bioanalysis, First Edition.
Edited by Neil Spooner, Emily Ehrenfeld, Joe Siple, and Mike S. Lee.

b

e

f

g

h

t

u

v

W

X

Printed and bound by CPI Group (UK) Ltd, Croydon, CR0 4YY

29/11/2023

08198509-0001